T0418803

Nanocarriers
for Organ-Specific and
Localized Drug Delivery

Nanocarriers for Organ-Specific and Localized Drug Delivery

Muhammad Raza Shah

HEJ Research Institute of Chemistry, International Center for Chemical and Biological Sciences, University of Karachi, Karachi, Pakistan

Muhammad Imran

HEJ Research Institute of Chemistry, International Center for Chemical and Biological Sciences, University of Karachi, Karachi, Pakistan

Shafi Ullah

HEJ Research Institute of Chemistry, International Center for Chemical and Biological Sciences, University of Karachi, Karachi, Pakistan

ELSEVIER

Elsevier
Radarweg 29, PO Box 211, 1000 AE Amsterdam, Netherlands
The Boulevard, Langford Lane, Kidlington, Oxford OX5 1GB, United Kingdom
50 Hampshire Street, 5th Floor, Cambridge, MA 02139, United States

Notices

Knowledge and best practice in this field are constantly changing. As new research and experience broaden our understanding, changes in research methods, professional practices, or medical treatment may become necessary.

Practitioners and researchers must always rely on their own experience and knowledge in evaluating and using any information, methods, compounds, or experiments described herein. In using such information or methods they should be mindful of their own safety and the safety of others, including parties for whom they have a professional responsibility.

To the fullest extent of the law, neither the Publisher nor the authors, contributors, or editors, assume any liability for any injury and/or damage to persons or property as a matter of products liability, negligence or otherwise, or from any use or operation of any methods, products, instructions, or ideas contained in the material herein.

ISBN: 978-0-12-821093-2

For information on all Elsevier publications
visit our website at https://www.elsevier.com/books-and-journals

Publisher: Matthew Deans
Acquisitions Editor: Sabrina Webber
Editorial Project Manager: Hilary Carr
Production Project Manager: Anitha Sivaraj
Cover Designer: Greg Harris

Typeset by STRAIVE, India

Working together to grow libraries in developing countries

www.elsevier.com • www.bookaid.org

Contents

Nanocarriers in drug delivery: Classification, properties, and targeted drug delivery applications

Introduction

Literally, the prefix "nano" comes from the Greek word "nanos," meaning very small in size. With regard to the definition given by NNI (*National Nanotechnology Initiative*), nanoparticles are small structures, ranging in size from 1 to 100 nm in at least one dimension. However, it may also be used for the particles comprising the size up to several hundreds of nanometers. Currently, a great revolution has been observed in the effective applications of nanostructures in various fields of science worldwide, which is profoundly attributed to nanotechnology and more specifically to the interventions of nanoparticles. The technology that frequently utilizes the matter at the nanometer scale is referred to as nanotechnology (Mansoori & Soelaiman, 2005; Suri, Fenniri, & Singh, 2007) and the concept of this technology dating back to the early 1970s, when an American physicist Richard Feynman (Nobel Prize laureate) during his lecture introduced the world to the new concept of nanotechnology for the first time in 1959 at Caltech (California Institute of Technology). For this reason, he is considered the founding father of modern nanotechnology. However, in 1974, a Japanese scientist Norio Taniguchi used and defined the term "nanotechnology" (Bayda, Adeel, Tuccinardi, Cordani, & Rizzolio, 2020). The scientific applications of this technology are widespread and grabbing nearly every field of science including chemistry, medicines, physics, biology, electronics, engineering, and much more (Bayda et al., 2020). Particularly, in medicines, the concept of novel strategies for the development of nanoformulations (i.e., *nanocarriers*) has rapidly emerged over the last few decades. In fact, nanocarriers could be successfully employed as drug delivery tools due to their optimized biological and physicochemical properties and a wide range of practical applications in the field of medicines (Sun et al., 2014). Numerous critical issues of conventional drug delivery systems including drug resistance, poor water solubility, drug toxicity, poor specificity, and low bioavailability are potentially associated with the decreased therapeutic potentiality of various drug systems. On the contrary, nanocarrier-based systems consisting of

1

colloidal nanoparticles (e.g., <500 nm) offer more advantages in terms of active drug transportation due to the high surface area to volume ratio. Hence, these systems of nanostructured materials provided a greater opportunity to effectively deliver active drugs into the targeted cells or tissues. The ultimate goal of nanocarriers' employment in drug delivery systems is to achieve the desired therapeutic outcomes by maximizing drug efficacy and minimizing its side effects during the course of treating a particular disease (Aslan, Ozpolat, Sood, & Lopez-Berestein, 2013; Yu, Trase, et al., 2016; Yu, Yang, Zhu, Guo, & Gan, 2016). Nanocarriers with tailored biochemical and physical properties are easily internalized by cells as compared to larger molecules, which makes them more acceptable delivery tools for presently available drugs and biologically active compounds. Among them, liposomes, polymers, carbon or silicon materials, solid lipids nanoparticles, magnetic nanoparticles, and dendrimers are examples of paramount importance that have been successfully employed in various drug delivery systems. Interestingly, these customized nanostructured materials have considerably reduced the dosage frequency by targeting the specific tissue's sites in a (spatial/temporal) guided manner to eliminate the side effects associated with conventional therapies. Particularly, they allow redressing the basic issues associated with already existing pharmaceutical treatment practices, for instance, nonspecific distribution, low bioavailability, rapid clearance, and uncontrolled release of drugs. Ultimately, this will lead to a sensitive and rational reduction in adverse reactions and toxicities as well (Lombardo, Kiselev, & Caccamo, 2019; Yu, Trase, et al., 2016; Yu, Yang, et al., 2016). However, despite the spectacular developments in establishing novel strategies, the majority of these nanocarriers' action is considerably linked with a large number of undesirable side effects and thereby diminishing their intended and most wanted applications in nanomedicines. This reveals some serious issues in the fabrication and engineering of nanocarrier-based systems for therapeutic applications, which are frequently attributed to the complex environment and various established interactions within a particular biological media (Siegler, Kim, & Wang, 2016; Werner et al., 2018).

Classification of nanocarriers

In many recent investigations, nanostructured systems have been extensively differentiated into two distinct categories of organic and inorganic nanocarriers, however, their physicochemical properties could be greatly optimized by modifying their compositions (organic, inorganic, or hybrid), shapes (rod, multilamellar, or sphere-shaped structure), surface characteristics (surface charge, coatings, incorporation of targeting moieties, PEGylation, or functional groups), dimensions (large or small sizes). Moreover, a great number of nanocarriers-based systems have been approved for the management of different diseases such as multiple tumor types and so on, however, several others still remain under observation in various phases of clinical trials (Tran, DeGiovanni, Piel, & Rai, 2017; Ventola, 2017).

Organic and polymer-based nanocarriers

Organic nanocarriers are considered to be one of the most contemplated groups of nanocarriers that are carbon-based nanomaterials. This category of nanocarriers has been widely found with high biocompatibility and enhanced drug loading properties. They allow the incorporation and transportation of a wide combination of various (hydrophobic/hydrophilic) drugs due to their colloidal stability and relatively large size. They also permit greater control of both chemical composition and morphological features (Fattal, Hillaireau, Mura, Nicolas, & Tsapis, 2012; Souery & Bishop, 2018). Based on preparation techniques, organic nanocarriers can be subdivided into two main groups such as amphiphilic systems that follow the process of self-assembly and those that are obtained by employing specific synthesis procedures (e.g., carbon nanotubes, dendrimers, chemical nanogels, and hyperbranched polymers). Notably, a proper combination of two methods is frequently adopted for the construction of a new generation of nanocarriers by considering a supramolecular approach (Akerman, Chan, Laakkonen, Bhatia, & Ruoslahti, 2002; Lombardo, Kiselev, Magazù, & Calandra, 2015; Ma & Zhao, 2015).

Polymer-based amphiphilic nanocarriers obtained by self-assembly processes

Several nanocarriers-based drug delivery systems are comprised of basic building blocks that self-assemble under the influence of certain driving interactions including hydrogen bonding, coordination bonding, solvation, hydration and electrostatic forces, van der Waals forces, $\pi - \pi$ staking interactions, and hydrophobic effects (Lombardo et al., 2015; Ma & Zhao, 2015). There is mounting evidence regarding the implications of amphiphilic macromolecules in formulating new materials for utilization in drug delivery systems. The amphiphilic macromolecules are composed of both a lipophilic part that usually contains hydrocarbon chains thereby minimizing its contact with water and a hydrophilic part, which could be charged or uncharged (cationic, anionic, or zwitterionic). During hydration, a microphase separation along with aggregates formation in aqueous solutions occurs due to hydrating hydrophilic and collapsing hydrophobic moieties. Particularly, when the concentration of the surfactants exceeds a specific concentration known as critical micelle concentration (CMC) (Kahlweit, 1985). Proper customization of amphiphiles' shapes (e.g., changing the critical packing factor parameter (Cpp) Fig. 1.1) would possibly allow for the desired manipulation in the development of

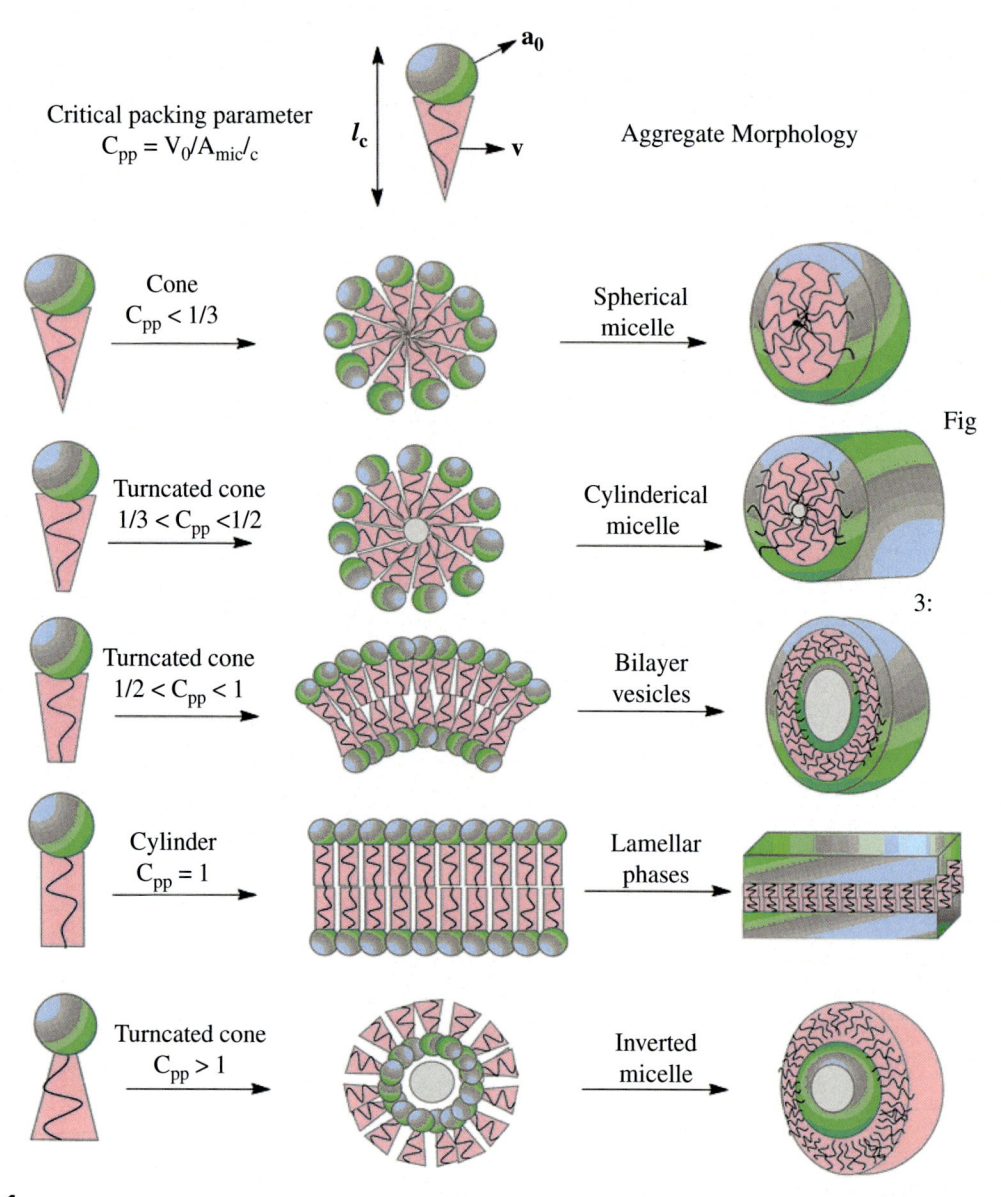

FIG. 1.1

Various self-assembled structures predicted from critical packing parameter C_{pp}. *From Lombardo, D., Kiselev, M. A., Magazù, S., & Calandra, P. (2015). Amphiphiles self-assembly: Basic concepts and future perspectives of supramolecular approaches.* Advances in Condensed Matter Physics, 2015. *https://doi.org/10.1155/2015/151683. Copyright 2015, Hindawi Publishing Corporation.*

nanostructures' designs including lamellar structures ($Cpp = 1$), spherical micelles ($Cpp \leq 1/3$), vesicles ($1/2 \leq Cpp \leq 1$) and cylindrical micelles ($1/3 \leq Cpp \leq 1/2$). However, for higher values ($Cpp > 1$), the amphiphiles will lump together into inversed phases (Kahlweit, 1985; Lombardo et al., 2015). Pertaining to their characteristic structure, liposomes and micelles offer promising safeguard mechanisms against distortion, also, providing broader opportunities for combined therapy and targeted delivery system (Allen & Cullis, 2004; Lombardo et al., 2015).

Micelle and vesicle nanocarriers from polymer-based amphiphiles

Polymers are extensively employed in numerous drug delivery systems due to numerous useful properties including ease in their construction and preparation, effective delivery of therapeutically active agents to the target cells or tissues, biodegradability, and biocompatibility. Various polymers possess specific properties depending on the physicochemical characteristics of their structural unit, whereas the versatility found in their chemical groups' modification has been widely applied for the drug conjugation and functionalization of several polymer-based nanostructures (Masood, 2016).

Interestingly, it is possible to get multiple morphologies and shapes of amphiphilic polymers' nanostructures in aqueous solutions by regulating the hydrophobic and hydrophilic equilibrium (via modulating the weight fraction (FW) of hydrophilic portion). For instance, vesicles ($FW = 20\%{-}40\%$), spherical micelles ($FW = 55\%{-}70\%$) and spherical vesicles ($FW = 45\%{-}55\%$) (Won, Brannan, Davis, & Bates, 2002; Yin, Chen, Zhang, & Han, 2016). A great amount of interest has been vested in the self-assembling amphiphilic polymer-based systems (e.g., micelles) in drug delivery designs (Mikhail & Allen, 2009). The hydrophobic core of micelles produces a microenvironment for the encapsulation of lipophilic drugs, usually resulting in significantly improved solubility of non-polar drugs and, thereby enhanced its bioavailability. On the other hand, the hydrophilic covering creates a stabilizing interface between the aqueous medium and hydrophobic core with the purpose to arrest aggregation, undesired linkages with other constituents and, promoting colloidal stability as well. Additionally, vesicles obtained from amphiphilic polymers (e.g., polymersomes) display a specific bilayer arrangement, consisting of an internal aqueous core, capable of encapsulating hydrophilic substances along with the integrating facility of hydrophobic drug molecules inside the bilayer structure. In some instances, certain responsive polymersomes have been identified with numerous multifunctional, self-assembling morphologies and structures (Siegler et al., 2016; Yin et al., 2016). Some specific tumor (micro) environments could be used for the construction of reactive copolymer incorporated nanosystems, which promote cargo release and imaging sensitivity. The basic endogenous stimuli involved in the activation mechanism are weakly acidic pH, redox species, temperature gradients, and a wide range of highly expressed enzymes (Siegler et al., 2016; Yin et al., 2016). Over the last few years, the natural polymers-based (i.e., dextran, hyaluronan, and chitosan) drug delivery designs have been profoundly contributed to the field of medicines. However, recently adopted protocols for the construction of synthetic polymers in different nanostructured delivery systems paved the way for novel discoveries in the field of nanomedicines (Elsabahy & Wooley, 2012). Polypeptides, polyesters, and polycarbonates are some of the most dedicated synthetic polymers.

Liposome nanocarriers

Even though polymer-based nanostructures possess outstanding properties for both in vitro and in vivo utilization, however, lipid-based nanocarriers in multiple drug delivery systems still remain prominent in the market for clinical usages. Liposomes provided a broader platform for the development of the most acceptable and promising drug delivery designs in the form of nanocarriers (Allen & Cullis, 2013). They offer several benefits due to their greater potentials for multipurpose self-assembly (Sackmann, 1995). It is generally easy to construct lipid-based nanostructures as compared to biopolymers concerning their large-scale availability in the form of phospholipid compounds. Also, they have optimal control over the kinetics of drug release. Natural or synthetic phosphor-lipids are composed of single or more hydrophobic tails and a hydrophilic head. In aqueous solutions, they tend to make extremely flexible vesicles (i.e., bilayer) via self-assembly in such a way that their hydrophilic heads are turned toward the water. In addition, they allow undergoing several dynamic and conformational changes necessary for many biological activities. The structure of liposomes within water solutions is highly dependent on various factors associated with their preparation including stirring, microfluidification, extrusion, and sonication. However, their sizes are predominantly ranging from 50 to 500 nm, which might be made up of large unilamellar vesicles (LUVs 100–1000 nm), small unilamellar vesicles (SUVs >100 nm), and sometimes giant unilamellar vesicles (GUVs >1 µm). While, multilamellar vesicles (MLVs) have been found with concentric bilayer surfaces usually presenting hydrated multilayers that resemble the structure of an onion (Katsaras & Gutberlet, 2001; Sackmann, 1995). Another recently adopted lipid-based system of nanostructures so-called solid lipid nanoparticles (SLN) are specifically engineered for lipophilic drugs, which comprise a hydrophobic and solid core with the capacity to entrap the drug by dissolving it in a hard fat matrix, usually bounded by phospholipid (monolayer) covering, which promote the colloidal stability water solutions (Khosa, Reddi, & Saha, 2018). The cellular uptake and release mechanisms are greatly influenced by the fluidity of a lipid bilayer, which is mainly controlled by both temperature and its composition (Hafez & Cullis, 2001). When temperature increases, then, an extremely arranged and crystalline state of phospholipids (i.e., bilayer) changes to a highly mobile fluid form (Kiselev & Lombardo, 2017).

Liposomal nanocarriers are one of the most acknowledged categories of nano-based drug delivery systems that represent the largest fraternity of clinically accepted anticancer formulations due to increased biocompatibility, low toxicity, ease of their size modulation, and hydrophilic/hydrophobic character (Chang & Yeh, 2012). Furthermore, liposome-based formulations are mainly dedicated to the treatment of multiple cancers, which are primarily administered through intravenous routes to avoid their gastrointestinal degradation (Hafez & Cullis, 2001). Among the most documented anticancer drugs, for instance, vincristine, doxorubicin, cisplatin, and paclitaxel remain prominent for liposome-based drug delivery systems. A number of liposomal formulations of the aforementioned agents are currently available in clinical practices against cancer character (Bozzuto & Molinari, 2015; Chang & Yeh, 2012).

Dendrimers

Dendrimers have been identified as unique, highly branched, synthetic macro-molecules containing a central cavity that further gives rise to various covalently attached branching units (Newkome, Moorefield, & Vögtle, 2001). Dendrimers could be prepared from sugar molecules, amino acids, and nucleotides as well. As mentioned earlier, dendrimers are hyperbranched, multivalent with a distinct molecular weight, and having numerous peripheral groups, collectively making them an attractive source in the field of nanomedicines. Dendrimers are prepared with the help of specific synthesis procedures that are completely different from self-assembled systems described so far, consisting of a sequence of iterative step-by-step reaction that allows for a reasonable manipulation of several parameters related to molecular design (i.e., shape, size, surface, and internal chemistry). Consequently, an extremely monodispersed system of nanostructures is achieved. Moreover, this stepwise preparation of dendrimer enables us to generate a unique and well-organized branching pattern (Basu, Sandanaraj, & Thayumanavan, 2002; Yang & Kao, 2006). Surface modification, generation, temperature, spacer length, and pH are among the various factors that can directly influence the structural features of dendrimers within solutions (Ballauff & Likos, 2004).

The active moieties of drugs could be entrapped inside the central cavities with the help of hydrogen bonds, hydrophobic bonds, or other certain chemical interactions. The drug molecules could also be joined through covalent bonds to the active groups of ending peaks. As was previously reported, dendrimers contained a distinct structure, which is more likely to allow the entrapment of numerous drugs like rifampicin that serve as first-line treatments against tuberculosis (Mignani et al., 2018). However, the disassociation of linked molecules might occur in the case of a single generation's dendrimers groot

(De Groot, Albrecht, Koekkoek, Beusker, & Scheeren, 2003). The formation of a chemical or physical bond serves as a key mechanism during the interactive phases of drugs and dendrimers. Thus, dendrimers can be successfully employed in drug delivery, magnetic resonance imaging scanning, vaccine delivery, gene delivery, and so on (Stiriba, Frey, & Haag, 2002). A well-documented application of dendrimers has been manifested in the conjugation of an appropriate chemical entity into their surface. Hence, this technique adds fuel to the development of novel prototypes, which can further play a key role in the detection of ligands and imaging agents. However, the significant role of dendrimer in drug delivery for the in vitro transformation of genetic material into cells is noteworthy (Astruc, Boisselier, & Ornelas, 2010; Yang & Kao, 2006). They have also proved their usefulness in association with prodrugs. Also, various anticancer drugs including doxorubicin, cisplatin exhibited enhanced anticancer potentials in combination therapy, using dendrimers' technology (Bhadra, Bhadra, Jain, & Jain, 2003; Lai et al., 2007; Malik, Evagorou, & Duncan, 1999; Zhuo, Du, & Lu, 1999).

Solid lipid nanocarriers

Solid lipid nanocarriers (SLN) are primarily composed of solid lipids such as stabilized waxes, complex mixtures of glycerides, or triglycerides (Kovacevic, Savic, Vuleta, Müller, & Keck, 2011). SLN have appeared as widely accepted nanostructures since the 1990s, which are frequently used for the delivery of lipophilic drugs. The most documented technique to prepare SLN is to disperse the molten solid lipids in water followed by the addition of emulsifiers with the help of high-pressure homogenization or micro-emulsification (Malam, Loizidou, & Seifalian, 2009; Müller, Mäder, & Gohla, 2000). Particularly, the lipids that remain solid at room temperatures such as free fatty acids or alcohol, mono, di, or triglycerides, and waxes or steroids are usually used for the synthesis of SLN (Üner & Yener, 2007). The drugs could be embedded into the core, shell, or matrix of the solid lipid, depending on the composition and synthesis condition. The disadvantages associated with conventional chemotherapy can be overcome by the SLN delivery systems due to their flexible nature. In fact, the conventional SLN could easily be removed by Reticular Endothelial System (RES) and may also pose complications while developing sustained drug release, encapsulating ionic and hydrophilic drug molecules. However, recently developed SLN are known to incorporate hydrophilic as well as ionic anti-tumor drugs jointly with lipophilic drugs. For instance, a lipid-polymer combined nanocarrier system has been demonstrated as an effective candidate for oral drug delivery (Hallan, Kaur, Kaur, Mishra, & Vaidya, 2016). In certain nano-based systems, some novel generations of nanocarriers like lipid drug conjugates (hydrophobic molecules of carriers), nanoscale lipid carriers (combination of solid-lipid and liquid-lipid) are known to redress the complications associated with conventional SLN. Also, such nanocarriers could be employed for drug delivery through various routes of administration like oral, topical, and parenteral. Thus, SLN belong to the group of more ideally engineered nanocarriers that enable us to deliver every kind of drug molecule to the targeted site. The potential role of SLN as a vehicle has been extensively implicated in nucleic acids as well as in gene delivery systems during the course of ophthalmic diseases (Müller et al., 2000; Yoo, Lee, Chung, Kwon, & Jeong, 2005), also, in targeted drug delivery of anticancer agents (Bondì et al., 2007; Stella et al., 2018).

Inorganic nanocarriers

This is a group of inorganic nanostructured materials with significant docile properties, including carbon nanotubes, mesoporous silica, magnetic and gold nanocarriers, quantum dots, and much more. Inorganic nanocarriers can be successfully utilized in drug targeting, imaging, cell labeling, diagnostics and, also in biosensing. They are also known to produce synergetic effects in various therapies (Santos, Bimbo, Peltonen, & Hirvonen, 2014). Moreover, changing the size or structure of inorganic nanocarriers usually allow spectacular plasmonic, optical and magnetic features. However, bearing in mind that the use of heavy metals for the construction of inorganic nanocarriers might result in long-lasting health problems (Ma et al., 2015).

Carbon nanotubes

In drug delivery systems, carbon nanotubes hold an eminent position due to their peculiar physicochemical and biological properties. Carbon nanotubes were initially identified by Sumio Iijima in 1991, which are regarded as one of the most promising and ideal sources for drug delivery. These are made up of a single or several graphene sheets, rolled together at specific angles and, displaying a hollow tube-like single-walled or multi-walled structure (Bianco, 2004; Iijima, 1991). In the context of drug delivery, carbon nanotubes brought about several novel applications to the field of nanomedicines pertaining to their specialized characteristics including unique chemical, electrical, mechanical, and thermal properties, high aspect ratio, nanosized needle structure, increased surface area, and ultralight weight (Madani, Naderi, Dissanayake, Tan, & Seifalian, 2011; Ng, Loh, Muthoosamy, Sridewi, & Manickam, 2016). The needle-penetration of carbon nanotubes executes the event of endocytosis facilely and thereby enabling them to cross cell membrane or other such barriers (Pérez-Herrero & Fernández-Medarde, 2015). The manipulated nanotubes are soluble in the water and present a prolonged circulation period by the time they enter the serum. While the unmanipulated nanotubes are insoluble in water and toxic in nature. Their surface modification, stability of the structure, and flexibility make them suitable candidates for targeting tumor cells. Therefore, the customized carbon nanotubes are extensively used to load or combine with antitumor drugs such as Paclitaxel

(Lay, Liu, Tan, & Liu, 2010), Methotrexate, Doxorubicin (Das, Datir, Singh, & Jain, 2013), and Mitomycin C for the control of various cancer types (Levi-Polyachenko, Merkel, Jones, Carroll, & Stewart IV, 2009). In addition to biomedical uses, carbon nanotubes provide an ideal source for several industrial utilities due to the versatility found in their properties.

Gold nanocarriers

Being a member of noble metals, gold has brought a medical breakthrough in the field of nanotechnology, allowing for various practical applications in the form of gold nanoparticles including photoacoustic imaging (Wang et al., 2004), photothermal therapy (Lu et al., 2010), chemotherapy (Qian et al., 2008), and gene therapy (Cao, Jin, & Mirkin, 2002). To construct gold nanoparticles, both the bottom-up and top-down approaches could be easily employed. So far, various anisotropies of gold nanoparticles such as nanoshell, nanostar, nanocage, nanoprism and nanorod have been frequently identified. The optical property of gold nanocarriers is one of the most significant properties that attract their uses in the field of nanomedicines. This allows the binding of various biomolecules like carbohydrates, proteins, enzymes, fluorophores, and genes to gold nanoparticles. Also, it promotes effective transportation inside the cell by dominating the barriers found in it. Besides, these are successfully employed in the imaging of cancer cells, which is one of the major applications of gold nanoparticles (Huang, El-Sayed, Qian, & El-Sayed, 2006; Loo, Lowery, Halas, West, & Drezek, 2005). Gold nanocarriers are also employed for single-photon emission computed tomography, computed tomography analysis, and positron emission tomography (Von Maltzahn et al., 2009).

Magnetic nanocarriers

The magnetic nanocarriers usually contain a magnetic core. Metal nanoparticles commonly possess innate magnetic properties as compared to the nanoparticles obtained from metal oxide. Hence, magnetic nanoparticles find various applications in biosensing due to their magnetic nature and their manipulated features (Berry, 2009; Koo et al., 2011). The study reports that superparamagnetic nanoparticles depict increased sensitivity to the magnetic field as compared to paramagnetic nanoparticles. The superparamagnetic iron oxide nanoparticles that are coated with polymer have been extensively chosen for molecular imaging pertaining to their magnetic resonance and serving as a contrast agent in the process (Huang, Barua, Sharma, Dey, & Rege, 2011). Interestingly, it promoted cell penetration and particle clearance. The removal of the magnetic field greatly influences paramagnetism. Using supermagnetic iron oxides, the tumor cells are targeted via passive targeting (Barry, 2008). The magnetic nanoparticles with functionalized surfaces enable them to be utilized in various implant components in the form of magnetic resonance imaging dependent sensors (Toma et al., 2005). The well-known examples of magnetic nanotubes are nanoferrites, hematite, magnetite, and maghemite. While, the distinctive features are known to enable it to be used for drug targeting in nanomedicines, hyperthermia mediators, gene therapy (Kim, Kim, Kim, & Lee, 2009; Tang, Zhang, Cong, Wan, & Jin, 2008; van Landeghem et al., 2009), and as contrast agents (Babincova, Altanerova, Altaner, Čičmanec, & Babinec, 2004). In this regard, an attempt has been made to attach Epirubicin drug with ferrofluid and the resulted accumulation of drug at the target site was quite significant (Lübbe, Bergemann, Brock, & McClure, 1999). Magnetic nanocarriers are also involved in antisense and gene therapeutic by the way of magnetofection (Mykhaylyk, Antequera, Vlaskou, & Plank, 2007). Trojan Horse, with size manipulable capacity in combination with paclitaxel, has been appeared with increased penetration into cancer cells thereby, leading to controlled drug release at the target site with greater cytotoxicity (Lai, Chiang, Kao, & Chen, 2018). However, to successfully obtain the desired therapeutic benefits of magnetic nanocarriers is always been a challenging task and a number of difficulties appear in the way to achieve all the specified objectives. The basic problem always found with magnetic nanosystems is the lack of an appropriate magnetic system. For instance, magnetic nanoparticles are more likely to tend to cumulate into clusters of larger dimensions thereby losing the specific properties linked with their smaller dimensions. As a result, the magnetic force may not be able enough to dominate the force built-in by blood flow and to keep the magnetic drugs limited to specified target site (Neuberger, Schöpf, Hofmann, Hofmann, & Von Rechenberg, 2005).

Mesoporous silica

The mesoporous silica exhibits a large spongy honeycomb architecture that usually allows the incorporation of several drug molecules into its construction. It finds many ways of its applications to drug delivery systems due to large-scale availability and simplicity of structural features. It has the potentiality to entrap both hydrophilic and hydrophobic drugs which can be connected to a ligand molecule for a controlled drug delivery system (Li et al., 2017). Some of the most desirable properties of mesoporous silica include higher drug encapsulating capacity, biocompatibility, thermochemical stability, and a greater pore volume with increased surface area. During the course of cancer treatment, both aspects of active and passive targeting could be successfully achieved by employing mesoporous silica (Wang et al., 2015). For instance, using mesoporous silica, several anticancer drugs like methotrexate and camptothecin have been successfully delivered so far (Lebold, Jung, Michaelis, & Bräuchle, 2009; Rosenholm et al., 2010).

Quantum dots

This group of nanocarriers is comprised of atoms (i.e., Zn, Te, In, and P) belonging to II–VI or III–V element groups of elements in the periodic. They are generally categorized as energy donors and colloidal nanocrystals (Alivisatos, 1996). The variation generated by the size of a quantum dot in the emission of light occurs between UV-near IR regions, indicating that the larger quantum dots (~5 nm) produce red fluorescence whereas smaller quantum dots (~2 nm) produce blue fluorescence (Bruchez, Moronne, Gin, Weiss, & Alivisatos, 1998). Certain properties of quantum dots such as optical features, elongated light emission, and lower photobleaching make them superior as compared to other organic dyes. These properties enable quantum dots to be considered for application in cell imaging procedures. For instance, the conjugate of quantum dot and peptide is used in mice to target cancer vasculature in vivo (Akerman et al., 2002). The toxicity of cadmium in the CdSe quantum dot is generally masked by encapsulating it inside the ZnS shell to safeguard against toxicity. In turn, this increased the accumulation of nanoparticles in the targeted vascular site. The study reports that these quantum dots have proved their efficiency both in drug delivery and reporter system. For example, quantum dots surface modified with local tumor peptide successfully attached to the nucleolin of tumor cells and improved cellular uptake (Christian et al., 2003). Also, certain RNA-associated quantum dots have been found to promote the gene knockdown (Derfus, Chen, Min, Ruoslahti, & Bhatia, 2007).

Absorption mechanisms of nanocarriers-based drugs

The gastrointestinal tract (GIT) plays a vital role in drug disposition, particularly, small intestine is the major portion of GIT, where the absorption of nutrients and various nanocarriers-based drugs frequently take place. In fact, the active and complex structure of the intestine's wall allows to regulate the absorption of various drugs, interactions found between local microflora and, immunes system as well, thereby, preserving the intestinal equilibrium (Davitt & Lavelle, 2015). It is necessary for a nanocarrier-associated drug to cross the thick mucosal barrier via diffusion before entering the endothelial cells. The mucus layer is generally synthesized by goblet cells and the principal components of this layer are glycoproteins and lipids (Boegh & Nielsen, 2015). The passage of drugs across the mucosal layer depends on various factors including molecular size, charge, and viscosity. The study reports that whenever mucosal layer encounters large molecules or mucoadhesive hydrophobic substances this might result in their reduced permeability. Similarly, the translocation of nanocarriers-based drugs via gut endothelia could also be arrested, however, lipophilic components can easily cross the mucosal barrier when combined with phospholipids, bile salts, or free fatty acids (Gleeson, Ryan, & Brayden, 2016). Upon encapsulation in suitable nanocarriers (e.g., polyethylene glycol (PEG), chitosan, gelatin, and lectin), drugs may pass through or adhere to the mucosal layer, thereby allowing the uptake of drugs (Mansuri,

Kesharwani, Jain, Tekade, & Jain, 2016). Also, the mucus layer repels the negatively charged nanocarriers and leading to reduced cellular uptake due to their short residence time within epithelial cells. On the other hand, when drugs such as estradiol were entrapped inside the positively charged nanoparticles such as poly(lactic-co-glycolic acid) (PLGA), their intracellular uptake was higher as compared to negatively charged or neutral nanoparticles (Hariharan et al., 2006). Once the nanocarriers overcome this mucosal barrier then they could also pass through epithelial layer either through transcellular (including M-cell facilitated transport), or paracellular (via tight junctions) transportation pathways.

Paracellular route

The inter-cellular spaces of epithelial cells serve as key pathways in the passive transport of materials (e.g., drugs) through the process of passive diffusion within intestine (Daugherty & Mrsny, 1999). The inter-cellular junctions are frequently occupied by epithelial tight junctions (TJs) that usually display a complex structure due to the presence of various proteins (e.g., claudin and occluding) in their composition (Lerner & Matthias, 2015). These TJs are potentially involved in the regulation of paracellular transport, inter-cellular adhesion, permeability of intestine, the movement of molecules between lumen and lamina propria and restrict the propagation of microbes toward host cells and tissues (Bischoff et al., 2014). Small water-soluble and polar molecules like amino acids, peptides, water, sugars, and ions (i.e., molecular weight < 50 Da) can easily pass through such TJs (Maher, Mrsny, & Brayden, 2016). Some particular agents like polyphenols enhance the function of TJs, whereas, others like caprylic acids could also reduce its function and thereby promoting the uptake of small molecules (Bohn et al., 2015). The paracellular transportation is considerably responsible for the suppression of intracellular metabolism, which is extremely necessary for drugs. Also, the study reports that certain polyphenols including gallic acid, quercetin, caffeic acid, resveratrol, rutin, and chrysin are weakly transported via passive diffusion both in artificial membrane permeability and Caco-2 cells monolayer assays (Rastogi & Jana, 2016). Furthermore, TJs can only expand to a maximum of 20 nm, limiting nanocarrier transport across the intestinal epithelium via the paracellular route. As a result, most nanocarrier-based delivery techniques are hampered by this gap (Yu, Trase, et al., 2016; Yu, Yang, et al., 2016).

Nano-systems having a size of less than 20 nm can damage TJs and expose the payload to systemic circulation. Later on, the TJs return to their original function. Furthermore, because the negatively charged membrane surface attracts positively charged particles, they can easily be moved by paracellular transport. Nano-systems containing cationic chitosan can open TJs, allowing paracellular transport to occur. Encapsulating drugs in chitosan and poly(glycolic acid) (PGA)-based nanoparticles, for example, improves their administration and transport through the intestinal barrier via paracellular transport (Tang et al., 2013).

Transcellular route

The passive or active transport of substances across cells via endocytosis is required for transcellular absorption. The majority of drugs are thought to be absorbed via transcellular pathway without the involvement of a receptor or carrier (Renukuntla, Vadlapudi, Patel, Boddu, & Mitra, 2013). This procedure has been postulated for carotenoids and non-polar polyphenol aglycones (Bohn et al., 2015). For example, curcumin passage through passive diffusion in Caco-2 cells when incorporated in SLN has been previously reported (Guri, Gülseren, & Corredig, 2013). On the other hand, some polar and charged biomolecules bind to a particular membrane transporter (carrier facilitated transport) usually found in the apical cell membrane. Then, these molecules are moved against the concentration gradient (i.e., active transport) (Yun, Cho, & Park, 2013), otherwise, they may not be able to pass the cell membrane (Li, 2011). Thus, such carrier proteins and receptors are extremely important for the uptake of numerous drugs. For instance, vitamin C, certain peptides, and fatty acids are transported with the help of sodium vitamin C co-transporter, proton-coupled peptide transporters, and fatty acid-binding proteins, respectively (Gleeson et al., 2016). Furthermore, various nanocarriers have the ability to stimulate or inhibit membrane transporters, and the stimulation or inhibitory potential varies depending on the nanocarrier concentration, type, exposure period, and other factors (Bohn et al., 2015).

Certain molecules that usually bind to specialized carriers' systems at cell membrane receptors or apical cell membranes are transferred to the cell by the process of endocytosis (e.g., phagocytosis or pinocytosis) (Huwyler, Kettiger, Schipanski, & Wick, 2013). To get internalized into the M-cells of Peyer's patch (specified in antigen sampling) primarily depends on phagocytosis (M-cells mediated transport), suggesting a possible avenue for nanocarrier-based delivery systems (Des Rieux et al., 2007; Yun et al., 2013). When M-cell expression is less than 1% of the total intestinal area, then, transportation through these cells becomes extremely challenging (Acosta, 2009). Pinocytosis is a mechanism through which the nanocarriers could also attach to the corresponding cell surface receptors (e.g., $\alpha 5 \beta 1$ integrin, lectins, and lactoferrin) and thereby get internalized (Plapied, Duhem, des Rieux, & Préat, 2011). The study reports that certain nanocarriers constructed for various drugs use unique ligands on their surfaces for a particular receptor to get improved intracellular delivery both in enterocytes and M-cells (Plapied et al., 2011). Because of the presence of lactoferrin receptors, gambogic acid loaded in lactoferrin-based nanoparticles led to its improved transport through cell membrane (Zhang et al., 2013). It is worth noting that some nutraceuticals are discharged back into the GIT lumen after being absorbed by efflux pumps (efflux transporters) located in cell membrane lipid bilayers, thus limiting their bioavailability (Misaka, Müller, & Fromm, 2013). For the effective development of nanocarrier-based drugs delivery systems, knowledge and understanding of multiple transport processes across the GIT are essential.

Release mechanisms of nutraceuticals from nanocarrier

The development of regulated and customized nano-based delivery systems necessitates a thorough understanding of release mechanisms. Understanding release mechanisms allow one to predict strategies to improve the safety of the payload in the nanocarrier, as well as their absorption and release from the system based on good understanding of release processes (Wise, 2000). Depending on the nature of the loaded molecules, carrier composition, particle geometry, amount of encapsulated material, and release media, the entrapped materials can be released from the carrier in a variety of ways. Hence, the release of an encapsulated drug from the carrier can be attributed to one of the four basic methods listed here.

Diffusion—The drug molecules just diffuse out into the surrounding media from the intact non-biodegradable biopolymers. Diffusion can occur through a homogenous matrix, water-filled pores, or an exterior shell connected to an interior reservoir. The solubility of drugs in the matrix, the size and geometry of carrier along with its diffusion coefficient through the matrix, all influence the overall rate of mass transfer. Several environmental and particle characteristics, such as temperature (temperature $\propto$ diffusion coefficient), porosity (porosity $\propto$ diffusion coefficient), and tortuosity (tortuosity $1/\propto$ diffusion coefficient), consequently, influence the diffusion coefficient.

Erosion—Erosion is another method for releasing payload from nanocarriers. The entrapped drug is released into the media either by homogeneous (happens in the nanocarrier's bulk volume) or heterogeneous erosion (that usually happens on the nanocarrier surface). This event of erosion can be triggered by enzymes and/or chemicals processes. Bulk erosion is a phenomenon in which the external fluid enters the nanocarrier by breaking physical or chemical bonds while the size of nanocarriers almost remains constant. Surface erosion, on the other hand, is a phenomenon in which the size of a nanocarrier (usually a biopolymer) is slowly decreased by erosion at the outer surface (Zhang, Yang, Chow, & Wang, 2003). The rate of erosion is affected by several factors, including release medium, size (erosion $1/\propto$ size), polymer molecular weight (erosion 1/∝ molecular weight) as well as physicochemical stability (Chirico, Dalmoro, Lamberti, Russo, & Titomanlio, 2007).

Swelling-shrinkage mechanism—When the drugs or other payload materials' dimensions (like size) are greater than the pore size of the nanocarriers, they become imprisoned within the nanosystem. The nano-system conditions are then altered by various triggers (such as pH, temperature, and water activity), causing the nanosystem to swell, resulting in an increase in the pore size of the nanocarriers, thereby, leading the entrapped payload to be released. On the contrary, in a shrinkage-induced release procedure, the payload is captured in the nanosystem during its swelling and subsequently released during shrinkage by changing the solution conditions (Arifin, Lee, & Wang, 2006).

Fragmentation—In this situation, the entrapped drug materials are released into the medium by physically disrupting the nanocarrier. The nanocarriers may be physically disrupted by fracturing or fragmentation through compression or shear forces in the mouth and GIT environment, or during processing (Bealer et al., 2020).

It's worth noting that diffusion is always potentially linked with each of the mechanisms listed earlier. Mathematical modeling is also critical for the design of effective delivery systems with optimum loading efficiency and release profiles (Fathi, Martín, & McClements, 2014). For the selection of an acceptable release model, the basic understanding of the drug release mechanism is essential. Various aspects are considered for a complete knowledge of the release mechanisms, including solubility of loaded materials in the medium and nano matrix, porosity, concentration, nanocarrier size, pore size distribution, and effective diffusion coefficients. Regarding drug delivery systems, such parameters are frequently obtained; however, in recent years, relatively limited data for nutraceutical delivery methods have been published.

Conclusion

Several issues such as drug resistance, poor water solubility, drug toxicity, poor selectivity, and limited bioavailability have been associated with traditional drug delivery systems, thus reducing their therapeutic potentiality. Therefore, nanocarriers have got wider scientific attention for improving drugs' therapeutic efficacy due to their optimized biological and physicochemical properties and a wide range of practical applications in the field of medicines. Currently, diverse forms of nanocarrier systems have been developed ranging from micelles and inorganic nanoparticles to self-assembling supramolecular. These nanocarriers offer unique surface functionalities that can be modified with ligands for delivering drugs to the target sites, this achieving maximum therapeutic efficacy for the loaded drugs with minimum off-target side effects.

References

Acosta, E. (2009). Bioavailability of nanoparticles in nutrient and nutraceutical delivery. *Current Opinion in Colloid and Interface Science*, *14*(1), 3–15. https://doi.org/10.1016/j.cocis.2008.01.002.

Akerman, M. E., Chan, W. C. W., Laakkonen, P., Bhatia, S. N., & Ruoslahti, E. (2002). Nanocrystal targeting in vivo. *Proceedings of the National Academy of Sciences, 99*(20), 12617–12621. https://doi.org/10.1073/pnas.152463399.

Alivisatos, A. P. (1996). Semiconductor clusters, nanocrystals, and quantum dots. *Science, 271*(5251), 933–937. https://doi.org/10.1126/science.271.5251.933.

Allen, T. M., & Cullis, P. R. (2004). Drug delivery systems: Entering the mainstream. *Science, 303*(5665), 1818–1822. https://doi.org/10.1126/science.1095833.

Allen, T. M., & Cullis, P. R. (2013). Liposomal drug delivery systems: From concept to clinical applications. *Advanced Drug Delivery Reviews, 65*(1), 36–48. https://doi.org/10.1016/j.addr.2012.09.037.

Arifin, D. Y., Lee, L. Y., & Wang, C. H. (2006). Mathematical modeling and simulation of drug release from microspheres: Implications to drug delivery systems. *Advanced Drug Delivery Reviews, 58*(12–13), 1274–1325. https://doi.org/10.1016/j.addr.2006.09.007.

Aslan, B., Ozpolat, B., Sood, A. K., & Lopez-Berestein, G. (2013). Nanotechnology in cancer therapy. *Journal of Drug Targeting, 21*(10), 904–913. https://doi.org/10.3109/1061186X.2013.837469.

Astruc, D., Boisselier, E., & Ornelas, C. (2010). Dendrimers designed for functions: From physical, photophysical, and supramolecular properties to applications in sensing, catalysis, molecular electronics, photonics, and nanomedicine. *Chemical Reviews, 110*(4), 1857–1959. https://doi.org/10.1021/cr900327d.

Babincova, M., Altanerova, V., Altaner, Č., Čičmanec, P., & Babinec, P. (2004). In vivo heating of magnetic nanoparticles in alternating magnetic field. *Medical Physics, 8*, 2219–2221.

Ballauff, M., & Likos, C. N. (2004). Dendrimers in solution: Insight from theory and simulation. *Angewandte Chemie, International Edition, 43*(23), 2998–3020. https://doi.org/10.1002/anie.200300602.

Barry, S. E. (2008). Challenges in the development of magnetic particles for therapeutic applications. *International Journal of Hyperthermia, 24*(6), 451–466. https://doi.org/10.1080/02656730802093679.

Basu, S., Sandanaraj, B. S., & Thayumanavan, S. (2002). Molecular recognition in dendrimers. In F. M. Herman (Ed.), *Encyclopedia of polymer science and technology* (pp. 385–424). New York: Wiley.

Bayda, S., Adeel, M., Tuccinardi, T., Cordani, M., & Rizzolio, F. (2020). The history of nanoscience and nanotechnology: From chemical–physical applications to nanomedicine. *Molecules, 25*(1).

Bealer, E. J., Onissema-Karimu, S., Rivera-Galletti, A., Francis, M., Wilkowski, J., de la Cruz, D. S., et al. (2020). Protein-polysaccharide composite materials: Fabrication and applications. *Polymers, 12*(2). https://doi.org/10.3390/polym12020464.

Berry, C. C. (2009). Progress in functionalization of magnetic nanoparticles for applications in biomedicine. *Journal of Physics D: Applied Physics, 42*(22). https://doi.org/10.1088/0022-3727/42/22/224003.

Bhadra, D., Bhadra, S., Jain, S., & Jain, N. K. (2003). A PEGylated dendritic nanoparticulate carrier of fluorouracil. *International Journal of Pharmaceutics, 257*(1–2), 111–124. https://doi.org/10.1016/S0378-5173(03)00132-7.

Bianco, A. (2004). Carbon nanotubes for the delivery of therapeutic molecules. *Expert Opinion on Drug Delivery, 1*(1), 57–65. https://doi.org/10.1517/17425247.1.1.57.

Bischoff, S. C., Barbara, G., Buurman, W., Ockhuizen, T., Schulzke, J.-D., Serino, M., et al. (2014). Intestinal permeability—A new target for disease prevention and therapy. *BMC Gastroenterology, 14*(1).

Boegh, M., & Nielsen, H. M. (2015). Mucus as a barrier to drug delivery—Understanding and mimicking the barrier properties. *Basic & Clinical Pharmacology & Toxicology, 116*(3), 179–186. https://doi.org/10.1111/bcpt.12342.

Bohn, T., Mcdougall, G. J., Alegría, A., Alminger, M., Arrigoni, E., Aura, A. M., et al. (2015). Mind the gap-deficits in our knowledge of aspects impacting the bioavailability of phytochemicals and their metabolites-a position paper focusing on carotenoids and polyphenols. *Molecular Nutrition & Food Research, 59*(7), 1307–1323. https://doi.org/10.1002/mnfr.201400745.

Bondì, M. L., Craparo, E. F., Giammona, G., Cervello, M., Azzolina, A., Diana, P., et al. (2007). Nanostructured lipid carriers-containing anticancer compounds: Preparation, characterization, and cytotoxicity studies. *Drug Delivery, 14*(2), 61–67. https://doi.org/10.1080/10717540600739914.

Bozzuto, G., & Molinari, A. (2015). Liposomes as nanomedical devices. *International Journal of Nanomedicine, 10*, 975. https://doi.org/10.2147/ijn.s68861.

Bruchez, M., Moronne, M., Gin, P., Weiss, S., & Alivisatos, A. P. (1998). Semiconductor nanocrystals as fluorescent biological labels. *Science, 281*(5385), 2013–2016.

Cao, Y. W. C., Jin, R., & Mirkin, C. A. (2002). Nanoparticles with Raman spectroscopic fingerprints for DNA and RNA detection. *Science, 297*(5586), 1536–1540. https://doi.org/10.1126/science.297.5586.1536.

Chang, H. I., & Yeh, M. K. (2012). Clinical development of liposome-based drugs: Formulation, characterization, and therapeutic efficacy. *International Journal of Nanomedicine, 7*, 49–60. https://www.dovepress.com/getfile.php?fileID=11737.

Chirico, S., Dalmoro, A., Lamberti, G., Russo, G., & Titomanlio, G. (2007). Analysis and modeling of swelling and erosion behavior for pure HPMC tablet. *Journal of Controlled Release, 122*(2), 181–188. https://doi.org/10.1016/j.jconrel.2007.07.001.

Christian, S., Pilch, J., Akerman, M. E., Porkka, K., Laakkonen, P., & Ruoslahti, E. (2003). Nucleolin expressed at the cell surface is a marker of endothelial cells in angiogenic blood vessels. *Journal of Cell Biology, 163*(4), 871–878. https://doi.org/10.1083/jcb.200304132.

Das, M., Datir, S. R., Singh, R. P., & Jain, S. (2013). Augmented anticancer activity of a targeted, intracellularly activatable, theranostic nanomedicine based on fluorescent and radiolabeled, methotrexate-folic acid-multiwalled carbon nanotube conjugate. *Molecular Pharmaceutics, 10*(7), 2543–2557. https://doi.org/10.1021/mp300701e.

Daugherty, A. L., & Mrsny, R. J. (1999). Transcellular uptake mechanisms of the intestinal epithelial barrier—Part one. *Pharmaceutical Science & Technology Today, 2*(4), 144–151. https://doi.org/10.1016/S1461-5347(99)00142-X.

Davitt, C. J. H., & Lavelle, E. C. (2015). Delivery strategies to enhance oral vaccination against enteric infections. *Advanced Drug Delivery Reviews, 91*, 52–69. https://doi.org/10.1016/j.addr.2015.03.007.

De Groot, F. M. H., Albrecht, C., Koekkoek, R., Beusker, P. H., & Scheeren, H. W. (2003). "Cascade-release dendrimers" liberate all end groups upon a single triggering event in the dendritic core. *Angewandte Chemie, International Edition, 42*(37), 4490–4494. https://doi.org/10.1002/anie.200351942.

Derfus, A. M., Chen, A. A., Min, D. H., Ruoslahti, E., & Bhatia, S. N. (2007). Targeted quantum dot conjugates for siRNA delivery. *Bioconjugate Chemistry, 18*(5), 1391–1396. https://doi.org/10.1021/bc060367e.

Des Rieux, A., Fievez, V., Théate, I., Mast, J., Préat, V., & Schneider, Y. J. (2007). An improved in vitro model of human intestinal follicle-associated epithelium to study nanoparticle transport by M cells. *European Journal of Pharmaceutical Sciences, 30*(5), 380–391. https://doi.org/10.1016/j.ejps.2006.12.006.

Elsabahy, M., & Wooley, K. L. (2012). Design of polymeric nanoparticles for biomedical delivery applications. *Chemical Society Reviews, 41*(7), 2545–2561. https://doi.org/10.1039/c2cs15327k.

Fathi, M., Martín, Á., & McClements, D. J. (2014). Nanoencapsulation of food ingredients using carbohydrate based delivery systems. *Trends in Food Science and Technology, 39*(1), 18–39. https://doi.org/10.1016/j.tifs.2014.06.007.

Fattal, E., Hillaireau, H., Mura, S., Nicolas, J., & Tsapis, N. (2012). Targeted delivery using biodegradable polymeric nanoparticles. In *Fundamentals and applications of controlled release drug delivery* (pp. 255–288). Springer US. https://doi.org/10.1007/978-1-4614-0881-9_10.

Gleeson, J. P., Ryan, S. M., & Brayden, D. J. (2016). Oral delivery strategies for nutraceuticals: Delivery vehicles and absorption enhancers. *Trends in Food Science and Technology, 53*, 90–101. https://doi.org/10.1016/j.tifs.2016.05.007.

Guri, A., Gülseren, I., & Corredig, M. (2013). Utilization of solid lipid nanoparticles for enhanced delivery of curcumin in cocultures of HT29-MTX and Caco-2 cells. *Food & Function, 4*(9), 1410–1419. https://doi.org/10.1039/c3fo60180c.

Hafez, I. M., & Cullis, P. R. (2001). Roles of lipid polymorphism in intracellular delivery. *Advanced Drug Delivery Reviews, 47*(2–3), 139–148. https://doi.org/10.1016/S0169-409X(01)00103-X.

Hallan, S. S., Kaur, P., Kaur, V., Mishra, N., & Vaidya, B. (2016). Lipid polymer hybrid as emerging tool in nanocarriers for oral drug delivery. *Artificial Cells, Nanomedicine, and Biotechnology, 44*(1), 334–349. https://doi.org/10.3109/21691401.2014.951721.

Hariharan, S., Bhardwaj, V., Bala, I., Sitterberg, J., Bakowsky, U., & Ravi Kumar, M. N. V. (2006). Design of estradiol loaded PLGA nanoparticulate formulations: A potential oral delivery system for hormone therapy. *Pharmaceutical Research, 23*(1), 184–195. https://doi.org/10.1007/s11095-005-8418-y.

Huang, H. C., Barua, S., Sharma, G., Dey, S. K., & Rege, K. (2011). Inorganic nanoparticles for cancer imaging and therapy. *Journal of Controlled Release, 155*(3), 344–357. https://doi.org/10.1016/j.jconrel.2011.06.004.

Huang, X., El-Sayed, I. H., Qian, W., & El-Sayed, M. A. (2006). Cancer cell imaging and photothermal therapy in the near-infrared region by using gold nanorods. *Journal of the American Chemical Society, 128*(6), 2115–2120. https://doi.org/10.1021/ja057254a.

Huwyler, J., Kettiger, H., Schipanski, A., & Wick, P. (2013). Engineered nanomaterial uptake and tissue distribution: From cell to organism. *International Journal of Nanomedicine, 8*, 3255. https://doi.org/10.2147/ijn.s49770.

Iijima, S. (1991). Helical microtubules of graphitic carbon. *Nature, 354*(6348), 56–58. https://doi.org/10.1038/354056a0.

Kahlweit. (1985). *Physics of amphiphiles, micelles, vesicles and microemulsions*. Amsterdam: North-Holland.

Katsaras, J., & Gutberlet, T. (2001). *Lipid bilayers: Structure and interactions*. Berlin: Springer-Verlag.

Khosa, A., Reddi, S., & Saha, R. N. (2018). Nanostructured lipid carriers for site-specific drug delivery. *Biomedicine and Pharmacotherapy, 103*, 598–613. https://doi.org/10.1016/j.biopha.2018.04.055.

Kim, D. H., Kim, K. N., Kim, K. M., & Lee, Y. K. (2009). Targeting to carcinoma cells with chitosan- and starch-coated magnetic nanoparticles for magnetic hyperthermia. *Journal of Biomedical Materials Research. Part A, 88*(1), 1–11. https://doi.org/10.1002/jbm.a.31775.

Kiselev, M. A., & Lombardo, D. (2017). Structural characterization in mixed lipid membrane systems by neutron and X-ray scattering. *Biochimica et Biophysica Acta - General Subjects, 1861*(1), 3700–3717. https://doi.org/10.1016/j.bbagen.2016.04.022.

Koo, H., Huh, M. S., Sun, I. C., Yuk, S. H., Choi, K., Kim, K., et al. (2011). In vivo targeted delivery of nanoparticles for theranosis. *Accounts of Chemical Research, 44*(10), 1018–1028. https://doi.org/10.1021/ar2000138.

Kovacevic, A., Savic, S., Vuleta, G., Müller, R. H., & Keck, C. M. (2011). Polyhydroxy surfactants for the formulation of lipid nanoparticles (SLN and NLC): Effects on size, physical stability and particle matrix structure. *International Journal of Pharmaceutics, 406*(1–2), 163–172. https://doi. org/10.1016/j.ijpharm.2010.12.036.

Lai, Y. H., Chiang, C. S., Kao, T. H., & Chen, S. Y. (2018). Dual-drug nanomedicine with hydrophilic F127-Modified magnetic nanocarriers assembled in amphiphilic gelatin for enhanced penetration and drug delivery in deep tumor tissue. *International Journal of Nanomedicine, 13*, 3011–3026. https://doi.org/10.2147/IJN.S161314.

Lai, P. S., Lou, P. J., Peng, C. L., Pai, C. L., Yen, W. N., Huang, M. Y., et al. (2007). Doxorubicin delivery by polyamidoamine dendrimer conjugation and photochemical internalization for cancer therapy. *Journal of Controlled Release, 122*(1), 39–46. https://doi.org/10.1016/j. jconrel.2007.06.012.

Lay, C. L., Liu, H. Q., Tan, H. R., & Liu, Y. (2010). Delivery of paclitaxel by physically loading onto poly(ethylene glycol) (PEG)-graftcarbon nanotubes for potent cancer therapeutics. *Nanotechnology, 21*(6). https://doi.org/10.1088/0957-4484/21/6/065101.

Lebold, T., Jung, C., Michaelis, J., & Bräuchle, C. (2009). Nanostructured silica materials as drug-delivery systems for doxorubicin: Single molecule and cellular studies. *Nano Letters, 9*(8), 2877–2883. https://doi.org/10.1021/nl9011112.

Lerner, A., & Matthias, T. (2015). Changes in intestinal tight junction permeability associated with industrial food additives explain the rising incidence of autoimmune disease. *Autoimmunity Reviews, 14*(6), 479–489. https://doi.org/10.1016/j.autrev.2015.01.009.

Levi-Polyachenko, N. H., Merkel, E. J., Jones, B. T., Carroll, D. L., & Stewart, J. H., IV. (2009). Rapid photothermal intracellular drug delivery using multiwalled carbon nanotubes. *Molecular Pharmaceutics, 6*(4), 1092–1099. https://doi.org/10.1021/mp800250e.

Li, X. (2011). *Oral bioavailability: Basic principles, advanced concepts, and applications. Vol. 16.* New York: Wiley.

Li, Y., Li, N., Pan, W., Yu, Z., Yang, L., & Tang, B. (2017). Hollow mesoporous silica nanoparticles with tunable structures for controlled drug delivery. *ACS Applied Materials and Interfaces, 9*(3), 2123–2129. https://doi.org/10.1021/acsami.6b13876.

Lombardo, D., Kiselev, M. A., & Caccamo, M. T. (2019). Smart nanoparticles for drug delivery application: Development of versatile nanocarrier platforms in biotechnology and nanomedicine. *Journal of Nanomaterials, 2019.* https://doi.org/10.1155/2019/3702518.

Lombardo, D., Kiselev, M. A., Magazù, S., & Calandra, P. (2015). Amphiphiles self-assembly: Basic concepts and future perspectives of supramolecular approaches. *Advances in Condensed Matter Physics, 2015.* https://doi.org/10.1155/2015/151683.

Loo, C., Lowery, A., Halas, N., West, J., & Drezek, R. (2005). Immunotargeted nanoshells for integrated cancer imaging and therapy. *Nano Letters, 5*(4), 709–711. https://doi.org/10.1021/ nl050127s.

Lu, W., Huang, Q., Ku, G., Wen, X., Zhou, M., Guzatov, D., et al. (2010). Photoacoustic imaging of living mouse brain vasculature using hollow gold nanospheres. *Biomaterials, 31*(9), 2617–2626. https://doi.org/10.1016/j.biomaterials.2009.12.007.

Lübbe, A. S., Bergemann, C., Brock, J., & McClure, D. G. (1999). Physiological aspects in magnetic drug-targeting. *Journal of Magnetism and Magnetic Materials, 194*(1), 149–155. https://doi.org/ 10.1016/S0304-8853(98)00574-5.

Ma, P., Xiao, H., Li, C., Dai, Y., Cheng, Z., Hou, Z., et al. (2015). Inorganic nanocarriers for platinum drug delivery. *Materials Today, 18*(10), 554–564. https://doi.org/10.1016/j. mattod.2015.05.017.

Ma, X., & Zhao, Y. (2015). Biomedical applications of supramolecular systems based on host-guest interactions. *Chemical Reviews, 115*(15), 7794–7839. https://doi.org/10.1021/cr500392w.

Madani, S. Y., Naderi, N., Dissanayake, O., Tan, A., & Seifalian, A. M. (2011). A new era of cancer treatment: Carbon nanotubes as drug delivery tools. *International Journal of Nanomedicine, 6.*

Maher, S., Mrsny, R. J., & Brayden, D. J. (2016). Intestinal permeation enhancers for oral peptide delivery. *Advanced Drug Delivery Reviews, 106,* 277–319. https://doi.org/10.1016/j.addr.2016.06.005.

Malam, Y., Loizidou, M., & Seifalian, A. M. (2009). Liposomes and nanoparticles: Nanosized vehicles for drug delivery in cancer. *Trends in Pharmacological Sciences, 30*(11), 592–599. https://doi.org/10.1016/j.tips.2009.08.004.

Malik, N., Evagorou, E. G., & Duncan, R. (1999). Dendrimer-platinate: A novel approach to cancer chemotherapy. *Anti-Cancer Drugs, 10*(8), 767–776. https://doi.org/10.1097/00001813-199909000-00010.

Mansoori, G. A., & Soelaiman, T. F. (2005). Nanotechnology—An introduction for the standards community. *Journal of ASTM International, 2*(6), 1–22.

Mansuri, S., Kesharwani, P., Jain, K., Tekade, R. K., & Jain, N. K. (2016). Mucoadhesion: A promising approach in drug delivery system. *Reactive and Functional Polymers, 100,* 151–172. https://doi.org/10.1016/j.reactfunctpolym.2016.01.011.

Masood, F. (2016). Polymeric nanoparticles for targeted drug delivery system for cancer therapy. *Materials Science and Engineering C, 60,* 569–578. https://doi.org/10.1016/j.msec.2015.11.067.

Mignani, S., Tripathi, R. P., Chen, L., Caminade, A.-M., Shi, X., & Majoral, J.-P. (2018). New ways to treat tuberculosis using dendrimers as nanocarriers. *Pharmaceutics, 10*(3).

Mikhail, A. S., & Allen, C. (2009). Block copolymer micelles for delivery of cancer therapy: Transport at the whole body, tissue and cellular levels. *Journal of Controlled Release, 138*(3), 214–223. https://doi.org/10.1016/j.jconrel.2009.04.010.

Misaka, S., Müller, F., & Fromm, M. F. (2013). Clinical relevance of drug efflux pumps in the gut. *Current Opinion in Pharmacology, 13*(6), 847–852. https://doi.org/10.1016/j.coph.2013.08.010.

Müller, R. H., Mäder, K., & Gohla, S. (2000). Solid lipid nanoparticles (SLN) for controlled drug delivery—A review of the state of the art. *European Journal of Pharmaceutics and Biopharmaceutics, 50*(1), 161–177. https://doi.org/10.1016/S0939-6411(00)00087-4.

Mykhaylyk, O., Antequera, Y. S., Vlaskou, D., & Plank, C. (2007). Generation of magnetic nonviral gene transfer agents and magnetofection in vitro. *Nature Protocols, 2*(10), 2391–2411. https://doi.org/10.1038/nprot.2007.352.

Neuberger, T., Schöpf, B., Hofmann, H., Hofmann, M., & Von Rechenberg, B. (2005). Superparamagnetic nanoparticles for biomedical applications: Possibilities and limitations of a new drug delivery system. *Journal of Magnetism and Magnetic Materials, 293*(1), 483–496. https://doi.org/10.1016/j.jmmm.2005.01.064.

Newkome, G. R., Moorefield, C. N., & Vögtle, F. (2001). *Dendrimers and dendrons: Concepts, syntheses, applications.* Weinheim/Chichester: Wiley-VCH.

Ng, C. M., Loh, H. S., Muthoosamy, K., Sridewi, N., & Manickam, S. (2016). Conjugation of insulin onto the sidewalls of single-walled carbon nanotubes through functionalization and diimide-activated amidation. *International Journal of Nanomedicine, 11,* 1607–1614. https://doi.org/10.2147/IJN.S98726.

Pérez-Herrero, E., & Fernández-Medarde, A. (2015). Advanced targeted therapies in cancer: Drug nanocarriers, the future of chemotherapy. *European Journal of Pharmaceutics and Biopharmaceutics, 93,* 52–79. https://doi.org/10.1016/j.ejpb.2015.03.018.

Plapied, L., Duhem, N., des Rieux, A., & Préat, V. (2011). Fate of polymeric nanocarriers for oral drug delivery. *Current Opinion in Colloid and Interface Science, 16*(3), 228–237. https://doi.org/10.1016/j.cocis.2010.12.005.

Qian, X., Peng, X. H., Ansari, D. O., Yin-Goen, Q., Chen, G. Z., Shin, D. M., et al. (2008). In vivo tumor targeting and spectroscopic detection with surface-enhanced Raman nanoparticle tags. *Nature Biotechnology, 26*(1), 83–90. https://doi.org/10.1038/nbt1377.

Rastogi, H., & Jana, S. (2016). Evaluation of physicochemical properties and intestinal permeability of six dietary polyphenols in human intestinal colon adenocarcinoma Caco-2 cells. *European Journal of Drug Metabolism and Pharmacokinetics, 41*(1), 33–43. https://doi.org/10.1007/s13318-014-0234-5.

Renukuntla, J., Vadlapudi, A. D., Patel, A., Boddu, S. H. S., & Mitra, A. K. (2013). Approaches for enhancing oral bioavailability of peptides and proteins. *International Journal of Pharmaceutics, 447*(1–2), 75–93. https://doi.org/10.1016/j.ijpharm.2013.02.030.

Rosenholm, J. M., Peuhu, E., Bate-Eya, L. T., Eriksson, J. E., Sahlgren, C., & Lindén, M. (2010). Cancer-cell-specific induction of apoptosis using mesoporous silica nanoparticles as drug-delivery vectors. *Small, 6*(11), 1234–1241. https://doi.org/10.1002/smll.200902355.

Sackmann, E. (1995). Physical basis of self-organization and function of membranes: Physics of vesicles. In R. Lipowsky, & E. Sackmann (Eds.), *Vol. 1A. Handbook of biological physics* (1st ed., pp. 213–304). Elsevier, ISBN:978-0-444-81975-8.

Santos, H. A., Bimbo, L. M., Peltonen, L., & Hirvonen, J. (2014). Inorganic nanoparticles in targeted drug delivery and imaging. *Targeted drug delivery: Concepts and design* (pp. 571–613). Springer Science and Business Media LLC.

Siegler, E. L., Kim, Y. J., & Wang, P. (2016). Nanomedicine targeting the tumor microenvironment: Therapeutic strategies to inhibit angiogenesis, remodel matrix, and modulate immune responses. *Journal of Cellular Immunotherapy*, 69–78. https://doi.org/10.1016/j.jocit.2016.08.002.

Souery, W. N., & Bishop, C. J. (2018). Clinically advancing and promising polymer-based therapeutics. *Acta Biomaterialia, 67*, 1–20. https://doi.org/10.1016/j.actbio.2017.11.044.

Stella, B., Peira, E., Dianzani, C., Gallarate, M., Battaglia, L., Gigliotti, C. L., et al. (2018). Development and characterization of solid lipid nanoparticles loaded with a highly active doxorubicin derivative. *Nanomaterials, 8*(2).

Stiriba, S. E., Frey, H., & Haag, R. (2002). Dendritic polymers in biomedical applications: From potential to clinical use in diagnostics and therapy. *Angewandte Chemie, International Edition, 41*(8), 1329–1334. https://doi.org/10.1002/1521-3773(20020415)41:8<1329::AID-ANIE1329>3.0.CO;2-P.

Sun, T., Zhang, Y. S., Pang, B., Hyun, D. C., Yang, M., & Xia, Y. (2014). Engineered nanoparticles for drug delivery in cancer therapy. *Angewandte Chemie, International Edition, 53*(46), 12320–12364. https://doi.org/10.1002/anie.201403036.

Suri, S. S., Fenniri, H., & Singh, B. (2007). Nanotechnology-based drug delivery systems. *Journal of Occupational Medicine and Toxicology, 2*(1). https://doi.org/10.1186/1745-6673-2-16.

Tang, D. W., Yu, S. H., Ho, Y. C., Huang, B. Q., Tsai, G. J., Hsieh, H. Y., et al. (2013). Characterization of tea catechins-loaded nanoparticles prepared from chitosan and an edible polypeptide. *Food Hydrocolloids, 30*(1), 33–41. https://doi.org/10.1016/j.foodhyd.2012.04.014.

Tang, Q.s., Zhang, D.s., Cong, X.m., Wan, M.l., & Jin, L.q. (2008). Using thermal energy produced by irradiation of Mn-Zn ferrite magnetic nanoparticles (MZF-NPs) for heat-inducible gene expression. *Biomaterials, 29*(17), 2673–2679. https://doi.org/10.1016/j.biomaterials.2008.01.038.

Toma, A., Otsuji, E., Kuriu, Y., Okamoto, K., Ichikawa, D., Hagiwara, A., et al. (2005). Monoclonal antibody A7-superparamagnetic iron oxide as contrast agent of MR imaging of rectal carcinoma. *British Journal of Cancer, 93*(1), 131–136. https://doi.org/10.1038/sj.bjc.6602668.

Tran, S., DeGiovanni, P., Piel, B., & Rai, P. (2017). Cancer nanomedicine: A review of recent success in drug delivery. *Clinical and Translational Medicine*. https://doi.org/10.1186/s40169-017-0175-0.

Üner, M., & Yener, G. (2007). Importance of solid lipid nanoparticles (SLN) in various administration routes and future perspective. *International Journal of Nanomedicine, 2*(3), 289–300.

van Landeghem, F. K. H., Maier-Hauff, K., Jordan, A., Hoffmann, K. T., Gneveckow, U., Scholz, R., et al. (2009). Post-mortem studies in glioblastoma patients treated with thermotherapy using magnetic nanoparticles. *Biomaterials, 30*(1), 52–57. https://doi.org/10.1016/j.biomaterials.2008.09.044.

Ventola, C. L. (2017). Progress in nanomedicine: Approved and investigational nanodrugs. *Pharmacy and Therapeutics, 42*(12), 742–755. https://www.ptcommunity.com/system/files/pdf/ptj4212742.pdf.

Von Maltzahn, G., Park, J. H., Agrawal, A., Bandaru, N. K., Das, S. K., Sailor, M. J., et al. (2009). Computationally guided photothermal tumor therapy using long-circulating gold nanorod antennas. *Cancer Research, 69*(9), 3892–3900. https://doi.org/10.1158/0008-5472.CAN-08-4242.

Wang, Y., Xie, X., Wang, X., Ku, G., Gill, K. L., O'Neal, D. P., et al. (2004). Photoacoustic tomography of a nanoshell contrast agent in the in vivo rat brain. *Nano Letters, 4*(9), 1689–1692. https://doi.org/10.1021/nl049126a.

Wang, Y., Zhao, Q., Han, N., Bai, L., Li, J., Liu, J., et al. (2015). Mesoporous silica nanoparticles in drug delivery and biomedical applications. *Nanomedicine: Nanotechnology, Biology, and Medicine, 11*(2), 313–327. https://doi.org/10.1016/j.nano.2014.09.014.

Werner, M., Auth, T., Beales, P. A., Fleury, J. B., Höök, F., Kress, H., et al. (2018). Nanomaterial interactions with biomembranes: Bridging the gap between soft matter models and biological context. *Biointerphases, 13*(2).

Wise, D. L. (2000). *Handbook of pharmaceutical controlled release technology*. New York: Marcel Dekker Inc.

Won, Y. Y., Brannan, A. K., Davis, H. T., & Bates, F. S. (2002). Cryogenic transmission electron microscopy (cryo-TEM) of micelles and vesicles formed in water by poly(ethylene oxide)-based block copolymers. *Journal of Physical Chemistry B, 106*(13), 3354–3364. https://doi.org/10.1021/jp013639d.

Yang, H., & Kao, W. J. (2006). Dendrimers for pharmaceutical and biomedical applications. *Journal of Biomaterials Science, Polymer Edition, 17*(1), 3–19. https://doi.org/10.1163/156856206774879171.

Yin, J., Chen, Y., Zhang, Z. H., & Han, X. (2016). Stimuli-responsive block copolymer-based assemblies for cargo delivery and theranostic applications. *Polymers, 8*(7). https://doi.org/10.3390/polym8070268.

Yoo, H. S., Lee, J. E., Chung, H., Kwon, I. C., & Jeong, S. Y. (2005). Self-assembled nanoparticles containing hydrophobically modified glycol chitosan for gene delivery. *Journal of Controlled Release, 103*(1), 235–243. https://doi.org/10.1016/j.jconrel.2004.11.033.

Yu, X., Trase, I., Ren, M., Duval, K., Guo, X., & Chen, Z. (2016). Design of nanoparticle-based carriers for targeted drug delivery. *Journal of Nanomaterials, 2016*. https://doi.org/10.1155/2016/1087250.

Yu, M., Yang, Y., Zhu, C., Guo, S., & Gan, Y. (2016). Advances in the transepithelial transport of nanoparticles. *Drug Discovery Today, 21*(7), 1155–1161. https://doi.org/10.1016/j.drudis.2016.05.007.

Yun, Y., Cho, Y. W., & Park, K. (2013). Nanoparticles for oral delivery: Targeted nanoparticles with peptidic ligands for oral protein delivery. *Advanced Drug Delivery Reviews, 65*(6), 822–832. https://doi.org/10.1016/j.addr.2012.10.007.

Zhang, Z.-H., Wang, X.-P., Ayman, W. Y., Munyendo, W. L. L., Lv, H.-X., & Zhou, J.-P. (2013). Studies on lactoferrin nanoparticles of gambogic acid for oral delivery. *Drug Delivery*, 86–93. https://doi.org/10.3109/10717544.2013.766781.

Zhang, M., Yang, Z., Chow, L. L., & Wang, C. H. (2003). Simulation of drug release from biodegradable polymeric microspheres with bulk and surface erosions. *Journal of Pharmaceutical Sciences, 92*(10), 2040–2056. https://doi.org/10.1002/jps.10463.

Zhuo, R. X., Du, B., & Lu, Z. R. (1999). In vitro release of 5-fluorouracil with cyclic core dendritic polymer. *Journal of Controlled Release, 57*(3), 249–257. https://doi.org/10.1016/S0168-3659(98)00120-5.

CHAPTER 2

Nanocarriers systems for brain targeted drug delivery and diagnosis

Introduction

Brain is an exceptional and extremely friable organ in the human body. It is set apart from the vascular system by the blood-brain barrier (BBB) which is comprised of zonulae occludentes (complex tight junctions). These zonulae occludentes are made of cerebral endothelial cells at the separating interface of the brain and blood. The main function of BBB is to monitor and selectively restrict the entrance of some molecules into the central nervous system (CNS) (Chen & Liu, 2012). The entrance of molecules into CNS through BBB depends on their chemical composition, surface properties, structure, and molecular weight. Low molecular weight lipophilic molecules enter the CNS with several times greater competency than large molecules. The BBB permits less than 5% of the available drugs to the brain at varying degrees. There are several drugs available in the market for different CNS-associated diseases, such as neuropeptides, antineoplastic agents, and antibiotics but BBB restricts their direct delivery to the CNS (Pardridge, 2003). Besides, even recently developed therapeutic molecules, for example, recombinant proteins, agents of gene therapy, and monoclonal antibodies (mAbs) are also incapable to cross BBB. Currently, investigators are concentrating on different strategies to deliver drugs/small therapeutic agents across the BBB and target the specific brain cells (P. Blumling III & A. Silva, 2012). In this perspective, nanocarriers-based delivery systems are being designed and prepared to meet the challenges associated with BBB.

The progressive and extensive research in nanotechnology has offered innovative applications of pre-designed nanomaterials in several fields of science owing to their exceptional and exclusive properties (Kuzum, Yu, & Philip Wong, 2013). The versatility of nanomaterials can be expanded by manipulating their chemical composition, surface, shape, and size. In the last two decades, controlling and directing the chemical properties of inorganic and organic materials at the nanoscale has aroused as the top research area. Nanomaterials offer a broad range of applications in different areas, like biosensing,

Nanocarriers for Organ-Specific and Localized Drug Delivery. https://doi.org/10.1016/B978-0-12-821093-2.00011-6

biological imaging, cancer therapy, gene delivery, and drug delivery (Cho, Wang, Nie, Chen, & Shin, 2008; He et al., 2018; Sharma et al., 2019). The drug delivery applications of nanocarriers-based systems have been widely studied for those drugs that have poor bioavailability issues. Keeping in mind the needs of commercial availability and clinical uses, it is obligatory to design smart nanocarriers or nanomedicines that can deliver the therapeutic agents to the desired site and can keep maintaining the functional features of therapeutic agents in the body. This scenario can be achieved only by addressing the harsh biological milieu that is formed by degrading chemicals and enzymes and protecting the therapeutic agents from modification in these harsh conditions (Petri et al., 2007). Delivery of drugs and other therapeutics to brain is extremely restricted due to the presence of sophisticated and delicate BBB.

There are three transport systems, i.e., receptor-mediated transporters, active efflux transporters, and carrier-mediated transporters which are involved in the transport of molecules across BBB. Receptor-mediated transporters are responsible for large molecules while active efflux and carrier-mediated transporters are mostly engaged with the transportation of small molecules (Gabathuler, 2010). So, the functionalization of nanocarriers with specific ligands can directly target endothelium cells of the brain and could further facilitate the internalization of drugs or small-sized molecules via transcytosis or endocytosis. Due to the nano-size and unique properties of nanocarriers, it becomes essential to well understand the interaction of nanocarriers with the brain cells and molecular instances happening at the cellular level. The process of nanocarriers' internalization is very crucial and must be understood completely to find out their outcome inside the cells, which may be either satisfactory or unfavorable. This chapter presents a comprehensive understanding of the developments made in nanocarriers based on effective drug delivery to the CNS, and what is the role of nanocarriers' surface modification in the uptake mechanisms.

The blood-brain barrier (BBB)

Human brain is extremely complex and the most vital organ of the body. Thereby, it is very imperative to protect it against any possible suffering that can lead to improper activation, inflammation, infection, and even the mortality of cerebral cells (Barchet & Amiji, 2009). The spinal cord and the brain combine make the CNS that are protected by two typical barriers, i.e., the blood-cerebrospinal fluid barrier (BCB) and BBB. These barriers of CNS are acting like selectively permeable membranes for different types of molecules and do not block all the incoming molecules (Habgood & Ek, 2010). The BBB is a typical structure and an interface between the cellular and acellular components in the CNS (Fig. 2.1). The basic and primary role of the BBB is to ensure a suitable environment for the functioning and interaction of the neurons,

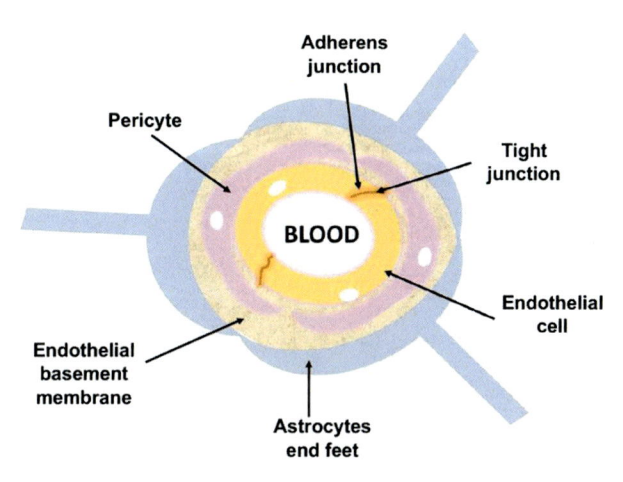

FIG. 2.1

Schematic representation of the BBB. *From Ahlawat, J., Guillama Barroso, G., Masoudi Asil, S., Alvarado, M., Armendariz, I., Bernal, J., et al. (2020). Nanocarriers as potential drug delivery candidates for overcoming the blood–brain barrier: Challenges and possibilities. ACS Omega, 5(22), 12583–12595. https://doi.org/10.1021/acsomega.0c01592.*

which is essential for protecting the CNS from pathogens, regulating influx and efflux, and maintaining homeostasis (Barchet & Amiji, 2009).

The BBB consists of an incessant film of endothelial cells linked via adherent junctions (AJs), tight junctions (TJs), and gap junctions (GJs). All these junctions have vital roles in the function of the BBB but the TJs are considered to be the main compositional component of the BBB and give greater transendothelial resistance (Zhou, Peng, Seven, & Leblanc, 2018). The entrance of drug or small molecules into CNS through the BBB is controlled by tying the adjoining cells so tightly that close up all the intercellular space among them (Habgood & Ek, 2010; Zhou et al., 2018). The special cells that exist in the BBB are adjacent neurons, microglia, astrocytes, and pericytes (Fig. 2.1). Pericytes are multiple functioning coating cells that wrap around the endothelial cells and regulate transcytosis, AJs, and TJs across the BBB (He et al., 2018). The main function of these cells is to control the blood flow in brain capillaries through contraction and relaxation of the blood capillaries' walls (Zhou et al., 2018). On the other hand, the astrocytes regulate neurotransmitters, remove the debris from cerebrospinal fluid (CSF), and maintain the redox potential that leads to maintaining the homeostasis by neurons. The interactions of astrocytes with the endothelial cells develop connections between the adjacent neurons and the brain capillary via projecting the capillaries. Moreover, microglia (the brain immune cells) can be activated through trauma or systemic inflammation. The primary function of microglia is to remove cellular or neuronal debris through phagocytosis (Redzic, 2011). While the basement membrane is composed of endothelial cells and pericytes which represent the surface of more than 99% of the CNS capillaries. The basement membrane

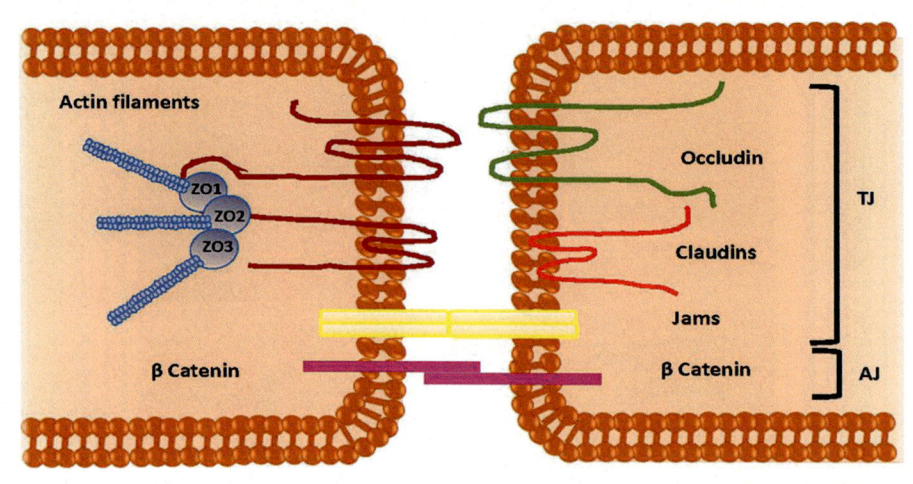

FIG. 2.2

Schematic representation of tight junctions and adherence junctions between endothelial cells. *From Ahlawat, J., Guillama Barroso, G., Masoudi Asil, S., Alvarado, M., Armendariz, I., Bernal, J., et al. (2020). Nanocarriers as potential drug delivery candidates for overcoming the blood–brain barrier: Challenges and possibilities. ACS Omega, 5(22), 12583–12595. https://doi.org/10.1021/acsomega.0c01592.*

is also rich in proteins, i.e., laminin and collagen type IV, and proteoglycans, i.e., heparan sulfate (Zhou et al., 2018) (Figs. 2.1 and 2.2).

Barriers to brain drug delivery

The existence of the intercellular clefts is almost nil in the brain capillaries. Therefore, transportation to the brain must take place transcellular and lipophilic molecules can easily diffuse via the endothelial membrane, leading to the free transportation of lipid-soluble solutes across the BBB (Serlin, Shelef, Knyazer, & Friedman, 2015). It makes compromised the therapeutic effectiveness of those promising drugs that cannot cross the BBB. That is why mostly lipophilic drugs are preferred for therapeutic purposes to address CNS-related disorders. However, high solubility in lipid does not guarantee the accumulation of the drug molecules in the brain because highly lipophilic compounds can also be deposited to other nontarget sites in the body (Habgood & Ek, 2010; Serlin et al., 2015).

There are some additional barriers also along with BBB. For example, when solutes passing through the mitochondria rich cerebral endothelial membrane are vulnerable to degrading enzymes, i.e., insulin-degrading, enkephalin, and neprilysin (Serlin et al., 2015). The BBB also has the potential to recognize neuropeptides and instantly degrade these peptides. The P-glycoproteins (Pgp) concentration also reinforces the BBB in the luminal membrane. This efflux transporter knocks out several drugs from the endothelial cytoplasm prior to their crossing the parenchyma (Savolainen, Edwards, Morgan, McNamara, & Anderson, 2002). If some drug molecules manage to escape Pgp barrier, then phase I and II metabolisms (the second line of defense) export out these molecules from the brain parenchyma. Phase I metabolism involves the unmasking or the addition of a reactive polar group to the drug molecules while phase II metabolism involves the addition of an anionic group to the drug molecules to make them more water-soluble and mark them for expulsion.

The choroid plexus is formed by a cluster of ependymal cells. It separates the CSF from blood and function as a barrier in the brain. The ependymal cells and the brain ventricles secrete the CSF (Redzic, 2011). The ependymal cells make a double layer on the brain surface by folding over onto themselves which is named as arachnoid membrane. This membrane works as a selectively permeable membrane to hydrophilic molecules. The arachnoid membrane and the choroid plexus jointly work at the barrier between the CSF and the blood. The BCB is another set of the barrier that actively contributes to CNS (Redzic, 2011). In the choroid plexus, BCB is also backed by a transport system that can pull out organic molecules into blood that are produced in CSF. This efflux system actively removes the methotrexate and penicillin from the CSF and inhibits their diffusion into the parenchyma (Redzic, 2011). These drugs show low solubility in lipid and are blocked by the TJs within the endothelial cells. To cross the BBB, these types of drugs require additional support for their transportation. Though these barriers (BCB and BBB) are extremely important for the brain to function normally, sometimes it is also found that they work like multidrug-resistant for the entry of drugs to the brain. So, these barriers make a big hindrance in the delivery of drugs into the CNS.

The drug nature (lipophilic or hydrophilic) is not the only factor that is required for the BBB permeability. There are some others factors also which affect the BBB permeability. Among them, the blood/plasma concentration of the therapeutic agent is very crucial which is achieved after a few hours of the drug administration. Some other processes such as degradation of the drug, losing lipophilic characteristic, combination with plasma protein, and efflux rate from the brain are also involved in the low blood/plasma concentration of the therapeutic agent in the brain along with the BBB.

Nanocarriers for brain drug delivery

Enhancing the transportation of drug molecules across the BBB has been one of the most challenging tasks for formulation chemists. With the advancement of nanomedicines, different types of nano-size tunable devices have been developed as an interesting approach potentially able to address the unmet challenge of enhancing transportation of therapeutic agents across the BBB (Guccione et al., 2017; Shaw et al., 2017). Nanocarriers technology among these devices is rapidly advancing. Nanocarriers are nano-size (1–100 nm) objects that are considered as complete units in terms of properties and transport. The types of nanocarriers based on the viral vectors, polymeric nanoparticles (NPs), liposomes, dendrimers, micelles, carbon nanotubes, carbon dots, or carbon nano-onions are more popular in biomedical sciences (Fig. 2.3). The high expectation from these nanocarriers is linked to the easy multi-functionalization of these nanocarriers along with their capability to carry payloads and cross the BBB. The drugs or therapeutic agents deliver to the brain across the BBB via

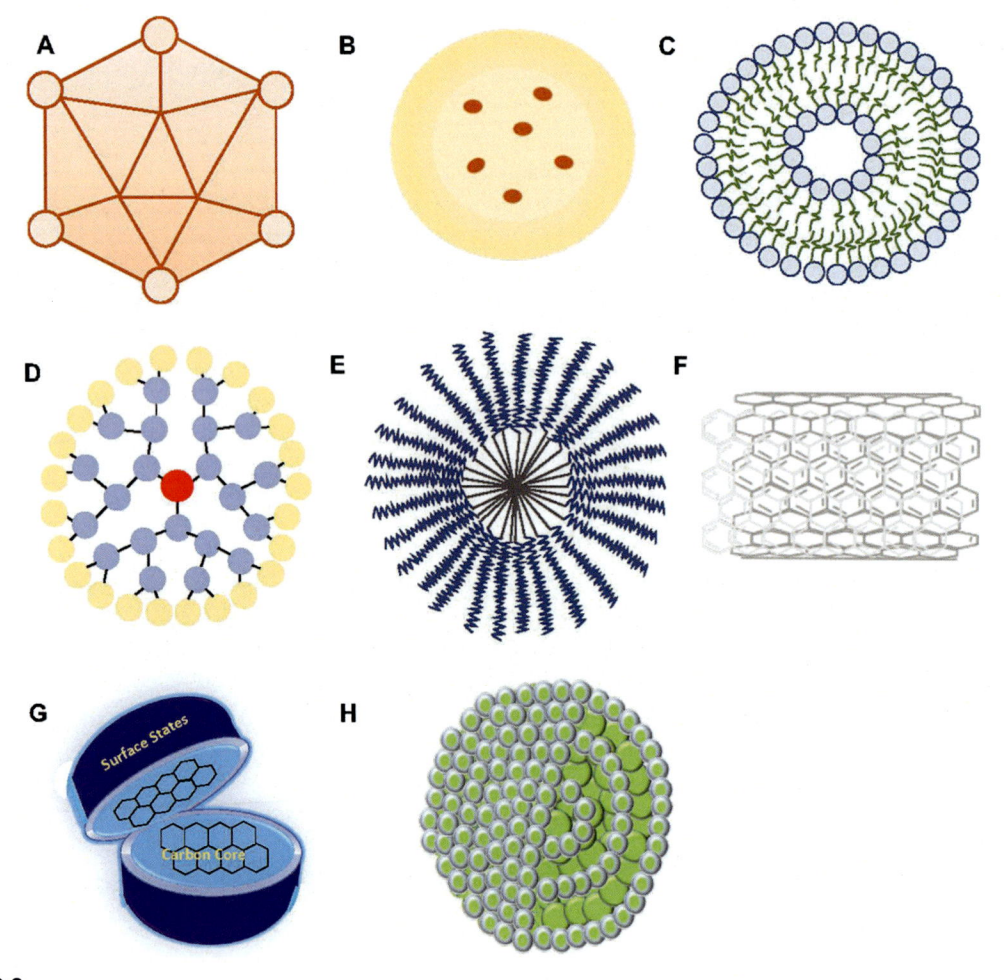

FIG. 2.3

Schematic representation of (A) viral vectors, (B) polymeric nanoparticles, (C) liposomes, (D), dendrimers, (E) micelles, (F) carbon nanotubes, (G) carbon dots, and (H) carbon nano-onions. *From Ahlawat, J., Guillama Barroso, G., Masoudi Asil, S., Alvarado, M., Armendariz, I., Bernal, J., et al. (2020). Nanocarriers as potential drug delivery candidates for overcoming the blood–brain barrier: Challenges and possibilities. ACS Omega, 5(22), 12583–12595. https://doi.org/10.1021/acsomega.0c01592.*

nanocarriers are wholly dependent on the biomimetic and physiochemical characteristics of the nanocarriers and do not depend on the physical or chemical properties of the drug which is encapsulated inside the nanocarriers.

Viral vectors

Viral vectors are the most practiced and highly successful approach to use for gene delivery to host cells. Currently, they get promising attention because of their capability to carry the desired genes into CNS and are used for addressing perilous neurological disorders. The DDSs based on viral vectors are widely used to carry nucleic acids to the parenchyma cells of the brain. The big advantage of administrating viral vectors in gene delivery to the brain cells is their efficient transfection efficiency (up to 80%) and expression of transgenes in the non-dividing cells for long period. The viral vectors such as adeno-associated virus (AAV), adenovirus (AdV), herpes simplex virus (HSV), and lentivirus have been used for drug delivery into CNS and have exhibited exciting results. However, there are some limitations also relating to the use of viral vectors in drugs and gene delivery applications, i.e., difficulties related to manufacturing, highly costly, and immunogenicity.

Shukla et al. conducted a study to assess the biodistribution and clearance of viruses in normal and cancer-induced mice. For this purpose, they introduced potato virus X and found that 30% of viral particles signals in mammary glands, the colon, and the brain tumor site of the mice. The remaining 70% of viral particles were passed through the reticuloendothelial system by the liver and spleen. Whereas the bile and kidney also showed slower clearance of them (Shukla et al., 2014). The AAV2 vector was also used for the glutamic acid decarboxylase gene delivery in Parkinson's disease patients. After 12 months of continuous treatment, a significant reduction was observed in levodopa-induced dyskinesias compared with the control (Niethammer et al., 2017). Brain K Kaspar group has reported for the first time an intravascular AAV9 which was introduced intravenously, bypassing the BBB, and efficiently targeted adult astrocytes and neonatal neurons in mice (Foust et al., 2009). Since most viruses cannot cross the BBB, a few administration routes, for example, direct injection into the CSF or stereotaxic injection have been practiced bypassing the BBB. Therefore, extensive research is still needed to be conducted on viral vectors' immunogenicity, safety, and use before going to clinical practice from preclinical studies.

Liposomes

Liposomes are small artificial lipid bilayer spherical shape vehicles that are composed of cholesterol and amphiphilic lipid or natural phospholipid. They can hold a variety of molecules including therapeutic agents such as proteins, nucleic acids, drugs, and vaccines. They are mostly biocompatible, amphiphilic in nature, can be easily modified, and enhance circulation time of the therapeutics in blood. These characteristics make them exceptional and efficient carriers for the delivery of both types of drugs whether lipophilic or hydrophilic (Ahlawat, Henriquez, & Narayan, 2018). Liposomes are mostly used for targeted drug delivery in CNS because of their ability to be easily modified, easily bypass the parenchyma of the brain, and deliver drugs to the desired sites (Barchet & Amiji, 2009). In spite of the unique properties of the liposomes, there are some limitations also associated with them such as unstable to be stored for long periods, leakage of the held drugs during storage, short half-life, batch-to-batch variation, low solubility, and high cost of production in some cases. Some safety issues such as hepatotoxicity and immunogenic response are raised upon liposome introduction (Wolfram et al., 2015). They also induce the overexpression of pro-inflammatory cytokines leading to inflammation.

Docetaxel, an anticancer medication for different types of cancers, cannot cross the BBB due to its physicochemical characteristics. Shaw et al. designed a liposomal nano-system encapsulating docetaxel to target solid tumors in the brain. This nano-formulation exhibited effective cellular uptake in the C6 glioma cells when assessed in an in vitro study. Whereas in vivo study on rats showed improved uptake of this nano-formulation by the brain and exhibited better pharmacokinetic as compared to the free docetaxel. The results suggest the competency of liposomes to be used as carriers for docetaxel targeting the glioma cells in the brain (Shaw et al., 2017). Rehman et al. developed a specialized lipid-based nanoparticulate system for the delivery of paclitaxel bearing thermo-responsive features as well. This formulation exhibited dual functions and not only improved the permeability of the drug across the BBB but also effectively targeted the glioblastoma cells. The formulation has the capability to show a sustained release of the drug at 37°C (physiological temperature), whereas an abrupt or rapid release could be observed at 39°C (Rehman et al., 2017). Sonkar et al. reported gold fabricated liposomes encapsulated glutathione gold nanoparticles and docetaxel for brain targeted imaging and drug delivery. The liposomes were decorated with transferrin to target the transferrin receptor. The gold fabricated liposomes effectively delivered docetaxel to the specific sites of the brain suggesting that co-loaded liposomes would be a promising platform for nano theranostic and cancer therapeutics for brain targeting (Sonkar et al., 2021).

Solid lipid nanoparticles

Solid lipid nanoparticles (SLNs) contain a solid core made of hydrophobic lipid and its primary function is to hold drugs in the dispersed or dissolved form (Kaur, Bhandari, Bhandari, & Kakkar, 2008). They are composed of biocompatible lipids, i.e., waxes, fatty acids, and triglycerides. The small size of

SLNs (40–200 nm) provides them the advantage of escaping from the reticulo-endothelial system and crossing the BBB (Pardeshi et al., 2012). In the manufacturing process, the drug is first mixed in melted lipid and then dispersed in the aqueous surfactants by micro-emulsification or high-pressure homogenization. The SLNs have advantages over other NPs systems as they are biocompatible, exhibit high drug entrapment efficiency, capable to provide an uninterrupted release of the drug even for weeks, and can be stored for long periods (Mishra, Patel, & Tiwari, 2010). Moreover, their composition can be modified and controlled to incorporate special properties to their surface to limit their uptake via the reticuloendothelial system and target the brain cells (Blasi, Giovagnoli, Schoubben, Ricci, & Rossi, 2007).

There are several reports that showed the enhanced drug delivery to the CNS mediated by SLNs. Neves et al. designed an SLN system functionalized with apolipoprotein E encapsulating resveratrol for targeted delivery to the brain (Neves, Queiroz, & Reis, 2016). The results concluded that apolipoprotein E functionalized SLNs showed better permeability of the resveratrol as compared to non-functionalized SLNs. The resveratrol-loaded SLNs would be a better approach to protect the drug from the breakdown in the bloodstream and effectively deliver it to the brain. Arduino et al. reported PEGylated SLNs encapsulating lipophilic kiteplatin Pt (IV) prodrugs for brain delivery to treat glioblastoma multiforme (Arduino et al., 2020). This formulation exhibited incredible results when subjected to evaluation via in vitro study using the glioblastoma cell line. Similarly, Wang et al. enhanced the access of 5-fluoro-2-deoxyuridine by synthesizing 3′,5-dioctanoyl-5-fluoro-2-deoxyuridine and incorporating it into SLNs (Wang, Sun, & Zhang, 2002). The results showed that the brain targeting efficiency of the SLNs incorporated functionalized drug was twofold higher than the free drug. They concluded that this SLN system could enhance the permeability of the drug across the BBB and might be a promising nano-system for the treatment of the CNS ailments.

Polymeric nanoparticles

Polymeric NPs are small-sized particles (60–200 nm) composed of a polymer core and can be loaded with therapeutic agents adsorbed onto the surface or entrapped within the polymeric core (Zhang et al., 2013). They have demonstrated great potentials in the field of targeted drug delivery, nano-formulations, and nanomedicines for the treatment of different types of diseases. The polymeric NPs have some exceptional advantages to be used as drug carriers such as protecting the drug molecules from the harsh environments (chemical modification and enzymatic degradation), enhanced drug solubility, controlled and sustained release of the drug, and increasing the therapeutic index of the drug by improving the bioavailability of drug (Cano et al., 2019, 2020). Polymeric NPs are synthesized in one of the two morphological forms, i.e., nanospheres and nanocapsules (Schaffazick, Pohlmann, Dalla-Costa, & Guterres, 2003). Nanospheres are made of a continuous polymeric framework in which the drug molecule can be adsorbed onto their surface or retained within the framework. Whereas nanocapsules possess an oily core which provides a medium to dissolve drugs and the oily core is surrounded by a polymeric shell controlling the drug release from the core (Crucho & Barros, 2017; Szczęch & Szczepanowicz, 2020).

Ginkgolide B is a derivative of *Ginkgo biloba* and is considered ideal in the treatment of Parkinson's disease because of its function as neuroprotective agent. It shows potential therapeutic properties, but the only issue associated with this drug is its poor bioavailability after administering in oral dosage forms. Zhao et al. developed polymeric NPs of PEG-*co*-poly(ε-caprolactone) entrapping Ginkgolide B (Zhao et al., 2020). In in-vitro study, it was observed that MDCK cell line could uptake these NPs via multiple nonspecific mechanisms such as caveolae-mediated/lipid raft endocytosis, clathrin-dependent endocytosis, and micropinocytosis. In the zebrafish model, these NPs easily crossed the blood-retinal, gastrointestinal, chorion, and blood-brain barriers. In in-vivo study after oral administration, the Ginkgolide B encapsulated polymeric NPs exhibited more exciting and desirable pharmacokinetics as compared to the free Ginkgolide B. Bukchin et al. reported amphiphilic polymeric NPs functionalized with retro-*enantio* peptide shuttle to improve their brain transport (Bukchin et al., 2020). Their results exhibited that after functionalization with peptide shuttle the cell compatibility and permeability of these NPs improved and internalized via the clathrin-mediated pathway. In in-vivo study, the peptide-functionalized NPs showed fourfold high accumulation in the brain parenchyma and blood vessels.

Polymeric micelles

The amphiphilic copolymers produce polymeric micelles and form aggregated structures upon exposure to aqueous environments. Polymeric micelles are spherical nanostructures composed of a hydrophobic core and a hydrophilic shell and highly stable in suspension form (Mishra et al., 2010). The stability of the polymeric micelles' system can be further improved by crosslinking the core or the shell chains. The core of the polymeric micelles is used as a reservoir for drug molecules holding or encapsulation. Furthermore, the polymeric micelles can be decorated with special tunable features to make them responsive to stimuli, i.e., temperature, light, pH, ultrasound, hypoxia, overexpression of a specific enzyme, etc., and triggering a sustain and control release of encapsulated drug molecules (Kabanov et al., 1992). Pluronic type block copolymer composed of propylene oxide and ethylene oxide is one of the most used polymers in preparing the polymeric micelles. The probable use of these nanocarriers in brain drug delivery has been evaluated by several research groups. For example, Pluronic-based nanocarriers decorated with chitosan and rabies virus glycoprotein (RVG29; target peptide specific for the brain) exhibited effective accumulation in the brain after i.v. injection in mice (Kim, Choi, Kim, & Tae, 2013). Ping sun et al. synthesized a transferrin receptor functionalized PEG-PLA copolymer to carry paclitaxel for chemotherapy of glioblastoma multiforme (Sun et al., 2020). The modified polymeric micelles were found to cross the BBB and absorbed quickly by tumor cells. The paclitaxel-loaded modified polymeric micelles exhibited improved anti-glioma effect and prolonged survival of mice bearing glioma as compared to the unmodified copolymer.

Dendrimers

Dendrimers are well-ordered, tree-shaped multiple branched homogenous synthetic molecules with a central core (Ahlawat et al., 2018). Their unique features such as rigidity, low polydispersity, predictable molecular weight, monodispersed phase, nanosized formulation, and easy modification render them excellent candidates to be used for developing nanocarriers for brain drug delivery (Zhang et al., 2017; Zhao et al., 2015). They are mostly used to carry hydrophobic drugs to the specific area of the brain because of their potential of accommodating hydrophobic molecules within their nanostructures. Despite the excellent properties of the dendrimers, there are some limitations also associated with them such as complications in accomplishing targeted delivery, batch-to-batch variations, and high cost of production. Some safety issues associated with cationic dendrimers, i.e., positively charged nanocarriers strongly interact with the negatively charged cell surfaces resulting in high toxicity. These interactions create nanopores in the cell membrane and provoke the probable leakage of cellular content, resulting in cell death (Janaszewska, Lazniewska, Trzepiński, Marcinkowska, & Klajnert-Maculewicz, 2019).

Polyamidation types of amine-terminated dendrimers have been used for delivery of drugs at the tumor site to target gliomas in the brain (Ahlawat et al., 2018; Zhao et al., 2015). A nano theranostic composed of a dendrimer (generation 5 poly(amidoamine)) multi-functionalized with 3-(4′-hydroxyphenyl) propionic acid-OSu, chlorotoxin (an anticancer drug), and PEG was designed. This nano theranostic was then radiolabeled with Iodine-131 to make it radioactive theranostic agent. The results showed that this theranostic agent was very effective against cancer cells and exhibited specificity toward those cancer cells which overexpressed the metallopeptidase 2. Therefore, from these significant results, it could be concluded that multi-functionalized dendrimeric nanocarriers possess great potentials to be utilized in the human brain against gliomas. The effect of the generation 6 poly(amidoamine) dendrimers was evaluated as nanocarrier in the canine model of the brain injury induced by the hypothermic circulatory arrest (Zhang et al., 2017). The results showed that the nanocarriers of generation 6 dendrimers revealed a prolonged circulation time in the bloodstream and an increased accumulation in the brain as compared to other dendrimers (generation 4, etc.). These interesting results could be attributed to the size of generation 6 dendrimers, their increased uptake in CSF, and reduced clearance by the liver and kidneys.

Micelles

Micelles are monolayered nanosized spherical aggregates of amphiphilic molecules and have been used as nanocarriers for entrapping, delivering, and controlling release of the drugs. These nanosized aggregates offer an effective and enhanced approach for the hydrophobic drugs to cross the BBB in treating brain-related disorders (Shiraishi, Wang, Kokuryo, Aoki, & Yokoyama, 2017). They have already shown their effectiveness as nanocarriers for many therapeutic molecules, i.e., peptides and small molecules. Currently, they have been shown to owe promising potentials in imaging brain stokes and inflammation via magnetic resonance as well as in treating Alzheimer's disease. However, there are some limitations also associated with micelles such as highly unstable to be kept stored for long periods, leakage of the drugs molecules during storing for a long time, being vulnerable to disassembly and deformation, short half-life, batch-to-batch variations, and high cost of production in some cases. Some safety issues are also associated with micelles, i.e., the beginning immunogenic response in the case of Taxol encapsulated micelles (Wolfram et al., 2015).

Several reports showed the enhanced drug delivery to the CNS mediated by micellar systems. For example, Shiraishi et al. reported an MRI contrast agent based on gadolinium-micelles (Shiraishi et al., 2017). They injected the contrast agent into rats and got clear contrast images after half an hour. The contrast images showed increased distribution of the micelles in the ischemic hemisphere and enhanced permeability of the BBB. Therefore, this micellar system can be a promising MRI contrast agent for the clinical diagnosis and to find out probable intracranial hemorrhage risks. Whereas Yin et al. developed a micellar system composed of poly (lactic-glycolic acid) lyso-GM1 and loaded with doxorubicin up to 61% encapsulation. The results showed that this nano-system could readily pass through the BBB and gather in the brain parenchyma via lysosomal and micropinocytosis pathways. This micellar system exhibited an effective anti-glioma effect in the rats and could be used as promising delivery tool for effectively delivering anti-glioma drugs.

Carbon nanotubes

Carbon nanotubes (CNTs) are single-walled or multi-walled cylindrical molecules of graphene sheets. These nanotubes have shown excellent applications in several scientific fields specifically in nanotechnology. Currently, CNTs have arisen as promising theranostic and nanocarrier systems in several brain-specific therapies. Such excellent activities of CNTs are mainly attributed to their easy surface modification and functionalization with specific chemical ligands resulting in unique physical, chemical, and biological properties (Ma et al., 2018). It has been observed that polymer-coated CNTs can readily pass through the BBB in several in-vitro and in-vivo studies. The polymer-coated CNTs make an effective interface with neurons leading to increase uptake of the CNTs. However, some limitations are also associated with CNTs such as no control over chirality and length of CNTs, high polydispersity, batch-to-batch variation, and high cost of production (Wolfram et al., 2015). Some safety issues are also associated with CNTs, i.e., a study reported that a dose of 3 mg/mouse of CNTs exhibited a carcinogenic effect (Zhang et al., 2017).

Berberine is an isoquinoline derivative and has been effectively used as an anti-Alzheimer's agent. Lohan et al. synthesized berberine incorporated multi-walled CNTs and then further functionalized their surface with polysorbate and phospholipids to make them biocompatible (Ma et al., 2018). These modified CNTs exhibited no significant toxicity when subjected to treatment with SHSY-5y cell lines. When this nano-formulation was subjected to experiment in-vivo, the recovery of memory was observed in rats after 2–3 weeks of formulation administration. The enhanced drug concentration in the plasma and brain tissues of rats indicates the effective absorption of this nano-formulation and its crossing of the BBB.

Carbon nano-onions

Carbon nano-onions (CNOs) are multilayered zero-dimensional fullerenes. These carbonaceous nanostructures possess unique structural and electronic features that depend on the approach used for their preparation. Their structural features such as thermal stability, larger surface area, accepting multiple electrons reversibly, wide absorption band, low cytotoxicity, and high biocompatibility make them promising candidates for various applications in nanomedicine field (Plonska-Brzezinska, 2019). CNOs cannot be utilized in their impeccable form due to their poor solubility in inorganic and organic solvents. Therefore, they must be passed through the process of surface functionalization to make them more dispersible and avoid their aggregation in different solvents. These carbonaceous nanostructures are generally prepared by thermal annealing of nanodiamonds in helium or nitrogen atmosphere. This process requires very high temperature and pressure and yields CNOs that are on the average composed of six to eight graphite shells and 5–6 nm in size. Despite the excellent properties of the CNOs, there are some limitations also associated with them such as difficulties in the preparation of well-dispersed CNOs in aqueous medium, complications in accomplishing targeted delivery, batch-to-batch variation, high cost of production, and inability to cross the BBB easily because of their high molecular weights (Plonska-Brzezinska, 2019).

Sarkar et al. reported a CNOs based drug livery system for the transportation of an anti-Alzheimer's drug to the brain (Pakhira, Ghosh, Allam, & Sarkar, 2016). They synthesized water-soluble CNOs were 25–50 nm in size and appeared like a Trojan horse in shape. An acetylcholinesterase inhibitor was effectively loaded to these nanostructures. The inhibitor showed sustained release from these nanostructures at physiological pH. They also assessed the capability of these nanostructures to deliver inhibitors into the brain across the BBB and target the glioma.

Carbon dots

Carbon dots (CDs) are zero-dimensional fluorescent nanocarbons usually possessing spherical shapes and are amorphous in nature. They were unintentionally isolated during the purification of single-walled CNTs initially. Now several methods or approaches have been practiced for the synthesis of CDs. Among them, laser ablation is the most practiced approach employed to produce CDs due to its flexibility and green synthesis aspect (e.g., the stabilizing agent is not used which is appropriate for biomedical applications). CDs are chemically modifiable and can be doped with heteroatoms such as B, N, S, and P to introduce some additional functionalities. They exhibit enhanced properties as compared to typical nanoclusters of noble metals or inorganic quantum dots, making them excellent luminescent probes and nanocarriers. The most exciting characteristics of CDs are their tunable fluorescence and the ability to easy modification for desired applications. These prevailing tools are used for developing biosensors or chemo-sensors based on their luminescence features. CDs are now preferred in the biomedical field due to their unique physical, optical, and biological properties, i.e., water-solubility, adjustable stability, feasible fluorescence, low toxicity, and high biocompatibility. Despite the excellent properties of CDs, there are some limitations also associated with them such as unstable on storing for long times, high polydispersity, particle to particle aggregation, and photobleaching (Yao et al., 2018).

Chung et al. has reported multifunctional CDs that possess nitrogenous polyaromatic functionalities (o-phenylenediamine) on their surface (Chung, Lee, &

Park, 2019). The modified CDs exhibited high potency toward the chelation with Cu (II) ions, inhibit the aggregation of cerebral β-amyloid (Aβ) peptides, and photo-oxygenate Aβ peptides. The irregular self-assembly of Aβ peptides results in toxic aggregates which are considered as a neuropathological trait of Alzheimer's disease. The effective photo-oxygenation of Aβ peptides and disturbing the β-sheet-rich structure led to suppressing the coordination of Aβ peptides with Cu (II) ions and self-assembly of toxic aggregation. Therefore, amyloid-β-related toxicity can be alleviated successfully by using multifunctional CDs.

Conclusions and outlook

The CNS disorders specifically brain diseases lack appropriate treatments due to the inability of corresponding therapeutic agents to pass through the BBB. For a therapeutic agent to exhibit its impact on brain diseases, it first needs to cross this gate and enter the CNS. Researchers across the globe are still being engaged in developing and testing small drug molecules that can cross the BBB and target CNS disorders. Brain-associated diseases, e.g., Alzheimer's disease and brain cancers could not be properly treated because of the inability of the drugs to pass through the BBB. Therefore, the role of nanotechnology is noteworthy in CNS-associated disorders when addressing the delivery of small drug molecules to the brain. Several DDSs based on the viral vectors, polymeric nanoparticles (NPs), liposomes, dendrimers, micelles, carbon nanotubes, carbon dots, or carbon nano-onions have already shown their significance in biomedical sciences. These nanocarriers are capable to protect drugs from enzymatic degradation, minimize drug immunogenicity, enhance drug solubility, improve drug plasma stability, deliver drugs at the target site, release the drug in a controlled manner, avoid reticuloendothelial clearance, and are small enough to pass through the highly selective barriers. Comprehensive knowledge about the function and mechanism of nanocarriers will help to design and develop novel strategies for treating brain-related diseases that are currently limited due to any reasons. Moreover, functionalization of the nanocarriers for targeting the cerebral circulation will lead to enhance uptake of the drug by the brain and minimizing toxicity to other tissues of the body. An effective nanocarrier for brain targeting should be stable in the blood, capable to bypass plasma protein binding and renal clearance, able to cross the BBB, and capable to knock out active efflux mechanisms. Though currently some of these problems have been addressed individually, the ultimate and ideal candidate for addressing the CNS disorders would be the one that addresses all these problems at once.

References

Ahlawat, J., Henriquez, G., & Narayan, M. (2018). Enhancing the delivery of chemotherapeutics: Role of biodegradable polymeric nanoparticles. *Molecules*, *23*(9), 2157. https://doi.org/10.3390/molecules23092157.

Arduino, I., Depalo, N., Re, F., Dal Magro, R., Panniello, A., Margiotta, N., et al. (2020). PEGylated solid lipid nanoparticles for brain delivery of lipophilic kiteplatin Pt(IV) prodrugs: An in vitro study. *International Journal of Pharmaceutics*, *583*, 119351. https://doi.org/10.1016/j.ijpharm.2020.119351.

Barchet, T. M., & Amiji, M. M. (2009). Challenges and opportunities in CNS delivery of therapeutics for neurodegenerative diseases. *Expert Opinion on Drug Delivery*, *6*(3), 211–225. https://doi.org/10.1517/17425240902758188.

Blasi, P., Giovagnoli, S., Schoubben, A., Ricci, M., & Rossi, C. (2007). Solid lipid nanoparticles for targeted brain drug delivery. *Advanced Drug Delivery Reviews*, *59*(6), 454–477. https://doi.org/10.1016/j.addr.2007.04.011.

Bukchin, A., Sanchez-Navarro, M., Carrera, A., Teixidó, M., Carcaboso, A. M., Giralt, E., et al. (2020). Amphiphilic polymeric nanoparticles modified with a retro-enantio peptide shuttle target the brain of mice. *Chemistry of Materials*, *32*(18), 7679–7693. https://doi.org/10.1021/acs.chemmater.0c01696.

Cano, A., Ettcheto, M., Chang, J. H., Barroso, E., Espina, M., Kühne, B. A., et al. (2019). Dual-drug loaded nanoparticles of Epigallocatechin-3-gallate (EGCG)/Ascorbic acid enhance therapeutic efficacy of EGCG in a APPswe/PS1dE9 Alzheimer's disease mice model. *Journal of Controlled Release*, *301*, 62–75. https://doi.org/10.1016/j.jconrel.2019.03.010.

Cano, A., Sánchez-López, E., Ettcheto, M., López-Machado, A., Espina, M., Souto, E. B., et al. (2020). Current advances in the development of novel polymeric nanoparticles for the treatment of neurodegenerative diseases. *Nanomedicine*, *15*(12), 1239–1261. https://doi.org/10.2217/nnm-2019-0443.

Chen, Y., & Liu, L. (2012). Modern methods for delivery of drugs across the blood-brain barrier. *Advanced Drug Delivery Reviews*, *64*(7), 640–665. https://doi.org/10.1016/j.addr.2011.11.010.

Cho, K., Wang, X., Nie, S., Chen, Z., & Shin, D. M. (2008). Therapeutic nanoparticles for drug delivery in cancer. *Clinical Cancer Research*, *14*(5), 1310–1316. https://doi.org/10.1158/1078-0432.CCR-07-1441.

Chung, Y. J., Lee, B. I., & Park, C. B. (2019). Multifunctional carbon dots as a therapeutic nanoagent for modulating Cu(ii)-mediated β-amyloid aggregation. *Nanoscale*, *11*(13), 6297–6306. https://doi.org/10.1039/c9nr00473d.

Crucho, C. I. C., & Barros, M. T. (2017). Polymeric nanoparticles: A study on the preparation variables and characterization methods. *Materials Science and Engineering C*, *80*, 771–784. https://doi.org/10.1016/j.msec.2017.06.004.

Foust, K. D., Nurre, E., Montgomery, C. L., Hernandez, A., Chan, C. M., & Kaspar, B. K. (2009). Intravascular AAV9 preferentially targets neonatal neurons and adult astrocytes. *Nature Biotechnology*, *27*(1), 59–65. https://doi.org/10.1038/nbt.1515.

Gabathuler, R. (2010). Approaches to transport therapeutic drugs across the blood-brain barrier to treat brain diseases. *Neurobiology of Disease*, *37*(1), 48–57. https://doi.org/10.1016/j.nbd.2009.07.028.

Guccione, C., Oufir, M., Piazzini, V., Eigenmann, D. E., Jähne, E. A., Zabela, V., et al. (2017). Andrographolide-loaded nanoparticles for brain delivery: Formulation, characterisation and in vitro permeability using hCMEC/D3 cell line. *European Journal of Pharmaceutics and Biopharmaceutics*, *119*, 253–263. https://doi.org/10.1016/j.ejpb.2017.06.018.

Habgood, M., & Ek, J. (2010). Delivering drugs into the brain: Barriers and possibilities. *Therapeutic Delivery*, *1*(4), 483–488. https://doi.org/10.4155/tde.10.58.

He, Q., Liu, J., Liang, J., Liu, X., Li, W., Liu, Z., et al. (2018). Towards improvements for penetrating the blood–brain barrier—Recent progress from a material and pharmaceutical perspective. *Cell*, *7*(4).

Janaszewska, A., Lazniewska, J., Trzepiński, P., Marcinkowska, M., & Klajnert-Maculewicz, B. (2019). Cytotoxicity of dendrimers. *Biomolecules*, *9*.

Kabanov, A. V., Batrakova, E. V., Melik-Nubarov, N. S., Fedoseev, N. A., Dorodnich, T. Y., Alakhov, V. Y., et al. (1992). A new class of drug carriers: Micelles of poly(oxyethylene)-poly (oxypropylene) block copolymers as microcontainers for drug targeting from blood in brain. *Journal of Controlled Release*, *22*(2), 141–157. https://doi.org/10.1016/0168-3659(92)90199-2.

Kaur, I. P., Bhandari, R., Bhandari, S., & Kakkar, V. (2008). Potential of solid lipid nanoparticles in brain targeting. *Journal of Controlled Release*, *127*(2), 97–109. https://doi.org/10.1016/j.jconrel.2007.12.018.

Kim, J. Y., Choi, W. I., Kim, Y. H., & Tae, G. (2013). Brain-targeted delivery of protein using chitosan- and RVG peptide-conjugated, pluronic-based nano-carrier. *Biomaterials*, *34*(4), 1170–1178. https://doi.org/10.1016/j.biomaterials.2012.09.047.

Kuzum, D., Yu, S., & Philip Wong, H.-S. (2013). Synaptic electronics: Materials, devices and applications. *Nanotechnology*, 382001. https://doi.org/10.1088/0957-4484/24/38/382001.

Ma, X., Zhong, L., Guo, H., Wang, Y., Gong, N., Wang, Y., et al. (2018). Multiwalled carbon nanotubes induced hypotension by regulating the central nervous system. *Advanced Functional Materials*, *28*.

Mishra, B., Patel, B. B., & Tiwari, S. (2010). Colloidal nanocarriers: A review on formulation technology, types and applications toward targeted drug delivery. *Nanomedicine: Nanotechnology, Biology, and Medicine*, *6*(1), 9–24. https://doi.org/10.1016/j.nano.2009.04.008.

Neves, A. R., Queiroz, J. F., & Reis, S. (2016). Brain-targeted delivery of resveratrol using solid lipid nanoparticles functionalized with apolipoprotein E. *Journal of Nanobiotechnology*, *14*(1). https://doi.org/10.1186/s12951-016-0177-x.

Niethammer, M., Tang, C. C., LeWitt, P. A., Rezai, A. R., Leehey, M. A., Ojemann, S. G., et al. (2017). Long-term follow-up of a randomized AAV2-GAD gene therapy trial for Parkinson's disease. *JCI Insight*, *2*(7), e90133. https://doi.org/10.1172/jci.insight.90133.

P. Blumling, J., III, & A. Silva, G. (2012). Targeting the brain: Advances in drug delivery. *Current Pharmaceutical Biotechnology*, *13*(12), 2417–2426. https://doi.org/10.2174/138920112803341833.

Pakhira, B., Ghosh, M., Allam, A., & Sarkar, S. (2016). Carbon nano onions cross the blood brain barrier. *RSC Advances*, *6*(35), 29779–29782. https://doi.org/10.1039/C5RA23534K.

Pardeshi, C., Rajput, P., Belgamwar, V., Tekade, A., Patil, G., Chaudhary, K., et al. (2012). Solid lipid based nanocarriers: An overview. *Acta Pharmaceutica*, *62*(4), 433–472. https://doi.org/10.2478/v10007-012-0040-z.

Pardridge, W. M. (2003). Blood-brain barrier drug targeting: The future of brain drug development. *Molecular Interventions*, *3*(2), 51–90. https://doi.org/10.1124/mi.3.2.90.

Petri, B., Bootz, A., Khalansky, A., Hekmatara, T., Müller, R., Uhl, R., et al. (2007). Chemotherapy of brain tumour using doxorubicin bound to surfactant-coated poly(butyl cyanoacrylate) nanoparticles: Revisiting the role of surfactants. *Journal of Controlled Release*, *117*(1), 51–58. https://doi.org/10.1016/j.jconrel.2006.10.015.

Plonska-Brzezinska, M. E. (2019). Carbon nano-onions: A review of recent progress in synthesis and applications. *ChemNanoMat*, *5*(5), 568–580. https://doi.org/10.1002/cnma.201800583.

Redzic, Z. (2011). Molecular biology of the blood-brain and the blood-cerebrospinal fluid barriers: Similarities and differences. *Fluids and Barriers of the CNS, 8*(1). https://doi.org/10.1186/2045-8118-8-3.

Rehman, M., Madni, A., Shi, D., Ihsan, A., Tahir, N., Chang, K. R., et al. (2017). Enhanced blood brain barrier permeability and glioblastoma cell targeting via thermoresponsive lipid nanoparticles. *Nanoscale, 9*(40), 15434–15440. https://doi.org/10.1039/c7nr05216b.

Savolainen, J., Edwards, J. E., Morgan, M. E., McNamara, P. J., & Anderson, B. D. (2002). Effects of a P-glycoprotein inhibitor on brain and plasma concentrations of anti-human immunodeficiency virus drugs administered in combination in rats. *Drug Metabolism and Disposition, 30*(5), 479–482. https://doi.org/10.1124/dmd.30.5.479.

Schaffazick, S. R., Pohlmann, A. R., Dalla-Costa, T., & Guterres, S. S. (2003). Freeze-drying polymeric colloidal suspensions: Nanocapsules, nanospheres and nanodispersion. A comparative study. *European Journal of Pharmaceutics and Biopharmaceutics, 56*(3), 501–505. https://doi.org/10.1016/S0939-6411(03)00139-5.

Serlin, Y., Shelef, I., Knyazer, B., & Friedman, A. (2015). Anatomy and physiology of the blood–brain barrier. *Seminars in Cell & Developmental Biology, 38*, 2–6.

Sharma, G., Sharma, A. R., Lee, S. S., Bhattacharya, M., Nam, J. S., & Chakraborty, C. (2019). Advances in nanocarriers enabled brain targeted drug delivery across blood brain barrier. *International Journal of Pharmaceutics, 559*, 360–372. https://doi.org/10.1016/j.ijpharm.2019.01.056.

Shaw, T. K., Mandal, D., Dey, G., Pal, M. M., Paul, P., Chakraborty, S., et al. (2017). Successful delivery of docetaxel to rat brain using experimentally developed nanoliposome: A treatment strategy for brain tumor. *Drug Delivery, 24*(1), 346–357. https://doi.org/10.1080/10717544.2016.1253798.

Shiraishi, K., Wang, Z., Kokuryo, D., Aoki, I., & Yokoyama, M. (2017). A polymeric micelle magnetic resonance imaging (MRI) contrast agent reveals blood–brain barrier (BBB) permeability for macromolecules in cerebral ischemia-reperfusion injury. *Journal of Controlled Release, 253*, 165–171. https://doi.org/10.1016/j.jconrel.2017.03.020.

Shukla, S., Wen, A. M., Ayat, N. R., Commandeur, U., Gopalkrishnan, R., Broome, A. M., et al. (2014). Biodistribution and clearance of a filamentous plant virus in healthy and tumor-bearing mice. *Nanomedicine, 9*(2), 221–235. https://doi.org/10.2217/nnm.13.75.

Sonkar, R., Sonali, Jha, A., Viswanadh, M. K., Burande, A. S., Narendra, et al. (2021). Gold liposomes for brain-targeted drug delivery: Formulation and brain distribution kinetics. *Materials Science and Engineering C, 120*. https://doi.org/10.1016/j.msec.2020.111652.

Sun, P., Xiao, Y., Di, Q., Ma, W., Ma, X., Wang, Q., et al. (2020). Transferrin receptor-targeted PEG-PLA polymeric micelles for chemotherapy against glioblastoma multiforme. *International Journal of Nanomedicine, 15*.

Szczęch, M., & Szczepanowicz, K. (2020). Polymeric core-shell nanoparticles prepared by spontaneous emulsification solvent evaporation and functionalized by the layer-by-layer method. *Nanomaterials, 10*(3), 496. https://doi.org/10.3390/nano10030496.

Wang, J. X., Sun, X., & Zhang, Z. R. (2002). Enhanced brain targeting by synthesis of 3′,5′-dioctanoyl-5-fluoro-2′-deoxyuridine and incorporation into solid lipid nanoparticles. *European Journal of Pharmaceutics and Biopharmaceutics, 54*(3), 285–290. https://doi.org/10.1016/S0939-6411(02)00083-8.

Wolfram, J., Zhu, M., Yang, Y., Shen, J., Gentile, E., Paolino, D., et al. (2015). Safety of nanoparticles in medicine. *Current Drug Targets, 16*(14), 1671–1681. https://doi.org/10.2174/1389450115666140804124808.

Yao, K., Lv, X., Zheng, G., Chen, Z., Jiang, Y., Zhu, X., et al. (2018). Effects of carbon quantum dots on aquatic environments: Comparison of toxicity to organisms at different trophic levels.

Environmental Science and Technology, 52(24), 14445–14451. https://doi.org/10.1021/acs.est.8b04235.

Zhang, X., Chen, G., Wen, L., Yang, F., Shao, A. L., Li, X., et al. (2013). Novel multiple agents loaded PLGA nanoparticles for brain delivery via inner ear administration: In vitro and in vivo evaluation. *European Journal of Pharmaceutical Sciences, 48*(4–5), 595–603. https://doi.org/10.1016/j.ejps.2013.01.007.

Zhang, F., Trent Magruder, J., Lin, Y. A., Crawford, T. C., Grimm, J. C., Sciortino, C. M., et al. (2017). Generation-6 hydroxyl PAMAM dendrimers improve CNS penetration from intravenous administration in a large animal brain injury model. *Journal of Controlled Release, 249*, 173–182. https://doi.org/10.1016/j.jconrel.2017.01.032.

Zhao, Y., Xiong, S., Liu, P., Liu, W., Wang, Q., Liu, Y., et al. (2020). Polymeric nanoparticles-based brain delivery with improved therapeutic efficacy of Ginkgolide B in Parkinson's disease. *International Journal of Nanomedicine, 15*.

Zhao, L., Zhu, J., Cheng, Y., Xiong, Z., Tang, Y., Guo, L., et al. (2015). Chlorotoxin-conjugated multifunctional dendrimers labeled with radionuclide 131I for single photon emission computed tomography imaging and radiotherapy of gliomas. *ACS Applied Materials and Interfaces, 7*(35), 19798–19808. https://doi.org/10.1021/acsami.5b05836.

Zhou, Y., Peng, Z., Seven, E. S., & Leblanc, R. M. (2018). Crossing the blood-brain barrier with nanoparticles. *Journal of Controlled Release, 270*, 290–303. https://doi.org/10.1016/j.jconrel.2017.12.015.

CHAPTER 3

Designing of nanocarriers for liver targeted drug delivery and diagnosis

Introduction

Liver is the key organ in human body responsible for numerous biological processes including protein synthesis, detoxification, and the production of necessary biochemicals for life sustenance. Viral hepatitis, liver cirrhosis, and hepatocellular carcinoma (HCC) are some of the chronic liver diseases that need immediate attention for sustainable life of patients with such chronic liver diseases. Among these diseases, HCC is the most prevalent liver malignancy and is a major cause of cancer-related deaths worldwide. HCC is the ninth leading cause of cancer-related deaths in the United States (Ward, Watson, Momin, & Richardson, 2001). HCC is invasive cancer that often occurs as a result of cirrhosis and generally appears in its advanced stages. By using proper measures such as hepatitis B vaccination, worldwide screening of blood products, observing safety measures for injection practices, spreading public awareness, treatment for intravenous drugs users and alcoholics, prevention of HCC can be assured. Gradual progress in surgical and nonsurgical strategies has shown benefits in overall survival of HCC patients. The rate of HCC occurrence and its mortality rate are continuously rising besides advancement in prevention applications, screening, and new technologies for diagnosis and treatment of HCC (Balogh et al., 2016).

The risk factor of HCC increases with chronic medical issues like obesity and diabetes mellitus. Sixty percent of patients above 50 years are considered to have advanced fibrosis with non-alcohol-induced steatohepatitis (NASH), diabetes mellitus, or obesity (Rinella, 2015). The liver is directly affected by diabetes due to its vital role in glycogenesis which results in chronic hepatitis, hepatic steatosis, liver failure, and cirrhosis (Gao, Fang, Zhao, Li, & Yao, 2013; Wang et al., 2012). HCC has a long latent period, that's why when it is diagnosed, most of the patients are at the middle or advanced stage so the only way of treatment is chemotherapy (Carrilho et al., 2015). Although many times chemotherapy showed less response rate and serious side effects such as; sorafenib an approved first-line drug for HCC, exhibits a 3-month median

45

Nanocarriers for Organ-Specific and Localized Drug Delivery. https://doi.org/10.1016/B978-0-12-821093-2.00006-2

survival benefit but is associated with drug-related adverse events including toxicity, hypertension, nausea, mucositis, alopecia, and hand-foot skin reaction (Llovet et al., 2008). These severe effects are resulted due to the limitations of conventional chemotherapy and also due to the HCC characteristics. Traditional anticancer agents owe less selectivity and high toxicity toward cancer tumors which results in damage to the normal cell proliferation process too (Chidambaram, Manavalan, & Kathiresan, 2011). Less accumulation of active therapeutic agents in HCC area results in very low drug concentration which is usually ineffective for HCC treatment (Needham & Dewhirst, 2001). Moreover, in chemotherapy often inherent and acquired multidrug resistance (MDR) occurs, which results in the recurrence of cancer cells leading to poor survival and/or poor quality of life (Han & Park, 2008). In recent years, we've seen the flourishing of nano-science at an extraordinary pace. It deals with the materials at nanometer scale where the properties of the conventional materials change extraordinarily. Growing interest in nanotechnology's potential applications in medicine has spawned a new area known as nanomedicine. Nanotechnology-based medicines are expected to lead to substantial advancement in disease prevention, diagnosis, and treatment. Nanocarriers-based drug delivery strategies have the potential to overcome the limitations of conventional drug delivery such as toxicity to normal cells, low bioavailability, and non-specific distribution of drugs. However, in dept understanding of the pathophysiologic foundation of the disease is vital for construction of effective nanocarriers-based delivery system to handle it (Mehta, Guvva, & Patil, 2008). In the following sections, we first describe the physicochemical characteristics of HCC, the changes that occur in HCC, and then we will describe different processes that can be targeted using nanocarriers-based drug delivery for the treatment of HCC.

Physiochemical and biological characteristics of HCC

It is very important to understand the biology of HCC and normal liver to find out the different targets that can be targeted for developing some possible targeted DDSs for the treatment of HCC. Here we discuss changes in cells, vascularity, proteins, and other factors especially those that are present on cell surface in HCC and give better presentation for developing nanocarriers-based drug delivery systems.

Morphology and HCC structure

The properties and metastases of HCC are very related to its structure especially extracellular matrix (ECM), which contains collagens, proteoglycans, and glycoproteins (Liotta, 1986; Rojkind & Ponce-Noyola, 1982). Biologically HCC is controlled by non-cellular tumor stroma components by affecting cancer cell's signaling pathways or tumor metastasis and invasion. Basically tumor stroma consists of extracellular matrix proteins, proteolytic enzymes, growth factors, and also inflammatory cytokines (Yang, Nakamura, & Roberts, 2011). All these components are facilitating tumor growth by exerting different functions on supporting cancer cell's abnormal structure and can be modified by HCC (Wu, Chen, & Xie, 2006). In ECM collagens is the most abundant protein. At sub-endothelial spaces of sinusoids collagens type I and III are present and also at intercellular spaces of the tumor cells to maintain the structure, and at the basal side fibronectin and laminin are present on endothelial cells. In HCC, a replacement occurs between collagen fibers-IV and VI for collagen fibers-I and III (fibrotic fibers), causing an impairment in nutrient exchange to activate the immune system. At the sub-endothelial space of sinusoids, some intercellular spaces help to contribute to the composition of baseline in cancer cells membrane (Torimura et al., 1994). Collagen XVIII is overexpressed at normal adult hepatocytes cells (Torimura et al., 1994).

Two different types of collagen XVII variants, i.e., SHORT and LONG represent specific tissue expression. The SHORT type is the basic major constituent of ECM in HCC tumor and the LONG type is present specifically and abundantly in HCC tissues as a liver-exclusive regulatory element (Musso et al., 2001). Additionally, proteoglycan is another basic constituent for the structural framework of cancer tumors (Leonardi et al., 2012). Three main categories of proteoglycans are present in ECM which include keratan sulfate, chondroitin sulfate, and heparan sulfate. As compared to normal human liver cells, HCC contains 24 times more chondroitin sulfate (Kovalszky et al., 1990). The unusual appearance of chondroitin sulfate plays a vital function in progression and metastasis processes in HCC and it can be potential marker target for developing drug delivery to target HCC (Jia et al., 2012).

Vascular changes

In normal liver hepatocytes, 75% of blood is supplied from portal vein while the remaining 25% from hepatic artery (Lautt & Greenway, 1987). But in HCC tissues (specifically in advanced HCC), blood supply occurs mainly through the arterial system due to de-differentiation and arterialization (Yang & Poon, 2008). Blood supply, angioarchitecture of the blood vessels, microvasculature and their functions are all altered in HCC. At initial state, well differentiated HCC normally receive blood through preferential portal vain however poorly and moderately differentiated HCCs at their advanced stage receive blood through arterial vain due to its high requirement for oxygen and nutrition (Yamamoto et al., 2001).

In HCC tissues, unpaired vessels with numerous arteries in the parenchyma cell of the tumor contain an unusual structure that forms irregular structure, abnormal diameter and density, lacking normal muscle layer and branching pattern (Semela & Dufour, 2004). Additionally, microvessels are unevenly distributed at the inner side of HCC, and microvascular density (numerous positively stained blood vessels per unit area) is indirectly correlated with tumor size in HCC (Ng et al., 2001). Another important factor related to increased permeability is the endothelial cells in hepatic sinusoids contain large fenestrations with a diameter of 150–175 nm (Ballet, 1990). Same as the endothelial fenestrae present on many other cancer cells with a diameter of approximately 50–80 nm (Roberts & Palade, 1997). Moreover, the removal and absence of mural pericytes and/or smooth vascular muscles in HCC also contribute to the leaking of cancer vessels (Jain & Booth, 2003).

HCC-specific membrane receptors

There are several proteins and receptors overexpressed on the cell surface of HCC tissues and also in normal healthy liver cells. In this part, we focus only on proteins and receptors that are expressed with specific ligands on HCC cells as represented in Fig. 3.1.

Asialoglycoprotein receptor (ASGP-R)

Asialoglycoprotein receptor (ASGP-R) was first isolated and characterized by Baenziger and Maynard in late 1970s, which is a glycoprotein that is overexpressed specifically on liver cells (Baenziger & Maynard, 1980). It contains a particular subunit with a molecular weight of 41,000 and has the ability to bind with the terminal part of galactose and galactosamines residues after identifying them (Weigel & Oka, 1982). Another important feature of this receptor is that it plays a vital role in the removal of desalinated proteins from blood serum via endocytosis and lysosomal degeneration (Ashwell & Harford, 1982). Normally ASGP-R ligands consist of numerous galactosamines and/or galactose such as asialofetuin, asialotransferrin, and asialoorosomucoids (Shi, Abrams, & Sepp-Lorenzino, 2013). Binding of asialoorosomucoid to ASGP-R has been found to be dependent on time, saturation, and dissociation and also sensitivity toward

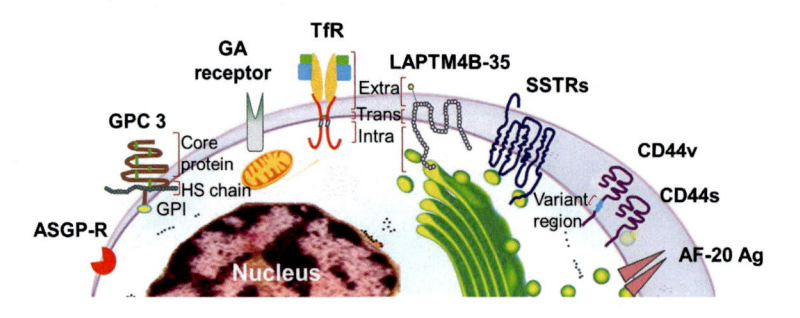

FIG. 3.1

Receptors and proteins expressed on HCC membrane. *From Zhang, X., Ng, H. L. H., Lu, A., Lin, C., Zhou, L., Lin, G., et al. (2016). Drug delivery system targeting advanced hepatocellular carcinoma: Current and future.* Nanomedicine: Nanotechnology, Biology, and Medicine, *12(4), 853–869. https://doi.org/10.1016/j.nano.2015.12.381.*

neuraminidase enzymes in hepatic cells of rats (Steer & Ashwell, 1980). ASGP-R also shows specific binding with targeted calcium receptors (Li, Huang, Diakur, & Wiebe, 2008). ASGPRs are overexpressed on the surfaces of HCC and hepatocyte cells which make them an excellent targeting receptor for designing anticancer drug delivery to liver for the treatment of liver cancer (Kaneo, Tanaka, Nakano, & Yamaguchi, 2001).

Glypican-3 (GPC3)

Glypican-3 (GPC3, also known as, GTR2-2, SGBS, OCI-5, DGSX, and MXR7) is a part of membrane-bound proteoglycans heparin sulfate. The principal structure of GPC3 contains glycosyl-phosphatidylinositol (GPI), heparin sulfate, and core protein. GPC3 protein is attached to the surface of a hepatic cell through a covalent bond with GPI. A prominent negative charge present on the sulfate sugar chains is beneficial for the attraction to a positive-charge protein like heparin-binding growth factor. GPC3 belongs to the glypican family, shows an N-terminal secretory signal peptide as well as a hydrophobic domain required for the addition of the GPI anchor at the C-terminus (Filmus & Selleck, 2001). GPC3 is over-expressed in several cancer types including HCC and its expression in HCC contributes to the growth and metastasis of HCC.

Glycyrrhetinic acid receptor

Glycyrrhetinic acid has a specific binding site for hepatic membrane in rats and was first reported in 1991 by Negishi and co-workers (Negishi, Irie, Nagata, & Ichikawa, 1991). This was further confirmed by Ismair et al., who observed cellular uptake of glycyrrhizin in hepatic cells through a carrier-facilitated pathway (Ismair et al., 2003). Glycyrrhetinic acid is an aglycone of glycyrrhizic acid containing a steroid structure from plant Glycyrrhiza root. The difference in structure between glycyrrhizin and glycyrrhetinic acid is the attachment of hydroxyl group at the C3 position, thus it initially binds through hydroxyl and then with glycosyl group. But as compared to glycyrrhetinic acid, glycyrrhizin has been found with lower affinity toward the binding of glycyrrhetinic acid receptor. It indicates that C3-hydroxyl group is important for targeting membrane of hepatic cells' receptor. In a recent study, it has been found that the β-configuration of hydrogen atom at C-18 of GA plays an important role in the targeting effect while the carbonyl at C-11 and hydroxyl group at C-3 of GA have little influence on targeting action to HCC. The overexpression of glycyrrhetinic acid receptors in HCC could be used as potential target for developing nanocarriers-based drug delivery system for targeting these receptors (Sun et al., 2018).

Transferrin receptor (TfR)

Transferrin receptor (TfR), also known as CD71, is a transmembrane glycoproteins receptor that was originally thought as a marker for proliferating cells due to its expression in cancer cells, however, some resting cells can also express it. TfR plays an important role in cell proliferation. It is homodimeric in nature that contains 90 kDa subunits connected by two disulfide intermolecular bonds (Jing & Trowbridge, 1987). Individual subunit consists of 760 amino acids with a molecular weight of 90–95 kDa, and it is distributed into three domains where the first transmembrane domain (a short single-pass domain) contains 28 amino acids. The second N-terminal domain (intracellular domain) contains 61 amino acids and the third large extracellular C-terminal domain contains 671 amino acids (Daniels, Delgado, Rodriguez, Helguera, & Penichet, 2006). In normal healthy cells, low or less expression of TfR is present but over-expressed on a number of proliferative cells (Sutherland et al., 1981). Cancer cells show rapid proliferation process thus roughly express more than 100,000 TfR receptors per cell (Inoue, Cavanaugh, Steck, Brünner, & Nicolson, 1993). In humans, numerous cases (33/34 cases) confirmed the presence of TfRs on HCC cells by immuno-staining (Sciot et al., 1988). This extensive TfR expression in rapidly proliferating cells like HCC makes it a potential target for development of targeted nanocarriers-based drug delivery and antitumor therapy.

Somatostatin receptors (SSTRs)

Somatostatin receptors (SSTRs) are a superfamily of transmembrane receptors connected with G-proteins that exhibit different expression densities in various endocrine and exocrine glands, e.g., pancreas, thyroid, adrenals, and various other organs like kidneys, brain, gut, and immune cells (Patel, 1999). SSTRs are overexpressed in many cancer cells lines (Patel, 1999) and they have also been reported to have over-expression in the human diseased liver conditions like acute and chronic hepatitis, cirrhosis, and HCC (Kouroumalis et al., 1998). It has been reported that 41% of HCCs patients show over-expressed SSTRs (Reubi et al., 1999) while its expression in normal healthy liver hepatocytes and hepatic stellate cells shows negative results (Reynaert et al., 2004). In another study, the unequal expression of SSTR in both HepG2 and SMMC-7721 cells (HCC cell lines) was found, however, no expression was found in L-02 cells hepatocytes, indicating SSTRs can be used as a major target site for the treatment of HCC with minor side effects (Liu, Huo, & Wang, 2004).

Folate receptors

Folate receptor is a distinct cancer biomarker that interacts with B9 vitamin (folic acid). Folate receptor-α (alpha isoform) is overexpressed on 40% of cancer cells in comparison to its β isoform which is present on macrophage surfaces and also expressed on HCC malignant cells of hematopoietic origin (Low & Kularatne, 2009). Cancer cells require a high amount of folate and transfer it via cell membrane by using folate receptors. Membrane-associated reduced folate carriers are present on almost all cell surfaces but folate receptors are principally present only on activated macrophage and also on polarizing epithelial cells (Lu & Low, 2012). Folate conjugated drug carriers contain high efficacy and show excellent binding with cancer cells and are internalized through receptor facilitated endocytosis. Folate-containing nanocarriers deliver their loaded drug via receptors facilitated endocytosis. The intracellular transfer can be done in acidic media that promote drug release from the carrier and essentially released drug can be retained or delayed until reached to the cytoplasm or targeted site to suppress the efflux generated by multidrug resistance pumps (Kularatne & Low, 2010; Low, Henne, & Doorneweerd, 2008). Distribution of therapeutic drugs, diagnostic and imaging agents by folate conjugated nanocarriers proved that they have superiority without addition of other targeting ligands on nanocarriers surface (Ling, Yuen, Magosso, & Barker, 2009; Malhi et al., 2012). Because these folate receptors are over-expressed on macrophages which is, in most cases, the sign of inflammatory diseases including Crohn's disease, psoriasis, rheumatoid arthritis, and atherosclerosis. Thus, folate-conjugated nanocarriers can effectively be used for dealing with inflammatory diseases of the body and liver (Kularatne & Low, 2010).

Epidermal growth factor receptors (EGFR)

The ErbB family of receptor tyrosine kinases, which is consisted of epidermal growth factor receptor (EGFR), ErbB2, ErbB3, and ErbB4, regulates complex signaling network impacting different cellular processes, e.g., survival, proliferation, angiogenesis, and metastasis in a number of cancers and also works in management of cell damage in normal healthy cells (Lehtinen et al., 2012). EGFR (a 170 kDa glycoprotein and member of the ErbB family) is abundantly present on various solid cancer cells like breast cancer, lung cancer, colorectal, pancreas, prostate, neck, and kidney. Thus, due to this over-expression property of this receptor, it promotes target site-selective chemotherapeutic drug delivery via nanocarriers (Danhier, Feron, & Préat, 2010; Kim & Huang, 2012). Modification of therapeutic drug encapsulated nanocarriers can be achieved mostly with antibodies, proteins, peptides, epithelial growth factors, and aptamers. EGFR-targeted nanocarriers have been mostly decorated/functionalized with antibodies and/or antibody fragments because of the United States' Food and Drug Administration (US FDA) approval of these agents in immunotherapies such as trastuzumab and cetuximab. Functionalization of nanocarriers with these motifs is being utilized for multifunctional purposes including drug delivery, imaging, radiofrequency, and/or photothermal ablation (Gao et al., 2009; Kim & Huang, 2012; Master & Sen Gupta, 2012; Wong & Vij, 2009). The above-mentioned strategies for targeting HCC have been depicted in Fig. 3.2:

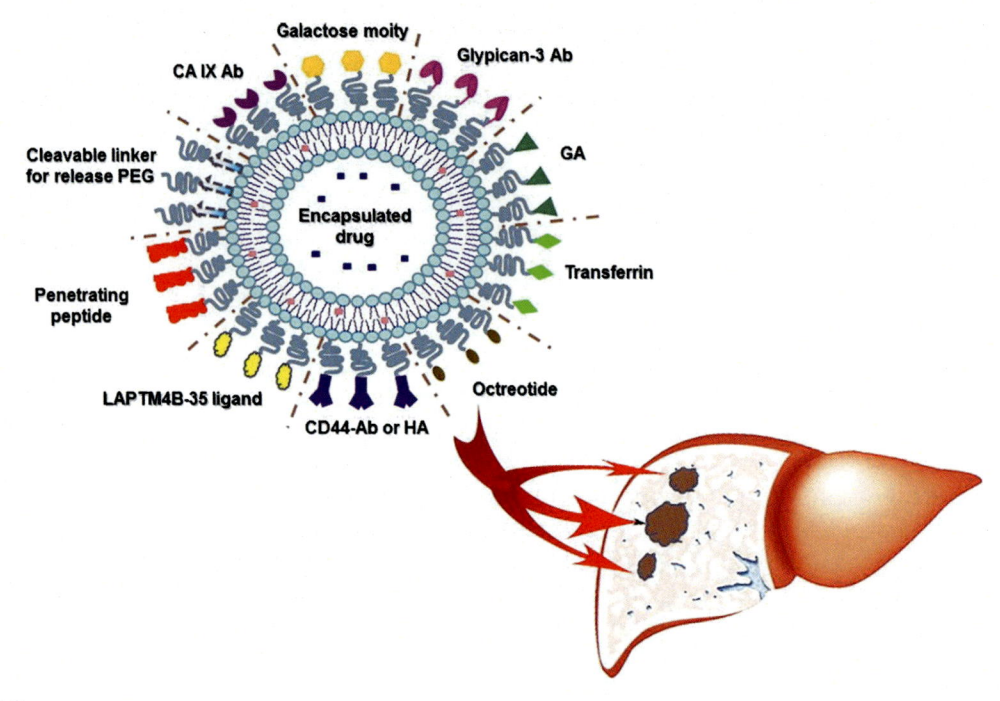

FIG. 3.2

Different strategies ligands based on targeting of HCC through nanocarriers. *Reproduced with permission from Zhang, X., Ng, H. L. H., Lu, A., Lin, C., Zhou, L., Lin, G., et al. (2016). Drug delivery system targeting advanced hepatocellular carcinoma: Current and future. Nanomedicine: Nanotechnology, Biology, and Medicine, 12(4), 853–869. https://doi.org/10.1016/j.nano.2015.12.381.*

Nanocarriers (NCs)

The traditional chemotherapy for liver cancer has many negative effects like high rate of drug clearance, severe side effects on healthy cells, multi-drug resistance (MDR), undesirable drug distribution to other parts of the body, and very less concentration of drug reaching to target site. Thus, it is important to develop unique strategies and NCs that carry therapeutic agents to targeted hepatic cancer cells in an acceptable amount and within a certain time duration. There are many advantages of nanocarriers-based therapeutic systems as compared to traditional chemotherapy like it has high drug loading efficacy, maximum cellular uptake, high drug release, and most important having very few side effects. These NCs contains minimum toxic effects on healthy cells and show maximum accumulation of drug in cancer cells (Ruman, Fakurazi, Masarudin, & Hussein, 2020).

Numerous macromolecular architectures like dendrimers, micelles, polymer-drug conjugates, liposomes, and nanoparticles have been designed to deliver therapeutic agents to the targeted site. Amphiphilic block-copolymers containing hydrophobic and hydrophilic segments can self-assemble into spherical shapes have been utilized for the construction of micelles (Aliabadi, Shahin, Brocks, & Lavasanifar, 2008; Matsumura, 2008; Yokoyama, 2005). Whereas, liposomes are made from amphiphilic natural and synthetic lipids that form bilayered spherical-shaped structures and encapsulate therapeutic agents in their central core or within the bilayer portion of the nanostructure (Yokoyama, 2005). In nanoparticles, normally nano-size colloidal particles are formed by a polymeric matrix that encapsulates drugs in them via secondary interaction between them or with chemical bond formation between them (Soussan, Cassel, Blanzat, & Rico-Lattes, 2009; Uhrich, Cannizzaro, Langer, & Shakesheff, 1999). Three most important properties of nanocarriers should be considered for designing targeted drug delivery systems. The first and foremost point to consider is the ideal size of nanocarrier. The size of the nanocarrier should be between 10 and 100 nm in most cases. The size of nanocarriers needs to be smaller than 400 nm for efficient extravasation (from the fenestrations in leaky vasculature). On the other hand, NCs size should be greater than 10 nm to prevent them from filtration (from kidneys). To avoid specific entrapments through liver, the nanocarrier should be less than 100 nm. The next point to consider is the surface charge of the nanocarrier. For effective evasion of the renal filtration, the surface charge of the nanosystem should be anionic and/or neutral. Thirdly, it is necessary to hide NCs from the reticuloendothelial system to prevent NCs material damage via opsonization and later by phagocytosis process (Gullotti & Yeo, 2009; Malam, Loizidou, & Seifalian, 2009). Liver chemotherapeutic agents are mostly inhibitors of tyrosine kinase (anti-angiogenesis). These chemotherapeutic agents block signals that are responsible for normal cell functions as well. Although initially, they restrict the proliferation of HCC, simultaneously they affect the normal growth of healthy cells like bone marrow, hair follicles, gastrointestinal tract (GIT) cells in the body (Chabner & Roberts, 2005). All these severe side effects of these chemotherapeutic agents can be avoided and controlled by utilizing nanocarriers for their encapsulation and then delivering to a specific site with high efficacy. Some of the most widely practiced NC's structures are given in Fig. 3.3 and their applications in HCC are discussed here:

Organic nanocarriers

A broad range of organic compound-based nanostructures have been used for so long in drug delivery systems for enhancing therapeutic efficacy and reducing the side effects associated with conventional drug delivery approaches (Chabner & Roberts, 2005). Liposomes are among the most widely used organic nanocarriers for drug delivery applications having spherical-shaped vesicles composed of self-assembling phospholipid bilayers surrounding an internal aqueous cavity. Other lipids-based organic nanocarriers that have been

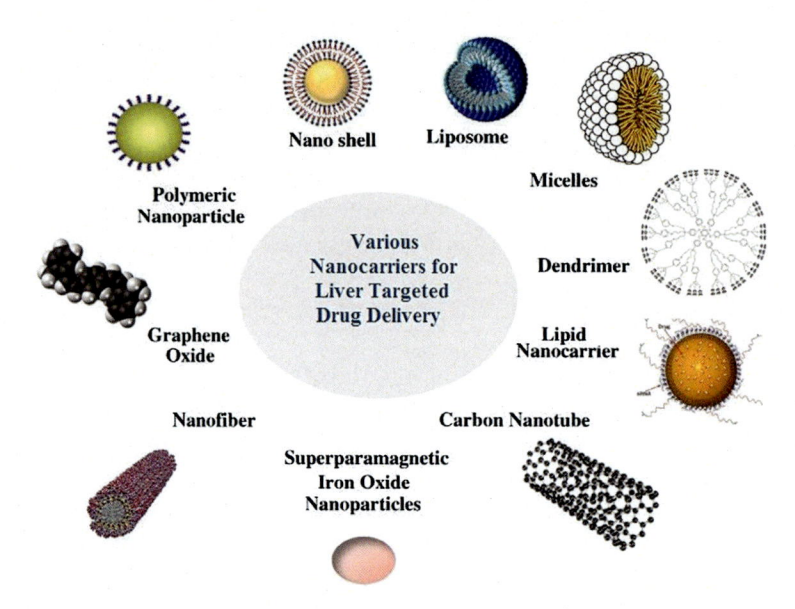

FIG. 3.3

Various nanocarriers for liver targeted drug delivery applications.

extensively explored for drug delivery purposes include solid lipid nanoparticles (SLNs), nanostructured lipid carriers (NLCs), oil emulsions including self-emulsifying drug delivery systems (SEDDs), polymer nanoparticles, micelles, dendrimers, and metal-organic frameworks (MOFs). Polymers are basically organic molecules with higher molecular weights and the NCs made from organic polymers thus come under organic nanocarriers. In HCC, polymers-based nanocarriers are among the most widely explored organic nanocarriers (Wang et al., 2010). Organic nanocarriers owe numerous good qualities such as they are biocompatible, biodegradable, nontoxic, safe for autogenous tissues, and also show long blood circulation times. Moreover, the surface functionalization possibilities of these organic nanosystems make them excellent delivery vehicles for targeting different sites in the body including brain, liver, and kidney and even blood components can also be targeted. Organic nanocarriers also possess flexible structures and thus the chemotherapeutic agents are easily encapsulated, adsorbed, and interact with the surface of organic nanocarriers (Masood, 2016). In the following section, we will discuss some of these organic nanocarriers with respect to their use in liver diseases.

Organic nanoparticles

Organic nanoparticles (NPs) are solid colloidal systems mostly made of polymers in which the therapeutic agents are dissolved, dispersed, or chemically bound with polymeric chains (Dufès, Uchegbu, & Schätzlein, 2005;

Uchegbu & Schatzlein, 2006). Different shaped and architectured NPs (nanospheres and nanocapsules) could be constructed because the shape and architecture of NPs depend on the applied preparation methods. In nanospheres, a matrix-like system is present in which drugs are dispersed or encapsulated by physical/secondary interactions with polymeric chains. In contrast, a vesicular system is formed in nanocapsules in which drugs are confined to a cavity surrounded by the polymeric membrane (Letchford & Burt, 2007). NPS based on carbohydrates (polysaccharides) received great attention for drug delivery in recent years because they are abundantly available in nature, cheaper, less toxic, biocompatible, biodegradable, and show very less immunogenicity (Liu, Jiao, Wang, Zhou, & Zhang, 2008; Seidi, Jenjob, Phakkeeree, & Crespy, 2018). Moreover, due to the presence of various available functional groups in these polymers, they can be functionalized with targeting ligands and other polysaccharides (Bodnar, Hartmann, & Borbely, 2005).

Many researchers utilized chitosan for the construction of polymer NPs and conducted targeted drug delivery to HCCs. Chitosan is a cationic polysaccharide polymer, biodegradable in nature and obtained by deacetylation of Chitin. El-Marakby et al. reported chitosan NPs' conjugates with valerate and enhanced their efficiency with glycyrrhizin for targeting glycyrrhetinic acid receptors in the liver and delivered ferulic acid to HCC. They found this targeted chitosan NPs system as a safe and efficient strategy for HCC treatment and targeting (El-Marakby, Hathout, Taha, Mansour, & Mortada, 2017). Liu et al. fabricated A54 peptide-modified chitosan-based NCs that showed higher HCC accumulation and maximum cancer tumor-suppressive effects in comparison with unmodified chitosan nanoparticles (Liu et al., 2016). In most cases, the surface of NPs is functionalized with ligands that help to reduce their interaction nonspecifically with macrophages and reticuloendothelial systems (RES) and thus enhance their accumulation in cancer tissues. The introduction of hepatoma-targeted ligands makes difficult the synthesis of NPs, thus, polysaccharides with their innate affinity toward HCCs is an alternate way to design hepatic targeting DDS (Li et al., 2014; Wang, Gu, Wang, Yang, & Mao, 2018; Yu et al., 2014). Ye et al. observed the effect of chitosan coating on doxorubicin entrapped NCs in HCC and found its enhanced efficacy in inhibition of HCC cells proliferation (Ye, Zheng, Ruan, Zheng, & Cai, 2018). Loutfy et al. prepared chitosan NPs and evaluated their efficacy in-vitro in HCCs model. They observed that the synthesized chitosan NPs were more cytotoxic against HCCs and proposed that these chitosan NPs were efficient for use in targeted drug delivery for HCCs (Loutfy et al., 2016). In another study, Lin et al. prepared lactobionic acid (LBA) grafted-pegylated-chitosan NPS that showed increased HepG2 cells transfection containing over-expressed asialoglycoprotein receptors compared to LBA free formulation (Lin, Chen, Liu, Chen, & Chang, 2011). Tsend-Ayush et al. also have recently fabricated LBA-conjugated TPGS NPs and evaluated enhancing therapeutic efficacy of etoposide against HCC. Their results suggested that the newly fabricated NCs could be potentially used as a favorable DDS for etoposide delivery to liver for treating HCC (Tsend-Ayush et al., 2017).

Micelles and dendrimers

Amphiphilic polymers having hydrophilic and lipophilic blocks can self-assemble in aqueous medium and form nano-sized core-shell micelles. Micelles have been utilized as carriers for a number of drugs, and they show efficient drug encapsulation and maximize blood circulation times of the encapsulated drugs, e.g., hydrophobic nature anticancer therapeutic agents (Peer et al., 2007; Tian et al., 2012; Torchilin, 2001). Polymeric micelles have long been utilized in HCCs as drug delivery carriers due to their core-shell morphology, small size, and uniform size distribution of the constructed nanostructures (Huang et al., 2010). Micelles have the ability to solubilize therapeutic agents thus enhancing their therapeutic efficacy and can be used for targeted delivery in HCC treatment (Blanco, Kessinger, Sumer, & Gao, 2009).

A micellar system was constructed from poly(ethylene glycol)-*b*-poly(γ-benzyl ʟ-glutamate) polymer and functionalized with Glycyrrhetinic acid (GA) for liver targeting. The GA-modified PEG-b-PBLG micelles were used for encapsulation of anticancer drug doxorubicin (DOX), and showed efficient drug loading, pH-dependent drug release, and fourfold higher cytotoxicity against hepatic carcinoma QGY-7703 compared to the free drug (Huang et al., 2011). Another GA modified micelles-based system of sulfated chitosan (GA-SCTS) was synthesized for the delivery of DOX and showed increased (2.18 times) cellular uptake by HCC compared with other liver cells confirming its effective use in HCC targeted delivery (Tian et al., 2012). Stearic acid was grafted in chitosan bounded GA in a study, and it was found to be excellent drug delivery carrier for HCC targeting because of its excellent properties like comparatively low critical aggregation concentration (CAC) with spherical structural morphology (Chen, Sun, et al., 2015). Huang et al. fabricated GA-conjugated micelles and Yang et al. prepared polyethylene glycol-conjugated GA micelles for encapsulation of DOX. Both micelles encapsulated higher amounts of DOX, showed excellent liver cells targeting, and showed enhanced tumor growth inhibition as compared to free DOX suggesting its potential use for HCC treatment and targeting (Huang et al., 2010; Yang et al., 2019).

Dendrimers

Dendrimers are hyper-branched, nano-sized, radially symmetrical macromolecules having well-defined and organized structures that are fabricated by the systematic addition of generations (layers) of molecules around a central core (Tomalia & Fréchet, 2001). In recent years, dendrimers have attracted researchers' attention for different applications in the biomedical field especially in drug delivery (Boas & Heegaard, 2004; Cheng, Wang, Rao, He, & Xu, 2008; Dufès et al., 2005). A number of desired characteristics of dendrimers like uniform distribution and size, circular design, highly branched structures, distinct molecular mass, and functionalizable surface make them excellent carriers for drug delivery (Cheng et al., 2008; Gillies & Frechet, 2005; Svenson & Tomalia, 2012). In dendrimers, the central core consists of monomers that contain at least two or more functionalities having the property of attaching moieties and thus form dendrimers with different structures by gradual increase in generations (layer). Normally two methods are utilized for the construction of dendrimers, one is divergent method (adding up repeating units around a central core) and the second one is convergent method where distinct fragments of dendrimers are formed and connected in the final step (Pearson, Sunoqrot, Hsu, Bae, & Hong, 2012).

Different types of dendrimers with diverse functional groups on surfaces and different generations can be obtained by these methods. Different polymeric dendrimers of poly(propylene imine) and poly(amidoamine) and polypeptide based and polyester frameworks have been developed and studied in biomedical field that show enhanced biodegradability and low toxicity (Boyd et al., 2006). LBA has been most widely used liver targeting ligand for targeted delivery to HCC as previously described. It has also been used in dendrimer-based systems for targeting liver cells and HCC. For example, Liu et al. have synthesized gold NPs encapsulated dendrimers that were conjugated with LBA for targeting HCC and evaluated for CT imaging of HCCs (Li et al., 2014). Samui et al. reported LBA conjugated multifunctional dendrimer-based DDS for targeted drug delivery to HCC (Samui, Pal, Karmakar, & Sahu, 2019). Yousef et al., synthesized nanoscale fourth generation (G4) polyamidoamine (PAMAM) dendrimers anchored to galactosamine (GAL, for targeting asialoglycoprotein receptors of HCC) and loaded with anticancer curcumin derivative (CDF) for targeted drug delivery to HCC in aggressive HCC xenograft mice model. Their results showed that the targeted dendrimer was able to achieve high and selective cellular uptake through ASGPR mediated endocytosis and enhanced the delivery of curcumin derivative into HCC cell line. Moreover, the fabricated targeted G4 dendrimer showed good bio-distribution and selective accumulation in liver cancer cells of the mice in-vivo. Medina et al. synthesized N-acetyl galactosamine (NAcGal)-grafted fifth generation (G5) poly(amidoamine) (PAMAM) dendrimers for targeted delivery of anticancer drugs to hepatic cancer cells. Their results demonstrated that the uptake of NAcGal-targeted G5 dendrimers occurs via ASGPR-mediated endocytosis in hepatic cancer cells and resulted in increased drug accumulation in HCC suggesting as an excellent targeted carrier hepatic cancer cells treatment (Medina et al., 2011).

Nano-gels

Nano-gels are interconnected nano-sized structures consisted of either hydrophilic or an arrangement of both hydrophilic and hydrophobic parts containing polymers exhibiting good properties to be used for drug delivery (Kabanov & Vinogradov, 2009). There are numerous advantages of these drug delivery carriers, e.g., they have the ability to convert from micro to nano-size, they possess excellent biocompatibility, can target a specific site, availability of maximum surface area for functionalization, and also respond to stimuli for controlled drug delivery (Hector et al., 1999; Yoo & Mok, 2015). Carbohydrate-containing nano-gels have been used as "smart" DDs due to their negligible toxicities and high biocompatibility normally linked with these polymers (Hector et al., 1999). Glycopolymers have also been widely used for targeted drug delivery because they contain functional groups at their extracellular surface facilitating interactions with cellular lectins, asialoglycoprotein, and Concavalin-A receptors for targeting (El-Sayed, Futaki, & Harashima, 2009). Quan et al. synthesized galactose-based thermo-responsive nano-gels for targeted delivery of Iodo-azomycin Arabino-furanoside (IAZA) for theranostic and therapeutic management of hypoxic HCC. These galactose-based nano-gels revealed maximum drug encapsulation and sustained release profile of the drug and were found to be an excellent drug delivery carrier for targeting hypoxic HCC (Quan, Wang, Zhou, Kumar, & Narain, 2015). Wang et al. introduced glycyrrhizin (GL) in alginate (ALG) and synthesized high potential DDs which inhibited rapid clearance of the drug by macrophages and increased the anticancer efficacy of the encapsulated doxo-rubicin (DOX). These synthesized nano-gels showed ionic and intermolecular hydrogen bond interactions with DOX and acted as multifunctional carriers. The response of immune cells and macrophage phagocytosis on this nano-gel was investigated against RAW 264.7 macrophages and also evaluated in-vitro and in in-vivo studies against HCC cells and cancer-containing mice model, respectively. Their findings revealed that DOX-loaded GL-conjugated ALG-nano-gels increased blood circulation time through inhibiting the activity of the inherent immune cells and the process of phagocytosis in macrophage and also increased the bioavailability of DOX significantly as compared to free drug. These results show a unique method of combination therapy combining anticancer drugs and natural compounds for cancer treatment, specifically HCC (Wang et al., 2019).

Another galactosyl ligand-based nano-gel system was synthesized by Wu et al. from interconnected polyphosphoester decorated with galactosylated polyethylene-glycol and loaded with DOX and evaluated for targeted delivery to hepatocytes and HCC. These nano-gels showed uniform size, zeta potential, and low PDI, and loaded higher amounts of DOX. In this work, DOX encap-sulated nano-gel containing galactosyl ligands showed excellent therapeutic effect against diethylnitrosamine-induced HCC and resulted in significantly enhanced bioavailability in-vivo via minimizing biological barriers and decreased the drug clearance from hepatocytes (Wu et al., 2013). A dual responsive nano-gel formed by free-radical emulsion polymerization from Poly (N-isopropyl acrylamide)-copoly (N,N-(dimethylamino) ethyl methacrylate) containing isopropylacrylamideas and N,N-(dimethylamino)ethyl methacrylate has also been reported. This system was heat and pH stimuli sensitive and methylene-bis-acrylamide was used as a cross-linker. Cisplatin loaded with magnetic iron oxide (Fe_3O_4) nanoparticles by secondary interaction was incorporated in the gel system. The chemotherapeutic activity of these nano-gels was evaluated against HCC, and it was found that high cisplatin release was observed at 40°C compared to 37°C. Moreover, this system released higher amount of the encapsulated drug at pH 5.7 compared with 7.4 revealing their response against temperature and pH, respectively. Cytotoxicity finding of this system revealed that empty nanogel (without drug) was biocompatible and was potentially suitable as a drug carrier against HCC (Salimi, Dilmaghani, Alizadeh, Akbarzadeh, & Davaran, 2018). These examples highlight previous work and reveal that nanogels have high potential to be used for targeted drug delivery to HCC.

Metal-organic frameworks (MOFs)

Metal-organic frameworks (MOFs) have great attraction for researchers due to their maximum surface area, adjustable pore size, and multiple functional groups with outstanding mechanical strength. MOFs also contain several desirable properties such as biocompatibility, homogeneity in size, small size, simple and easy functionalization, ability to encapsulate maximum drug with sustained and prolonged release profiles. They also have great potential to act as carriers for chemotherapeutic agents and can also be used for theranostic and targeting purposes after the incorporation of fluorescent materials and site-specific ligands (Horcajada et al., 2010; Rengaraj et al., 2017; Wu & Yang, 2017). As previously described, LBA has excellent potential to be used as a targeting ligand against HCC due to over-expressed receptors (asialoglycoprotein) in HCC (Zeng et al., 2014). For this purpose, LBA conjugated nano-sized MOFs have been synthesized by Samui et al., for targeting HCC. They designed and conjugated nano-size NH2-MIL-53(Al) MOFs with LBA and used them for site-specific drug delivery toward HCC. These LBA conjugated MOFs showed cellular imaging due to small size and natural fluorescent property although low drug encapsulating capacity in LBA conjugated NH2-MIL-53(Al) MOF was found, the cellular uptake in HCCs was enhanced. Furthermore, these nano-sized MOFs can also act as a nanocarrier for other chemotherapeutic agents against HCCs (Samui et al., 2019).

Wang et al. reported a one pot and solvent-free green synthesis of MOFs for tumor targeting from a biocompatible MIL-101 containing multifunctional frameworks because of covalently bonded benzoic imine and disulfide bonds which responded to pH and redox reactions, respectively. The synthesized biocompatible MOFs showed enhanced cancer cell uptake and intracellular drug release. In vitro results of these DDs revealed less cytotoxicity toward healthy cells and significant inhibition against tumor growth with the minimum side effect when surface modification was made in drug-loaded synthesized MOFs (Wang et al., 2015). Yawei et al. reported the synthesis of zeolitic imidazolate framework for encapsulation of Dihydroartemisinin (DHA) and showed around 77.2% drug loading efficiency, highly biocompatibility, mechanical stability, and sustain and controllable drug release profile in acidic environments of cancer tumor cells. DHA has potentially enhanced therapeutic activity against cancer cells, however, it has also shown limited inherent properties like poor aqueous solubility, rapid clearance, less selectivity. These synthesized DHA-loaded frameworks exhibited maximum anticancer activity against cancer cells in in-vitro and in-vivo experiments in comparison with free DHA drug (Li et al., 2020). All above-mentioned examples of MOFs highlight the excellent potential of MOFs and their feasibility for the construction of safe and stable carriers' system for HCC treatment.

Liposomes

Liposomes are lipid vesicles that aggregate spontaneously and form bilayered spherical-shaped vesicles in aqueous environments. The lipids used for the construction of liposomes are mostly phospholipids having high potential applications in DDs due to biocompatibility, high drug encapsulation, and controlled and sustained drug release profiles (Xia et al., 2015). Liposomes also show increased drug accumulation in cancer cells relative to normal healthy cells resulting in less systematic toxicities and side effects. For the enhancement of the anticancer activity and chemotherapeutic activity of anticancer drugs toward HCC, GA has been most widely used to modify the liposomes' surface to make it directed toward specific site of liver in HCC (Tian, Wang, Wang, & Ke, 2014). GA derivative (3-succinic-30-stearyl; Suc-GA) was introduced onto the surface of liposomes with a certain molar ratio by ethanol injection method (amphiphilic) and applied as a targeted carrier to interact with GA receptors present on HCC (Mao, Hou, Jin, Zhang, & Jiang, 2003). Suc-GA-modified liposomes were then loaded with Calcein and found to exhibit maximum cellular uptake through a site-specific receptor in HCC, thus, were considered to be an efficient model to target liver cells (Sheng-Jun et al., 2007). In another study, Oxaliplatin (OX) was more loaded in GA-modified liposomes via well-known thin-film hydration method with more than 90% entrapment efficiency and resulted in sustained release of OX. In addition, these GA-modified OX-loaded liposomes were not-toxic to epithelial cells and the formulation supplied maximum amount of OX specifically toward HCC revealing the effectiveness of GA conjugated liposomes in liver targeted drug delivery (Chen, Jiang, Wu, Li, & Gao, 2015). Similar to GA-modified OX liposomes, docetaxel

(DX) loaded in these liposomes with the same reported method have shown 2.28 times more cellular uptake in liver cells as compared to non-parenchymal cells. DX-loaded GA modified reveals efficient cancer inhibition (more than twofold) than DX liposomes and showed same pharmacokinetics (Li, Xu, Ke, & Tian, 2012).

Wogonin (WG) was also loaded in GA modified liposomes and found that upon liposomal delivery, it showed good cellular uptake with less IC50 value (1.46-fold) against cancer cell lines as compared to free Wogonin (Tian et al., 2014). These GA-decorated liposomes containing WG showed the property of liver targeting with no cytotoxicity and maximum cellular uptake of anticancer drug toward HCC with high efficacy as compared to simple WG liposomes and free WG. In addition, liposomes have also been conjugated with Hyaluronic acid (HA) for targeting hepatocytes and to increase the therapeutic efficacy of encapsulated drugs (Arpicco et al., 2013). HA- conjugated targeted liposomes (tHA-LIP) were synthesized for encapsulation of DOX. This formulation showed increased blood circulation time of DOX, high DOX loading capacity, low systematic toxicity, and enhanced therapeutic activity (Deng et al., 2017). In another study, HA liposomes were modified with different molecular ratios of phospholipid derivatized HA to form bilayered liposomes. These HA-modified liposomes showed high potential to interact with tumor tissues due to overexpressed CD44 and HA receptor for endocytosis (HARE). As compared to unmodified liposomes, the HA modified liposomes showed less IC50 value and also showed enhanced anticancer activity (Taetz et al., 2009). Thus, liposomes may be used as an excellent platform for drug delivery to liver because of their easy surface functionalization, ability to encapsulate both hydrophilic and lipophilic drugs and small size. Nevertheless, liposomes also show drawbacks including stability issues, high cost, and sterilization issues.

Inorganic nanocarriers

Inorganic NCs have also been extensively used for the therapy and diagnosis of different diseases. Gold nanoparticles (NPs), cerium oxide NPs (CeO_2NPs) as well as silver NPs are the examples of inorganic NPs commonly used for liver diseases. The structures of these inorganic NCs enable them to be easily modified with different drugs for the therapy of liver diseases especially liver cancer, e.g., DOX, capecitabine, and cisplatin. Inorganic NCs normally possess desirable physical properties like fluorescence (semiconductor, quantum dots), optical properties, and magnetic moments (iron oxides) thus they can be modified with potentially reactive groups and biomolecules to introduce biological functionality in them. In addition, most of these inorganic NCs are proven to be nontoxic (Tee, Peng, & Ho, 2019). Inorganic NCs have been employed for gene therapy and immunotherapy of different diseases, however, they have been less explored for liver diseases. Various reported inorganic NCs and their application against HCC targeted nano-drug delivery are discussed here:

Graphene oxide and iron oxide based nanocarriers

Inorganic nanocarriers based on graphene oxide exhibit excellent drug encapsulation efficiency. Graphene oxides owe the capability to transfer high electron density to a specific graphene sheet and make them potentially improve their efficacy in drug delivery (Gao et al., 2016). These graphene oxide-based nanocarriers have high surface area thus they offer to interact with numerous therapeutic agents and accommodate them (Vickers, 2017). Yang et al. have reported different targeted DDs of graphene oxide with various functional groups loaded with fluorescein isothiocyanate, carboxymethyl chitosan, lactobionic acid, and also chemotherapeutic agents like DOX. These carriers promoted cell death after 24 h incubation and revealed great biocompatibility toward liver tumor cells (Pan et al., 2016). Yuan et al. prepared different combinations of graphene NCs modified with monoclonal antibodies, folic acid, and gold NPs to promote site-specific drug delivery toward HCC and they showed controlled and sustained release properties (Yuan et al., 2015).

Iron-oxide nanoparticles contain Superparamagnetic properties and show excellent potentials in diagnosis and treatment of various diseases. Moreover, these superparamagnetic iron oxide nanoparticles (SPIONs) can be encapsulated in another delivery system with other drugs to synergize the therapeutic effects of the encapsulated drugs. Depalo et al. prepared micelles from polyethylene-glycol loaded with sorafenib and SPIONs and evaluated their inhibition effects against human hepatocellular carcinoma (HepG2) cells. Findings of their study revealed that these SPIONs containing micelles show improved anticancer activity due to the impact of the magnetic field of iron oxide (Depalo et al., 2017). Arias et al. and Unsoy et al. prepared Chitosan-Bound Fe_3O_4 Nanoparticles for encapsulation of different anticancer drugs and explored their anticancer efficiency. The Fe_3O_4 was selected as a magnetic nucleus due to its magnetic saturation and high magnetic susceptibility whereas chitosan was selected because of its biocompatible and non-toxic nature. These polymeric shells containing magnetic nanoparticles were applied for site-specific drug delivery of doxorubicin, ftorafur, epirubicin, bortezomib, and other chemotherapeutic agents and showed excellent activities (Arias, López-Viota, Sáez-Fernández, Ruiz, & Delgado, 2011; Unsoy et al., 2014).

Carbon nanotubes

Carbon nanotubes (CNTs) are cylindrical structures formed from graphene sheets having the ability to penetrate cells and deliver therapeutic agents. According to the number of graphene layers, CNTs can be divided into two different classes namely single-walled CNTs (SW-CNTs) and multi-walled CNTs (MW-CNTs) (Kostarelos et al., 2007). These carriers have been extensively used for different biomedical applications especially drug delivery due to their nanometer scale size, distinct shape and structure, low toxicity and also show interesting physical properties. CNTs are generally considered to be less toxic, having great biocompatibility and the encapsulated drugs in CNTs show high treatment efficiency with minimum drug dose in cancer-targeted drug delivery (Liu, Chen, et al., 2008). Moreover, they provide an extensive pore surface for encapsulation of a variety of drugs and functionalization with different ligands (Schipper et al., 2008). Qi et al. prepared galactosylated chitosan grafted oxidized multi-walled CNTs (O-CNTs-LCH-DOX) loaded with DOX and evaluated its delivery in mice model bearing H22 tumor. The prepared MW-CNTs showed good biocompatibility and low toxicity, pH-dependent drug release, exhibited excellent anticancer activity against HCC in-vitro, and higher antitumor activity in mice (Qi et al., 2015). In a similar example, Ji et al. prepared a highly effective and targeted SW-CNTs based drug delivery system modified with chitosan and folic acid for controlled loading and release of DOX (Fig. 3.4). The prepared DOX/FA/CHI/SW-CNTs not only effectively inhibited the growth of HCC cells (SMMC-7721 cell lines) but also showed much less toxicity compared to free DOX in-vivo (Ji et al., 2012). In a recent study, Elsayed et al., used functionalized CNTs for loading of Sorafenib (SFN) and assessing its in-vitro and in-vivo anticancer activity. The SFN was loaded on CNTs through physical adsorption to obtain drug-loaded CNTs (CNTs-SFN). The in-vitro anticancer activity of this formulation against HepG2 cells showed almost twofold more potency compared to free SFN and showed superiority to free SFN in all end points of the in-vivo study results in DENA-induced HCC rat model showed (Elsayed et al., 2019). Though CNTs functionalized with targeting ligands, biosensors, or other materials have shown excellent applications in the delivery of therapeutic agents and disease diagnosis as well as for the targeted delivery of drugs in different areas. However, there are still tremendous opportunities to be explored and significant challenges and risks to be solved for drug delivery to liver.

Nanoshells and nanofibers

Nanoshells and nanofibers have got great attention of the researchers in the past few decades as favorable vehicles for cancer treatment (Bardhan, Lal, Joshi, & Halas, 2011). They are nanoparticles with a dielectric (e.g., silica) core coated with a thin metal (usually gold) (Wang et al., 2014). One most important and useful example of this carrier system is gold nanoshells (AuNS). The drug is encapsulated or adsorbed on the shell's surface through electrostatic stabilization or specific functional groups. AuNS have been employed for the delivery of anticancer agents (e.g., paclitaxel, DOX, small interfering RNA (SiRNA), and single-stranded DNA) to cancer cells for the enhancement of their therapeutic efficacies. These nanoshells can also be modified with targeting ligands, e.g., aptamers, antibodies, and peptides for increasing their binding to the specific desired body targets. Liu et al. reported the construction of gold

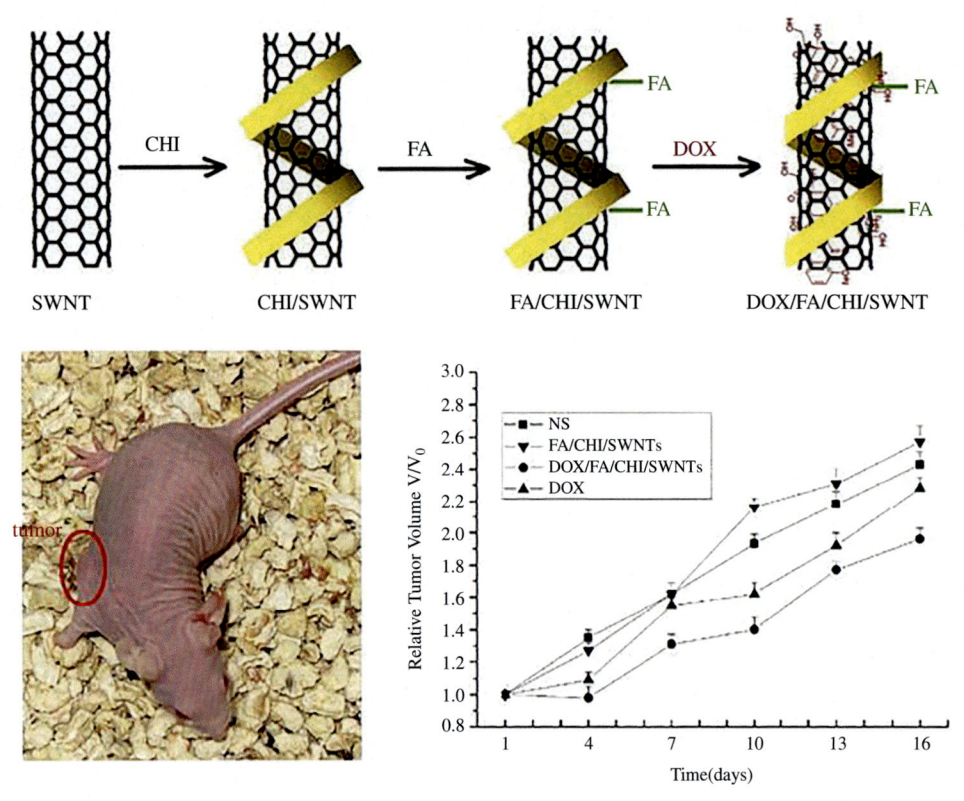

FIG. 3.4

Fabrication of single-walled CNTs modified with folic acid and chitosan for doxorubicin loading and in-vivo anticancer evaluation. *Reproduced with permission from Ji, Z., Lin, G., Lu, Q., Meng, L., Shen, X., Dong, L., et al. (2012). Targeted therapy of SMMC-7721 liver cancer in vitro and in vivo with carbon nanotubes based drug delivery system.* Journal of Colloid and Interface Science, 365*(1)*, *143–149. https://doi.org/10.1016/j.jcis.2011.09.013.*

nanoshells and found great targeting capability of this system toward cancer cells (BEL7404 and BEL-7402 cells) without affecting the normal healthy liver cells (HL-7702) (Liu, Liang, Gao, Luo, & Lu, 2010). Organic and inorganic nanofibers are the examples of other tremendous carrier systems for targeted drug delivery to liver cancer. Zhang et al. explore multilayered polylactide electrospun encapsulated cisplatin nanofiber against cancer tissue and observed enhanced cytotoxicity against tumor and sustained drug release from the nanofibers. The cisplatin-loaded multilayered polylactide electrospun nanofibers also inhibited the recurrence of cancer cells in a murine model. The formulation also prolonged survival time and less systemic toxicities were observed in comparison with other treatment groups (Zhang et al., 2014).

Metallic and magnetic nanoparticles

Metals-based NPs have long been used in various biomedical fields and particularly in drug delivery for site-specific and enhanced drug delivery (Goldman et al., 2004). Silver and gold have been the most prominent examples of metallic NPs due to their innate biocompatible nature, simple preparation methods, and owe electronic and optical properties that make them excellent systems for drug delivery and other biomedical applications. Gold NPs have been found to exhibit anti-angiogenic potentials since they counter the effect of heparin connecting growth factors like bFGF and VEGF165 and thus reduce the growth of tumors. Tomuleasa et al. have reported that the susceptibility of hepatic cancer cells to Dox, Cisplatin, and Capecitabine was enhanced upon encapsulation of these drugs in gold NPs. Similarly, gold NPs were found to be devoid of any cytotoxic or anti-propagative effects when evaluated against Hep3B HCC cell lines (Gannon, Patra, Bhattacharya, Mukherjee, & Curley, 2008). Gold nanoparticles can also be conjugated with ligands such as nucleotides enzymes, antibodies targeting to specific regions of the body (Loo et al., 2005). Ceramics are widely recognized as biomaterials and are made of metal oxides in most cases, e.g., titania (TiO_2), alumina (Al_2O_3), zinc oxide (ZnO), magnetite (Fe_3O_4), and silica (SiO_2). Ceramic NPs having innate porous or hollow structures have emerged as excellent counterparts to organic NCs for a variety of biomedical applications. Silica and magnetite NPs have been extensively used as NCs for drug delivery purposes, particularly for cancer treatment. Moreover, Silica has the ability to be functionalized with different functional groups that make it proficient to load a wide range of chemotherapeutics. For example, Ao et al. modified silica NPs with lipoprotein and encapsulated Docetaxel and Thalidomide for enhancing their cytotoxicity against HCC cell line and these NPs excellent anticancer activity and tumor targeting (Ao, Xiao, & Ao, 2018).

Conclusion

There has been an enormous increase in liver diseases over the last few years; hepatocellular carcinoma being the most prominent among these diseases. The diagnosis of HCC is often made in advanced stage of the disease thus, the available treatment modalities are no longer effective to significant extent. The increasing interest of researchers in cancer nanomedicine is encouraging and give a glimpse of hope for the creation of effective tools for both theranostic and therapeutics delivery to HCC. Dendrimers, micelles, nano-gels, metal-organic frameworks, liposomes, ceramics, and metallic NPs are some nanocarriers that are regarded as useful and effective drug delivery agents. Extensive studies are required to fully understand the exact mechanism of action of NCs, their drug delivery and drug clearance, their long-term effects on the body and to get a wider sense of their therapeutic potentials in comparison with the currently practiced treatments. Though the toxicity and biocompatibility studies have shown them safe and useful, nanocarriers have certain drawbacks due to their structure, nature, functional groups, composition, and so on. Despite all the limitations, research in nanocarriers is expanding and increasing day by day. These endeavors have also made possible the construction of different nanocarriers for a variety of diseases including liver cancers. Researcher has exploited the leaky vasculature property of cancer tissue that allows the nanocarriers to enter these cancerous tissues and deliver their payload. There is still a long way to go before these NCs can be successfully employed as therapeutic agents. Numerous clinical trial studies have to be conducted to eliminate any possibilities of adverse reactions, immunotoxicity, and side effects.

References

Aliabadi, H. M., Shahin, M., Brocks, D. R., & Lavasanifar, A. (2008). Disposition of drugs in block copolymer micelle delivery systems. *Clinical Pharmacokinetics, 47*(10), 619–634. https://doi.org/10.2165/00003088-200847100-00001.

Ao, M., Xiao, X., & Ao, Y. (2018). Low density lipoprotein modified silica nanoparticles loaded with docetaxel and thalidomide for effective chemotherapy of liver cancer. *Brazilian Journal of Medical and Biological Research, 51*(3). https://doi.org/10.1590/1414-431X20176650.

Arias, J. L., López-Viota, M., Sáez-Fernández, E., Ruiz, M. A., & Delgado, A. V. (2011). Engineering of an antitumor (core/shell) magnetic nanoformulation based on the chemotherapy agent ftorafur. *Colloids and Surfaces A: Physicochemical and Engineering Aspects, 384*(1–3), 157–163. https://doi.org/10.1016/j.colsurfa.2011.03.051.

Arpicco, S., Lerda, C., Dalla Pozza, E., Costanzo, C., Tsapis, N., Stella, B., et al. (2013). Hyaluronic acid-coated liposomes for active targeting of gemcitabine. *European Journal of Pharmaceutics and Biopharmaceutics, 85*(3), 373–380. https://doi.org/10.1016/j.ejpb.2013.06.003.

Ashwell, G., & Harford, J. (1982). Carbohydrate-specific receptors of the liver. *Annual Review of Biochemistry, 51*, 531–554. https://doi.org/10.1146/annurev.bi.51.070182.002531.

Baenziger, J. U., & Maynard, Y. (1980). Human hepatic lectin. Physiochemical properties and specificity. *Journal of Biological Chemistry, 255*(10), 4607–4613.

Ballet, F. (1990). Hepatic circulation: Potential for therapeutic intervention. *Pharmacology & Therapeutics, 47*(2), 281–328. https://doi.org/10.1016/0163-7258(90)90091-f.

Balogh, J., Victor, D., Asham, E. H., Burroughs, S. G., Boktour, M., Saharia, A., et al. (2016). Hepatocellular carcinoma: A review. *Journal of Hepatocellular Carcinoma*, 41–53. https://doi.org/10.2147/JHC.S61146.

Bardhan, R., Lal, S., Joshi, A., & Halas, N. J. (2011). Theranostic nanoshells: From probe design to imaging and treatment of cancer. *Accounts of Chemical Research, 44*(10), 936–946. https://doi.org/10.1021/ar200023x.

Blanco, E., Kessinger, C. W., Sumer, B. D., & Gao, J. (2009). Multifunctional micellar nanomedicine for cancer therapy. *Experimental Biology and Medicine, 234*(2), 123–131. https://doi.org/10.3181/0808-MR-250.

Boas, U., & Heegaard, P. M. H. (2004). Dendrimers in drug research. *Chemical Society Reviews, 33*(1), 43–63. https://doi.org/10.1039/b309043b.

Bodnar, M., Hartmann, J. F., & Borbely, J. (2005). Preparation and characterization of chitosan-based nanoparticles. *Biomacromolecules*, *6*(5), 2521–2527. https://doi.org/10.1021/bm0502258.

Boyd, B. J., Kaminskas, L. M., Karellas, P., Krippner, G., Lessene, R., & Porter, C. J. H. (2006). Cationic poly-L-lysine dendrimers: Pharmacokinetics, biodistribution, and evidence for metabolism and bioresorption after intravenous administration to rats. *Molecular Pharmaceutics*, *3*(5), 614–627. https://doi.org/10.1021/mp060032e.

Carrilho, F. J., de Mattos, A. A., Vianey, A. F., Vezozzo, D. C. P., Marinho, F., Souto, F. J., et al. (2015). Recomendações da Sociedade Brasileira de Hepatologia para diagnóstico e tratamento do carcinoma hepatocelular. *Arquivos de Gastroenterologia*, *52*, 2–14. https://doi.org/10.1590/S0004-28032015000500001.

Chabner, B. A., & Roberts, T. G. (2005). Chemotherapy and the war on cancer. *Nature Reviews Cancer*, *5*(1), 65–72. https://doi.org/10.1038/nrc1529.

Chen, J., Jiang, H., Wu, Y., Li, Y., & Gao, Y. (2015). A novel glycyrrhetinic acid-modified oxaliplatin liposome for liver-targeting and in vitro/vivo evaluation. *Drug Design, Development and Therapy*, *9*, 2265–2275. https://doi.org/10.2147/DDDT.S81722.

Chen, Q., Sun, Y., Wang, J., Yan, G., Cui, Z., Yin, H., et al. (2015). Preparation and characterization of glycyrrhetinic acid-modified stearic acid-grafted chitosan micelles. *Artificial Cells, Nanomedicine and Biotechnology*, *43*(4), 217–223. https://doi.org/10.3109/21691401.2013.845570.

Cheng, Y., Wang, J., Rao, T., He, X., & Xu, T. (2008). Pharmaceutical applications of dendrimers: Promising nanocarriers for drug delivery. *Frontiers in Bioscience*, *13*(4), 1447–1471. https://doi.org/10.2741/2774.

Chidambaram, M., Manavalan, R., & Kathiresan, K. (2011). Nanotherapeutics to overcome conventional cancer chemotherapy limitations. *Journal of Pharmacy & Pharmaceutical Sciences*, *14*(1), 67–77.

Danhier, F., Feron, O., & Préat, V. (2010). To exploit the tumor microenvironment: Passive and active tumor targeting of nanocarriers for anti-cancer drug delivery. *Journal of Controlled Release*, *148*(2), 135–146. https://doi.org/10.1016/j.jconrel.2010.08.027.

Daniels, T. R., Delgado, T., Rodriguez, J. A., Helguera, G., & Penichet, M. L. (2006). The transferrin receptor part I: Biology and targeting with cytotoxic antibodies for the treatment of cancer. *Clinical Immunology*, *121*(2), 144–158. https://doi.org/10.1016/j.clim.2006.06.010.

Deng, C., Zhang, Q., Fu, Y., Sun, X., Gong, T., & Zhang, Z. (2017). Coadministration of oligomeric hyaluronic acid-modified liposomes with tumor-penetrating peptide-iRGD enhances the antitumor efficacy of doxorubicin against melanoma. *ACS Applied Materials and Interfaces*, *9*(2), 1280–1292. https://doi.org/10.1021/acsami.6b13738.

Depalo, N., Iacobazzi, R. M., Valente, G., Arduino, I., Villa, S., Canepa, F., et al. (2017). Sorafenib delivery nanoplatform based on superparamagnetic iron oxide nanoparticles magnetically targets hepatocellular carcinoma. *Nano Research*, *10*(7), 2431–2448. https://doi.org/10.1007/s12274-017-1444-3.

Dufes, C., Uchegbu, I. F., & Schätzlein, A. G. (2005). Dendrimers in gene delivery. *Advanced Drug Delivery Reviews*, *57*(15), 2177–2202. https://doi.org/10.1016/j.addr.2005.09.017.

El-Marakby, E. M., Hathout, R. M., Taha, I., Mansour, S., & Mortada, N. D. (2017). A novel serum-stable liver targeted cytotoxic system using valerate-conjugated chitosan nanoparticles surface decorated with glycyrrhizin. *International Journal of Pharmaceutics*, *525*(1), 123–138. https://doi.org/10.1016/j.ijpharm.2017.03.081.

El-Sayed, A., Futaki, S., & Harashima, H. (2009). Delivery of macromolecules using arginine-rich cell-penetrating peptides: Ways to overcome endosomal entrapment. *The AAPS Journal*, *11*(1), 13–22. https://doi.org/10.1208/s12248-008-9071-2.

Elsayed, M. M. A., Mostafa, M. E., Alaaeldin, E., Sarhan, H. A. A., Shaykoon, M. S., Allam, S., et al. (2019). Design and characterisation of novel sorafenib-loaded carbon nanotubes with distinct tumour-suppressive activity in hepatocellular carcinoma. *International Journal of Nanomedicine*, *14*, 8445–8467. https://doi.org/10.2147/IJN.S223920.

Filmus, J., & Selleck, S. B. (2001). Glypicans: Proteoglycans with a surprise. *Journal of Clinical Investigation*, *108*(4), 497–501. https://doi.org/10.1172/JCI200113712.

Gannon, C. J., Patra, C. R., Bhattacharya, R., Mukherjee, P., & Curley, S. A. (2008). Intracellular gold nanoparticles enhance non-invasive radiofrequency thermal destruction of human gastrointestinal cancer cells. *Journal of Nanobiotechnology*, *6*. https://doi.org/10.1186/1477-3155-6-2.

Gao, C., Fang, L., Zhao, H. C., Li, J. T., & Yao, S. K. (2013). Potential role of diabetes mellitus in the progression of cirrhosis to hepatocellular carcinoma: A cross-sectional case-control study from Chinese patients with HBV infection. *Hepatobiliary & Pancreatic Diseases International*, *12*(4), 385–393. https://doi.org/10.1016/S1499-3872(13)60060-0.

Gao, P., Liu, M., Tian, J., Deng, F., Wang, K., Xu, D., et al. (2016). Improving the drug delivery characteristics of graphene oxide based polymer nanocomposites through the "one-pot" synthetic approach of single-electron-transfer living radical polymerization. *Applied Surface Science*, *378*, 22–29. https://doi.org/10.1016/j.apsusc.2016.03.207.

Gao, J., Zhong, W., He, J., Li, H., Zhang, H., Zhou, G., et al. (2009). Tumor-targeted PE38KDEL delivery via PEGylated anti-HER2 immunoliposomes. *International Journal of Pharmaceutics*, 145–152. https://doi.org/10.1016/j.ijpharm.2009.03.018.

Gillies, E., & Frechet, J. (2005). Dendrimers and dendritic polymers in drug delivery. *Drug Discovery Today*, *10*(1), 35–43. https://doi.org/10.1016/s1359-6446(04)03276-3.

Goldman, E. R., Clapp, A. R., Anderson, G. P., Uyeda, H. T., Mauro, J. M., Medintz, I. L., et al. (2004). Multiplexed toxin analysis using four colors of quantum dot fluororeagents. *Analytical Chemistry*, *76*(3), 684–688. https://doi.org/10.1021/ac035083r.

Gullotti, E., & Yeo, Y. (2009). Extracellularly activated nanocarriers: A new paradigm of tumor targeted drug delivery. *Molecular Pharmaceutics*, *6*(4), 1041–1051. https://doi.org/10.1021/mp900090z.

Han, K. H., & Park, J. Y. (2008). Chemotherapy for advanced hepatocellular carcinoma. *Journal of Gastroenterology and Hepatology*, *23*(5), 682–684. https://doi.org/10.1111/j.1440-1746.2008.05444.x.

Hector, A., Schmid, B., Beierkuhnlein, C., Caldeira, M. C., Diemer, M., Dimitrakopoulos, P. G., et al. (1999). Plant diversity and productivity experiments in European grasslands. *Science*, *286*(5442), 1123–1127. https://doi.org/10.1126/science.286.5442.1123.

Horcajada, P., Chalati, T., Serre, C., Gillet, B., Sebrie, C., Baati, T., et al. (2010). Porous metal-organic-framework nanoscale carriers as a potential platform for drug delivery and imaging. *Nature Materials*, *9*(2), 172–178. https://doi.org/10.1038/nmat2608.

Huang, W., Wang, W., Wang, P., Tian, Q., Zhang, C., Wang, C., et al. (2010). Glycyrrhetinic acid-modified poly(ethylene glycol)-b-poly(γ-benzyl l-glutamate) micelles for liver targeting therapy. *Acta Biomaterialia*, *6*(10), 3927–3935. https://doi.org/10.1016/j.actbio.2010.04.021.

Huang, W., Wang, W., Wang, P., Zhang, C. N., Tian, Q., Zhang, Y., et al. (2011). Glycyrrhetinic acid-functionalized degradable micelles as liver-targeted drug carrier. *Journal of Materials Science: Materials in Medicine*, *22*(4), 853–863. https://doi.org/10.1007/s10856-011-4262-2.

Inoue, T., Cavanaugh, P. G., Steck, P. A., Brünner, N., & Nicolson, G. L. (1993). Differences in transferrin response and numbers of transferrin receptors in rat and human mammary carcinoma lines of different metastatic potentials. *Journal of Cellular Physiology*, *156*(1), 212–217. https://doi.org/10.1002/jcp.1041560128.

Ismair, M. G., Stanca, C., Ha, H. R., Renner, E. L., Meier, P. J., & Kullak-Ublick, G. A. (2003). Interactions of glycyrrhizin with organic anion transporting polypeptides of rat and human liver. *Hepatology Research*, *26*(4), 343–347. https://doi.org/10.1016/S1386-6346(03)00154-2.

Jain, R. K., & Booth, M. F. (2003). What brings pericytes to tumor vessels? *Journal of Clinical Investigation, 112*(8), 1134–1136. https://doi.org/10.1172/JCI200320087.

Ji, Z., Lin, G., Lu, Q., Meng, L., Shen, X., Dong, L., et al. (2012). Targeted therapy of SMMC-7721 liver cancer in vitro and in vivo with carbon nanotubes based drug delivery system. *Journal of Colloid and Interface Science, 365*(1), 143–149. https://doi.org/10.1016/j.jcis.2011.09.013.

Jia, X.-L., Li, S.-Y., Dang, S.-S., Cheng, Y.-A., Zhang, X., Wang, W.-J., et al. (2012). Increased expression of chondroitin sulphate proteoglycans in rat hepatocellular carcinoma tissues. *World Journal of Gastroenterology: WJG, 18*(30).

Jing, S. Q., & Trowbridge, I. S. (1987). Identification of the intermolecular disulfide bonds of the human transferrin receptor and its lipid-attachment site. *The EMBO Journal, 6*(2), 327–331. https://doi.org/10.1002/j.1460-2075.1987.tb04758.x.

Kabanov, A. V., & Vinogradov, S. V. (2009). Nanogels as pharmaceutical carriers: Finite networks of infinite capabilities. *Angewandte Chemie International Edition, 48*(30), 5418–5429. https://doi.org/10.1002/anie.200900441.

Kaneo, Y., Tanaka, T., Nakano, T., & Yamaguchi, Y. (2001). Evidence for receptor-mediated hepatic uptake of pullulan in rats. *Journal of Controlled Release, 70*(3), 365–373. https://doi.org/10.1016/S0168-3659(00)00368-0.

Kim, S. K., & Huang, L. (2012). Nanoparticle delivery of a peptide targeting EGFR signaling. *Journal of Controlled Release, 157*(2), 279–286. https://doi.org/10.1016/j.jconrel.2011.08.014.

Kostarelos, K., Lacerda, L., Pastorin, G., Wu, W., Wieckowski, S., Luangsivilay, J., et al. (2007). Cellular uptake of functionalized carbon nanotubes is independent of functional group and cell type. *Nature Nanotechnology, 2*(2), 108–113. https://doi.org/10.1038/nnano.2006.209.

Kouroumalis, E., Skordilis, P., Thermos, K., Vasilaki, A., Moschandrea, J., & Manousos, O. N. (1998). Treatment of hepatocellular carcinoma with octreotide: A randomised controlled study. *Gut, 42*(3), 442–447. https://doi.org/10.1136/gut.42.3.442.

Kovalszky, I., Pogany, G., Molnar, G., Jeney, A., Lapis, K., Karacsonyi, S., et al. (1990). Altered glycosaminoglycan composition in reactive and neoplastic human liver. *Biochemical and Biophysical Research Communications, 167*(3), 883–890. https://doi.org/10.1016/0006-291X(90)90606-N.

Kularatne, S. A., & Low, P. S. (2010). Targeting of nanoparticles: Folate receptor. *Methods in Molecular Biology, 624*, 249–265. https://doi.org/10.1007/978-1-60761-609-2_17.

Lautt, W. W., & Greenway, C. V. (1987). Conceptual review of the hepatic vascular bed. *Hepatology, 7*(5), 952–963. https://doi.org/10.1002/hep.1840070527.

Lehtinen, J., Raki, M., Bergström, K. A., Uutela, P., Lehtinen, K., Hiltunen, A., et al. (2012). Pretargeting and direct immunotargeting of liposomal drug carriers to ovarian carcinoma. *PLoS One, 7*.

Leonardi, G. C., Candido, S., Cervello, M., Nicolosi, D., Raiti, F., Travali, S., et al. (2012). The tumor microenvironment in hepatocellular carcinoma (Review). *International Journal of Oncology, 40*(6), 1733–1747. https://doi.org/10.3892/ijo.2012.1408.

Letchford, K., & Burt, H. (2007). A review of the formation and classification of amphiphilic block copolymer nanoparticulate structures: Micelles, nanospheres, nanocapsules and polymersomes. *European Journal of Pharmaceutics and Biopharmaceutics, 65*(3), 259–269. https://doi.org/10.1016/j.ejpb.2006.11.009.

Li, H., Cui, Y., Liu, J., Bian, S., Liang, J., Fan, Y., et al. (2014). Reduction breakable cholesteryl pullulan nanoparticles for targeted hepatocellular carcinoma chemotherapy. *Journal of Materials Chemistry B, 2*(22), 3500–3510. https://doi.org/10.1039/c4tb00321g.

Li, Y., Huang, G., Diakur, J., & Wiebe, L. I. (2008). Targeted delivery of macromolecular drugs: Asialoglycoprotein receptor (ASGPR) expression by selected hepatoma cell lines used in antiviral drug development. *Current Drug Delivery, 5*(4), 299–302. https://doi.org/10.2174/156720108785915069.

Li, Y., Song, Y., Zhang, W., Xu, J., Hou, J., Feng, X., et al. (2020). MOF nanoparticles with encapsulated dihydroartemisinin as a controlled drug delivery system for enhanced cancer therapy and mechanism analysis. *Journal of Materials Chemistry B, 8*(33), 7382–7389. https://doi.org/10.1039/d0tb01330g.

Li, J., Xu, H., Ke, X., & Tian, J. (2012). The anti-tumor performance of docetaxel liposomes surface-modified with glycyrrhetinic acid. *Journal of Drug Targeting, 20*(5), 467–473. https://doi.org/10.3109/1061186X.2012.685475.

Lin, W. J., Chen, T. D., Liu, C. W., Chen, J. L., & Chang, F. H. (2011). Synthesis of lactobionic acid-grafted-pegylated-chitosan with enhanced HepG2 cells transfection. *Carbohydrate Polymers, 83*(2), 898–904. https://doi.org/10.1016/j.carbpol.2010.08.072.

Ling, S. S. N., Yuen, K. H., Magosso, E., & Barker, S. A. (2009). Oral bioavailability enhancement of a hydrophilic drug delivered via folic acid-coupled liposomes in rats. *Journal of Pharmacy and Pharmacology, 61*(4), 445–449. https://doi.org/10.1211/jpp.61.04.0005.

Liotta, L. A. (1986). Tumor invasion and metastases—Role of the extracellular matrix: Rhoads memorial award lecture. *Cancer Research, 46*(1), 1–7.

Liu, Z., Chen, K., Davis, C., Sherlock, S., Cao, Q., Chen, X., et al. (2008). Drug delivery with carbon nanotubes for in vivo cancer treatment. *Cancer Research, 68*(16), 6652–6660. https://doi.org/10.1158/0008-5472.CAN-08-1468.

Liu, H. L., Huo, L., & Wang, L. (2004). Octreotide inhibits proliferation and induces apoptosis of hepatocellular carcinoma cells. *Acta Pharmacologica Sinica, 25*(10), 1380–1386.

Liu, Z., Jiao, Y., Wang, Y., Zhou, C., & Zhang, Z. (2008). Polysaccharides-based nanoparticles as drug delivery systems. *Advanced Drug Delivery Reviews, 60*(15), 1650–1662. https://doi.org/10.1016/j.addr.2008.09.001.

Liu, S. Y., Liang, Z. S., Gao, F., Luo, S. F., & Lu, G. Q. (2010). In vitro photothermal study of gold nanoshells functionalized with small targeting peptides to liver cancer cells. *Journal of Materials Science: Materials in Medicine, 21*(2), 665–674. https://doi.org/10.1007/s10856-009-3895-x.

Liu, N., Tan, Y., Hu, Y., Meng, T., Wen, L., Liu, J., et al. (2016). A54 peptide modified and redox-responsive glucolipid conjugate micelles for intracellular delivery of doxorubicin in hepatocarcinoma therapy. *ACS Applied Materials and Interfaces, 8*(48), 33148–33156. https://doi.org/10.1021/acsami.6b09333.

Llovet, J. M., Ricci, S., Mazzaferro, V., Hilgard, P., Gane, E., Blanc, J. F., et al. (2008). Sorafenib in advanced hepatocellular carcinoma. *New England Journal of Medicine, 359*(4), 378–390. https://doi.org/10.1056/NEJMoa0708857.

Loo, C., Hirsch, L., Lee, M. H., Chang, E., West, J., Halas, N., et al. (2005). Gold nanoshell bioconjugates for molecular imaging in living cells. *Optics Letters, 30*(9), 1012–1014. https://doi.org/10.1364/OL.30.001012.

Loutfy, S. A., El-Din, H. M. A., Elberry, M. H., Allam, N. G., Hasanin, M., & Abdellah, A. M. (2016). Synthesis, characterization and cytotoxic evaluation of chitosan nanoparticles: In vitro liver cancer model. *Advances in Natural Sciences: Nanoscience and Nanotechnology, 7*(3).

Low, P. S., Henne, W. A., & Doorneweerd, D. D. (2008). Discovery and development of folic-acid-based receptor targeting for imaging and therapy of cancer and inflammatory diseases. *Accounts of Chemical Research, 41*(1), 120–129. https://doi.org/10.1021/ar7000815.

Low, P. S., & Kularatne, S. A. (2009). Folate-targeted therapeutic and imaging agents for cancer. *Current Opinion in Chemical Biology, 13*(3), 256–262. https://doi.org/10.1016/j.cbpa.2009.03.022.

Lu, Y., & Low, P. S. (2012). Folate-mediated delivery of macromolecular anticancer therapeutic agents. *Advanced Drug Delivery Reviews, 64*, 342–352. https://doi.org/10.1016/j.addr.2012.09.020.

Malam, Y., Loizidou, M., & Seifalian, A. M. (2009). Liposomes and nanoparticles: Nanosized vehicles for drug delivery in cancer. *Trends in Pharmacological Sciences, 30*(11), 592–599. https://doi.org/10.1016/j.tips.2009.08.004.

Malhi, S. S., Budhiraja, A., Arora, S., Chaudhari, K. R., Nepali, K., Kumar, R., et al. (2012). Intracellular delivery of redox cycler-doxorubicin to the mitochondria of cancer cell by folate receptor targeted mitocancerotropic liposomes. *International Journal of Pharmaceutics, 432*(1–2), 63–74. https://doi.org/10.1016/j.ijpharm.2012.04.030.

Mao, S. J., Hou, S. X., Jin, H., Zhang, L. K., & Jiang, B. (2003). Preparation of liposomes surface-modified with glycyrrhetinic acid targeting to hepatocytes. *Zhongguo Zhongyao Zazhi, 28*(4), 330–331.

Masood, F. (2016). Polymeric nanoparticles for targeted drug delivery system for cancer therapy. *Materials Science and Engineering C, 60*, 569–578. https://doi.org/10.1016/j.msec.2015.11.067.

Master, A. M., & Sen Gupta, A. (2012). EGF receptor-targeted nanocarriers for enhanced cancer treatment. *Nanomedicine, 7*(12), 1895–1906. https://doi.org/10.2217/nnm.12.160.

Matsumura, Y. (2008). Polymeric micellar delivery systems in oncology. *Japanese Journal of Clinical Oncology, 38*(12), 793–802. https://doi.org/10.1093/jjco/hyn116.

Medina, S. H., Tekumalla, V., Chevliakov, M. V., Shewach, D. S., Ensminger, W. D., & El-Sayed, M. E. (2011). N-acetylgalactosamine-functionalized dendrimers as hepatic cancer cell-targeted carriers. *Biomaterials, 32*(17), 4118–4129.

Mehta, D., Guvva, S., & Patil, M. (2008). Future impact of nanotechnology on medicine and dentistry. *Journal of Indian Society of Periodontology, 34*. https://doi.org/10.4103/0972-124X.44088.

Musso, O., Theret, N., Heljasvaara, R., Rehn, M., Marko, B., Campion, J. P., et al. (2001). Tumor hepatocytes and basement membrane-producing cells specifically express two different forms of the endostatin precursor, collagen XVIII, in human liver cancers. *Hepatology, 33*(4), 868–876. https://doi.org/10.1053/jhep.2001.23189.

Needham, D., & Dewhirst, M. W. (2001). The development and testing of a new temperature-sensitive drug delivery system for the treatment of solid tumors. *Advanced Drug Delivery Reviews, 53*(3), 285–305. https://doi.org/10.1016/S0169-409X(01)00233-2.

Negishi, M., Irie, A., Nagata, N., & Ichikawa, A. (1991). Specific binding of glycyrrhetinic acid to the rat liver membrane. *Biochimica et Biophysica Acta (BBA)-Biomembranes, 1066*(1), 77–82.

Ng, I. O. L., Poon, R. T. P., Lee, J. M. F., Fan, S. T., Ng, M., & Tso, W. K. (2001). Microvessel density, vascular endothelial growth factor and its receptors Flt-1 and Flk-1/KDR in hepatocellular carcinoma. *American Journal of Clinical Pathology, 116*(6), 838–845. https://doi.org/10.1309/FXNL-QTN1-94FH-AB3A.

Pan, Q., Lv, Y., Williams, G. R., Tao, L., Yang, H., Li, H., et al. (2016). Lactobionic acid and carboxymethyl chitosan functionalized graphene oxide nanocomposites as targeted anticancer drug delivery systems. *Carbohydrate Polymers, 151*, 812–820. https://doi.org/10.1016/j.carbpol.2016.06.024.

Patel, Y. C. (1999). Somatostatin and its receptor family. *Frontiers in Neuroendocrinology, 20*(3), 157–198. https://doi.org/10.1006/frne.1999.0183.

Pearson, R. M., Sunoqrot, S., Hsu, H. J., Bae, J. W., & Hong, S. (2012). Dendritic nanoparticles: The next generation of nanocarriers? *Therapeutic Delivery, 3*(8), 941–959. https://doi.org/10.4155/tde.12.76.

Peer, D., Karp, J. M., Hong, S., Farokhzad, O. C., Margalit, R., & Langer, R. (2007). Nanocarriers as an emerging platform for cancer therapy. *Nature Nanotechnology, 2*(12), 751–760. https://doi.org/10.1038/nnano.2007.387.

Qi, X., Rui, Y., Fan, Y., Chen, H., Ma, N., & Wu, Z. (2015). Galactosylated chitosan-grafted multiwall carbon nanotubes for pH-dependent sustained release and hepatic tumor-targeted delivery of doxorubicin in vivo. *Colloids and Surfaces B: Biointerfaces, 133*, 314–322. https://doi.org/10.1016/j.colsurfb.2015.06.003.

Quan, S., Wang, Y., Zhou, A., Kumar, P., & Narain, R. (2015). Galactose-based thermosensitive nanogels for targeted drug delivery of iodoazomycin arabinofuranoside (IAZA) for theranostic management of hypoxic hepatocellular carcinoma. *Biomacromolecules, 16*(7), 1978–1986. https://doi.org/10.1021/acs.biomac.5b00576.

Rengaraj, A., Puthiaraj, P., Heo, N. S., Lee, H., Hwang, S. K., Kwon, S., et al. (2017). Porous NH2-MIL-125 as an efficient nano-platform for drug delivery, imaging, and ROS therapy utilized Low-Intensity Visible light exposure system. *Colloids and Surfaces B: Biointerfaces, 160*, 1–10. https://doi.org/10.1016/j.colsurfb.2017.09.011.

Reubi, J. C., Zimmermann, A., Jonas, S., Waser, B., Neuhaus, P., Läderach, U., et al. (1999). Regulatory peptide receptors in human hepatocellular carcinomas. *Gut, 45*(5), 766–774. https://doi.org/10.1136/gut.45.5.766.

Reynaert, H., Rombouts, K., Vandermonde, A., Urbain, D., Kumar, U., Bioulac-Sage, P., et al. (2004). Expression of somatostatin receptors in normal and cirrhotic human liver and in hepatocellular carcinoma. *Gut, 53*(8), 1180–1189. https://doi.org/10.1136/gut.2003.036053.

Rinella, M. E. (2015). Nonalcoholic fatty liver disease a systematic review. *JAMA: The Journal of the American Medical Association, 313*(22), 2263–2273. https://doi.org/10.1001/jama.2015.5370.

Roberts, W. G., & Palade, G. E. (1997). Neovasculature induced by vascular endothelial growth factor is fenestrated. *Cancer Research, 57*(4), 765–772.

Rojkind, M., & Ponce-Noyola, P. (1982). The extracellular matrix of the liver. *Collagen and Related Research, 2*(2), 151–175. https://doi.org/10.1016/s0174-173x(82)80031-9.

Ruman, U., Fakurazi, S., Masarudin, M. J., & Hussein, M. Z. (2020). Nanocarrier-based therapeutics and theranostics drug delivery systems for next generation of liver cancer nanodrug modalities. *International Journal of Nanomedicine, 15*.

Salimi, F., Dilmaghani, K. A., Alizadeh, E., Akbarzadeh, A., & Davaran, S. (2018). Enhancing cisplatin delivery to hepatocellular carcinoma HepG2 cells using dual sensitive smart nanocomposite. *Artificial Cells, Nanomedicine and Biotechnology, 46*(5), 949–958. https://doi.org/10.1080/21691401.2017.1349777.

Samui, A., Pal, K., Karmakar, P., & Sahu, S. K. (2019). In situ synthesized lactobionic acid conjugated NMOFs, a smart material for imaging and targeted drug delivery in hepatocellular carcinoma. *Materials Science and Engineering C, 98*, 772–781. https://doi.org/10.1016/j.msec.2019.01.032.

Schipper, M. L., Nakayama-Ratchford, N., Davis, C. R., Kam, N. W. S., Chu, P., Liu, Z., et al. (2008). A pilot toxicology study of single-walled carbon nanotubes in a small sample of mice. *Nature Nanotechnology, 3*(4), 216–221. https://doi.org/10.1038/nnano.2008.68.

Sciot, R., Paterson, A. C., Van Eyken, P., Callea, F., Kew, M. C., & Desmet, V. J. (1988). Transferrin receptor expression in human hepatocellular carcinoma: An immunohistochemical study of 34 cases. *Histopathology, 12*(1), 53–63. https://doi.org/10.1111/j.1365-2559.1988.tb01916.x.

Seidi, F., Jenjob, R., Phakkeeree, T., & Crespy, D. (2018). Saccharides, oligosaccharides, and polysaccharides nanoparticles for biomedical applications. *Journal of Controlled Release, 284*, 188–212. https://doi.org/10.1016/j.jconrel.2018.06.026.

Semela, D., & Dufour, J. F. (2004). Angiogenesis and hepatocellular carcinoma. *Journal of Hepatology, 41*(5), 864–880. https://doi.org/10.1016/j.jhep.2004.09.006.

Sheng-Jun, M., Yue-Qi, B., Hui, J., Da-Peng, W., Ru, H., & Shi-Xiang, H. (2007). Preparation, characterization and uptake by primary cultured rat hepatocytes of liposomes surface-modified with glycyrrhetinic acid. *Die Pharmazie—An International Journal of Pharmaceutical Sciences, 62*(8), 614–619.

Shi, B., Abrams, M., & Sepp-Lorenzino, L. (2013). Expression of asialoglycoprotein receptor 1 in human hepatocellular carcinoma. *Journal of Histochemistry and Cytochemistry*, *61*(12), 901–909. https://doi.org/10.1369/0022155413503662.

Soussan, E., Cassel, S., Blanzat, M., & Rico-Lattes, I. (2009). Drug delivery by soft matter: Matrix and vesicular carriers. *Angewandte Chemie International Edition*, *48*(2), 274–288. https://doi.org/10.1002/anie.200802453.

Steer, C. J., & Ashwell, G. (1980). Studies on a mammalian hepatic binding protein specific for asia-loglycoproteins. Evidence for receptor recycling in isolated rat hepatocytes. *Journal of Biological Chemistry*, *255*(7), 3008–3013.

Sun, Y., Dai, C., Yin, M., Lu, J., Hu, H., & Chen, D. (2018). Hepatocellular carcinoma-targeted effect of configurations and groups of glycyrrhetinic acid by evaluation of its derivative-modified lipo-somes. *International Journal of Nanomedicine*, *13*, 1621–1632. https://doi.org/10.2147/IJN.S153944.

Sutherland, R., Delia, D., Schneider, C., Newman, R., Kemshead, J., & Greaves, M. (1981). Ubiq-uitous cell-surface glycoprotein on tumor cells is proliferation-associated receptor for transfer-rin. *Proceedings of the National Academy of Sciences*, *78*(7), 4515–4519. https://doi.org/10.1073/pnas.78.7.4515.

Svenson, S., & Tomalia, D. A. (2012). Dendrimers in biomedical applications-reflections on the field. *Advanced Drug Delivery Reviews*, *64*, 102–115. https://doi.org/10.1016/j.addr.2012.09.030.

Taetz, S., Bochot, A., Surace, C., Arpicco, S., Renoir, J.-M., Schaefer, U. F., et al. (2009). Hyaluronic acid-modified DOTAP/DOPE liposomes for the targeted delivery of anti-telomerase siRNA to CD44-expressing lung cancer cells. *Oligonucleotides*, 103–116. https://doi.org/10.1089/oli.2008.0168.

Tee, J. K., Peng, F., & Ho, H. K. (2019). Effects of inorganic nanoparticles on liver fibrosis: Optimiz-ing a double-edged sword for therapeutics. *Biochemical Pharmacology*, *160*, 24–33. https://doi.org/10.1016/j.bcp.2018.12.003.

Tian, J., Wang, L., Wang, L., & Ke, X. (2014). A wogonin-loaded glycyrrhetinic acid-modified lipo-some for hepatic targeting with anti-tumor effects. *Drug Delivery*, *21*(7), 553–559. https://doi.org/10.3109/10717544.2013.853850.

Tian, Q., Wang, X. H., Wang, W., Zhang, C. N., Wang, P., & Yuan, Z. (2012). Self-assembly and liver targeting of sulfated chitosan nanoparticles functionalized with glycyrrhetinic acid. *Nanomedi-cine: Nanotechnology, Biology, and Medicine*, *8*(6), 870–879. https://doi.org/10.1016/j.nano.2011.11.002.

Tomalia, D., & Fréchet, J. (2001). *Introduction to the dendritic state. Dendrimers and other dendritic poly-mers* (pp. 3–40). Wiley.

Torchilin, V. P. (2001). Structure and design of polymeric surfactant-based drug delivery systems. *Jour-nal of Controlled Release*, *73*(2–3), 137–172. https://doi.org/10.1016/S0168-3659(01)00299-1.

Torimura, T., Ueno, T., Inuzuka, S., Kin, M., Ohira, H., Kimura, Y., et al. (1994). The extracellular matrix in hepatocellular carcinoma shows different localization patterns depending on the dif-ferentiation and the histological pattern of tumors: Immunohistochemical analysis. *Journal of Hepatology*, *21*(1), 37–46. https://doi.org/10.1016/S0168-8278(94)80134-7.

Tsend-Ayush, A., Zhu, X., Ding, Y., Yao, J., Yin, L., Zhou, J., et al. (2017). Lactobionic acid-conjugated TPGS nanoparticles for enhancing therapeutic efficacy of etoposide against hepato-cellular carcinoma. *Nanotechnology*, *28*(19).

Uchegbu, I. F., & Schatzlein, A. G. (2006). *Polymers in drug delivery*. Taylor and Francis Group.

Uhrich, K. E., Cannizzaro, S. M., Langer, R. S., & Shakesheff, K. M. (1999). Polymeric systems for controlled drug release. *Chemical Reviews*, *99*(11), 3181–3198. https://doi.org/10.1021/cr940351u.

Unsoy, G., Yalcin, S., Khodadust, R., Mutlu, P., Onguru, O., & Gunduz, U. (2014). Chitosan magnetic nanoparticles for pH responsive Bortezomib release in cancer therapy. *Biomedicine and Pharmacotherapy*, *68*(5), 641–648. https://doi.org/10.1016/j.biopha.2014.04.003.

Vickers, N. J. (2017). Animal communication: When i'm calling you, will you answer too? *Current Biology*, *27*.

Wang, X. G., Dong, Z. Y., Cheng, H., Wan, S. S., Chen, W. H., Zou, M. Z., et al. (2015). A multifunctional metal-organic framework based tumor targeting drug delivery system for cancer therapy. *Nanoscale*, *7*(38), 16061–16070. https://doi.org/10.1039/c5nr04045k.

Wang, Q.-S., Gao, L.-N., Zhu, X.-N., Zhang, Y., Zhang, C.-N., Xu, D., et al. (2019). Co-delivery of glycyrrhizin and doxorubicin by alginate nanogel particles attenuates the activation of macrophage and enhances the therapeutic efficacy for hepatocellular carcinoma. *Theranostics*, *9*(21).

Wang, X., Gu, X., Wang, H., Yang, J., & Mao, S. (2018). Enhanced delivery of doxorubicin to the liver through self-assembled nanoparticles formed via conjugation of glycyrrhetinic acid to the hydroxyl group of hyaluronic acid. *Carbohydrate Polymers*, *195*, 170–179. https://doi.org/10.1016/j.carbpol.2018.04.052.

Wang, B., Qiao, W., Wang, Y., Yang, L., Zhang, Y., & Shao, P. (2010). Cancer therapy based on nanomaterials and nanocarrier systems. *Journal of Nanomaterials*, *2010*. https://doi.org/10.1155/2010/796303.

Wang, C., Wang, X., Gong, G., Ben, Q., Qiu, W., Chen, Y., et al. (2012). Increased risk of hepatocellular carcinoma in patients with diabetes mellitus: A systematic review and meta-analysis of cohort studies. *International Journal of Cancer*, *130*(7), 1639–1648. https://doi.org/10.1002/ijc.26165.

Wang, Y.-X. J., Zhu, X.-M., Liang, Q., Cheng, C. H. K., Wang, W., & Leung, K. C.-F. (2014). In vivo chemoembolization and magnetic resonance imaging of liver tumors by using iron oxide nanoshell/doxorubicin/poly(vinyl alcohol) hybrid composites. *Angewandte Chemie*, *126*(19), 4912–4915. https://doi.org/10.1002/ange.201402144.

Ward, J. W., Watson, M., Momin, B., & Richardson, L. C. (2001). Hepatocellular carcinoma—United States. *Morbidity and Mortality Weekly Report*, *59*(17), 517–520.

Weigel, P. H., & Oka, J. A. (1982). Endocytosis and degradation mediated by the asialoglycoprotein receptor in isolated rat hepatocytes. *Journal of Biological Chemistry*, *257*(3), 1201–1207.

Wong, A., & Vij, N. (2009). Modified epidermal growth factor receptor (EGFR)-bearing liposomes (MRBLs) are sensitive to EGF in solution. *PLoS One*, *4*(10), e7391. https://doi.org/10.1371/journal.pone.0007391.

Wu, X. Z., Chen, D., & Xie, G. R. (2006). Extracellular matrix remodeling in hepatocellular carcinoma: Effects of soil on seed? *Medical Hypotheses*, *66*(6), 1115–1120. https://doi.org/10.1016/j.mehy.2005.12.043.

Wu, J., Sun, T. M., Yang, X. Z., Zhu, J., Du, X. J., Yao, Y. D., et al. (2013). Enhanced drug delivery to hepatocellular carcinoma with a galactosylated core-shell polyphosphoester nanogel. *Biomaterials Science*, *1*(11), 1143–1150. https://doi.org/10.1039/c3bm60099h.

Wu, M. X., & Yang, Y. W. (2017). Metal–organic framework (MOF)-based drug/cargo delivery and cancer therapy. *Advanced Materials*, *29*(23).

Xia, S., Tan, C., Zhang, Y., Abbas, S., Feng, B., Zhang, X., et al. (2015). Modulating effect of lipid bilayer-carotenoid interactions on the property of liposome encapsulation. *Colloids and Surfaces B: Biointerfaces*, *128*, 172–180. https://doi.org/10.1016/j.colsurfb.2015.02.004.

Yamamoto, T., Hirohashi, K., Kaneda, K., Ikebe, T., Mikami, S., Uenishi, T., et al. (2001). Relationship of the microvascular type to the tumor size, arterialization and dedifferentiation of human hepatocellular carcinoma. *Japanese Journal of Cancer Research*, *92*(11), 1207–1213. https://doi.org/10.1111/j.1349-7006.2001.tb02141.x.

Yang, T., Lan, Y., Cao, M., Ma, X., Cao, A., Sun, Y., et al. (2019). Glycyrrhetinic acid-conjugated polymeric prodrug micelles co-delivered with doxorubicin as combination therapy treatment for liver cancer. *Colloids and Surfaces B: Biointerfaces*, *175*, 106–115. https://doi.org/10.1016/j.colsurfb.2018.11.082.

Yang, J. D., Nakamura, I., & Roberts, L. R. (2011). The tumor microenvironment in hepatocellular carcinoma: Current status and therapeutic targets. *Seminars in Cancer Biology*, *21*(1), 35–43. https://doi.org/10.1016/j.semcancer.2010.10.007.

Yang, Z. F., & Poon, R. T. P. (2008). Vascular changes in hepatocellular carcinoma. *The Anatomical Record: Advances in Integrative Anatomy and Evolutionary Biology*, *291*(6), 721–734. https://doi.org/10.1002/ar.20668.

Ye, B. L., Zheng, R., Ruan, X. J., Zheng, Z. H., & Cai, H. J. (2018). Chitosan-coated doxorubicin nano-particles drug delivery system inhibits cell growth of liver cancer via p53/PRC1 pathway. *Biochemical and Biophysical Research Communications*, *495*(1), 414–420. https://doi.org/10.1016/j.bbrc.2017.10.156.

Yokoyama, M. (2005). Drug targeting with nano-sized carrier systems. *Journal of Artificial Organs*, *8*(2), 77–84. https://doi.org/10.1007/s10047-005-0285-0.

Yoo, H., & Mok, H. (2015). Evaluation of multimeric siRNA conjugates for efficient protamine-based delivery into breast cancer cells. *Archives of Pharmacal Research*, *38*(1), 129–136. https://doi.org/10.1007/s12272-014-0359-8.

Yu, C. Y., Wang, Y. M., Li, N. M., Liu, G. S., Yang, S., Tang, G. T., et al. (2014). In vitro and in vivo evaluation of pectin-based nanoparticles for hepatocellular carcinoma drug chemotherapy. *Molecular Pharmaceutics*, *11*(2), 638–644. https://doi.org/10.1021/mp400412c.

Yuan, Y., Zhang, Y., Liu, B., Wu, H., Kang, Y., Li, M., et al. (2015). The effects of multifunctional MiR-122-loaded graphene-gold composites on drug-resistant liver cancer. *Journal of Nanobiotechnology*, *13*(1). https://doi.org/10.1186/s12951-015-0070-z.

Zeng, Y., Zhang, D., Wu, M., Liu, Y., Zhang, X., Li, L., et al. (2014). Lipid-AuNPs@PDA nanohybrid for MRI/CT imaging and photothermal therapy of hepatocellular carcinoma. *ACS Applied Materials and Interfaces*, *6*(16), 14266–14277. https://doi.org/10.1021/am503583s.

Zhang, Y., Liu, S., Wang, X., Zhang, Z. Y., Jing, X. B., Zhang, P., et al. (2014). Prevention of local liver cancer recurrence after surgery using multilayered cisplatin-loaded polylactide electrospun nanofibers. *Chinese Journal of Polymer Science (English Edition)*, *32*(8), 1111–1118. https://doi.org/10.1007/s10118-014-1491-0.

CHAPTER 4

Nanocarriers-based improved drug delivery for treatment and management of cardiovascular diseases

Introduction

Cardiovascular diseases (CVDs) refer to a variety of disease conditions that affect either the heart (from Greek *kardia* "heart") or the blood vessels (Latin vascularis, of or pertaining to vessels or tubes). In the past few decades, owing to major changes in people's lifestyle, including consumption of processed food and reduced activity levels, CVDs are becoming the leading cause of mortality worldwide and pose a widespread impact on the overall health status of the affected patients. CVDs have thus become a global public health concern and result in an enormous health and economic burden on the health systems of developed and developing countries around the world. In 2016, the death toll due to CVDs was estimated to be more than 17.9 million, representing more than one-third of the total deaths worldwide. Heart attack and stroke constitute 85% of these deaths (World Health Organization (WHO), 2017). Around 45% of total deaths in Europe are due to CVDs and it represents 3.9 million deaths per year due to CVDs (Timmis et al., 2018). Direct healthcare cost due to CVD treatment is over 100 billion Euros each year in Europe and these figures are unfortunately expected to rise over the next few years because of increased CVDs' risk factors including diabetes, obesity, and increase in the elderly population as well (Heidenreich, 2015). The available treatment strategies such as pharmacological, heart transplantation, coronary artery bypass surgery, and other ventricular assistant devices have greatly enhanced patients' longevity to date. However, less invasive and cost-effective treatments with high efficacy are still vital for future.

Anomalies in morphogenesis, muscle repair and function, and heart rhythm characterize human cardiac diseases. Congenital heart diseases, Cardiomyopathy, Heart attack or Myocardial Infarction, Coronary artery disease, Atherosclerosis, Deep vein thrombosis and pulmonary embolism, stroke, aortic disease, and heart arrhythmias are a few of the many Cardiovascular ailments.

The effective treatment of cardiovascular ailments however poses a number of challenges. While traditional formulations of a variety of drugs working

Nanocarriers for Organ-Specific and Localized Drug Delivery. https://doi.org/10.1016/B978-0-12-821093-2.00005-0

through different mechanisms of action for the treatment of CVDs can be found in the market, they are still far from ideal due to a slew of issues such as low biological efficacy, poor water solubility, lack of target specificity as well as drug resistance. Continuous progress in nanotechnology in recent decades has made a dramatic expansion in the potentials of nanomaterials in biomedical applications. Nanocarriers have been extensively employed in the imaging, diagnosis, and treatment of a variety of diseases, cancer being the prominent one. Nanomedicine for cancer treatment has been the main focus of research for decades and a number of nanotechnology-based products (e.g., Doxil, Abraxane) have been approved for human use in clinics by the US FDA and the EMA and a number of nanomedicines are in the advanced stages of clinical trials (Bobo, Robinson, Islam, Thurecht, & Corrie, 2016). Contrary to this, the field of Cardiovascular nanomedicine is still in its infancy and has recently started growing. Nonetheless, in recent years, promising lab-scale results and increased amount of research publications have been reported thus we hope that cardiovascular nanotechnology will reach its apex in coming decades. In this chapter, we highlight the efforts made for construction of nanocarriers-based biocompatible systems for cardiovascular drug delivery.

Nanostructure mediated drug delivery for cardiovascular diseases

A number of factors influence the success of nanocarriers in delivering drugs to target sites in cardiovascular diseases. The main points to be considered for developing cardiovascular targeted delivery systems include, (1): understanding of the basics of various heart disorders, such as their etiology, pathological manifestations, clinical characteristics, and desired therapies as identification of specific target sites pertaining to the relevant cardiovascular ailment, (2) choice of appropriate nano drug delivery carriers to deliver the drug efficiently, and (3): use of effective targeting ligands. Targeted drug delivery implies that a drug is deposited specifically at the target site, with minimal side effects on other organs following administration. Liposomes, dendrimers, microparticles, nanoparticles, polymer-drug conjugates, micelles, nanostents, and microbubbles (shown in Fig. 4.1) have all been used till now for designing and developing nanoscale drug delivery systems for cardiovascular diseases. In the following section, we will address recent studies that have used such nanocarriers to treat the respective cardiovascular diseases.

Nanocarriers in cardiac arrhythmias

Arrhythmia or atrial fibrillation is a disordered pattern of heartbeat in which the heart beats too quickly, too slowly, or at an abnormal pace. Arrhythmia

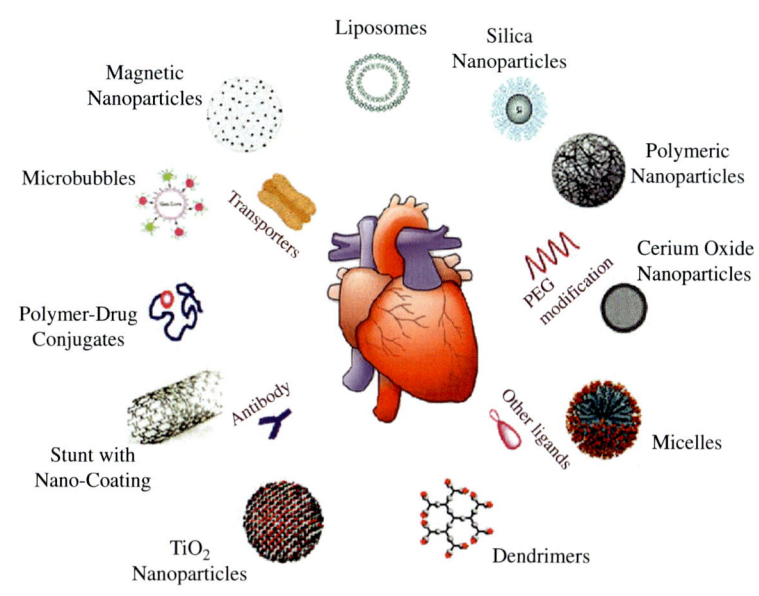

FIG. 4.1

Different nanocarriers for drug delivery in cardiovascular diseases. *No Permission Required.*

most often occurs when the cells responsible for producing electrical signals work abnormally, the electrical signals do not adequately propagate through the heart, or an electrical signal is produced by a random part of the heart (Fig. 4.2). Sodium channel blockers, calcium channel blockers, potassium channel blockers, and beta blockers are four types of therapeutic agents that can be used to treat arrhythmia. However, majority of these antiarrhythmic medications have poor bioavailability, undergo first-pass metabolism, have low water solubility, and have other negative side effects.

Amiodarone, the most widely employed antiarrhythmic drug is known to cause adverse side effects owing to its accumulation in liver, spleen, and lungs. The pharmaceutical compound also suffers from poor water solubility. Recently, nano-encapsulation has been extensively investigated and employed for the targeted delivery of Amiodarone and other antiarrhythmic agents. Lipid nanocapsules, nanoliposomes, nano niosomes, polymeric nanoparticles, solid lipid nanoparticles are examples of common nanocarriers used for the purpose.

Lamprecht and his colleagues synthesized and completely characterized amiodarone-loaded lipid nanocapsules (LNCs) (Lamprecht, Bouligand, & Benoit, 2002), which precisely refer to nanocarriers mimicking lipoprotein and ranging between 20 and 100 nm in size. The nonionic surfactant used was polyethylene glycol-660 hydroxystearate, and soybean lecithin was used to increase LNCs' stability. A liposomal formulation comprising amiodarone

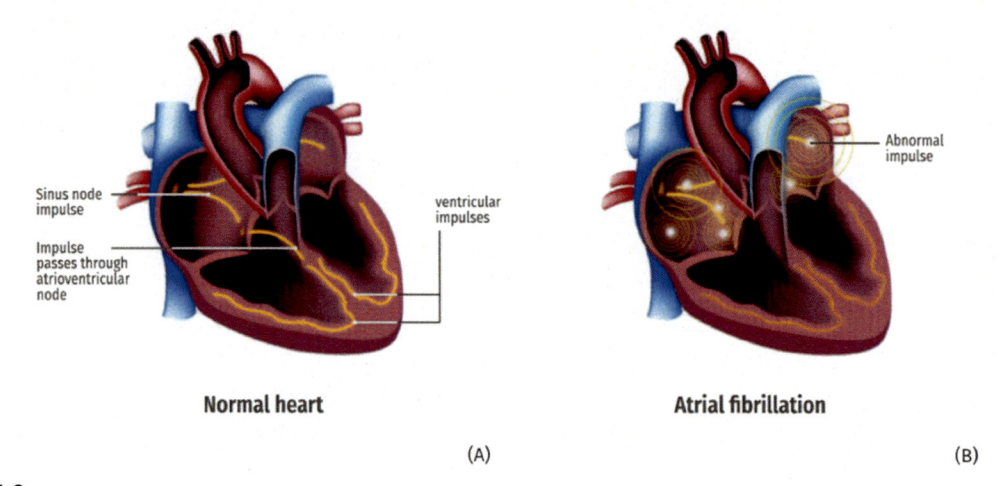

Normal heart   **Atrial fibrillation**

(A)   (B)

FIG. 4.2

Physiology of atrial fibrillation or cardiac arrhythmia. *From Hagiwara, Y., Fujita, H., Oh, S. L., Tan, J. H., Tan, R. S., Ciaccio, E. J., et al. (2018). Computer-aided diagnosis of atrial fibrillation based on ECG Signals: A review.* Information Sciences, 467, *99–114. https://doi.org/10.1016/j.ins.2018.07.063.*

hydrochloride was proposed by Zhuge et al. (2016). When compared to free amiodarone hydrochloride, intravenous administration of liposomal amiodarone to rats led to a more stable and sustained drug release as well as an improvement in drug circulation time. The liposomal amiodarone was found to be 4.1 times higher in the heart of cardiac radiofrequency ablation (CA) rat model treated with this formulation than in the sham-operated community treated in identical manner, according to the bio-distribution report. Wang and co-workers formulated Lidocaine hydrochloride, an antiarrhythmic agent that works by blocking sodium channels, into polymeric liposomes, which they thoroughly characterized in terms of morphology, particle size, polydispersity, drug-encapsulation strength, rate of drug release, and storage stability (Wang et al., 2013). Polymeric liposomes were generated using cholesterol and octadecyl-quaternized lysine modified chitosan (OQLCS). The synthesized polymeric liposome has also been conjugated with the TAT peptide, a cell-penetrating peptide. They were able to promote the cross membrane uptake of polymeric liposomes using TAT with minimal toxicity. In an ischemia/reperfusion(i/R) rat model, the Amiodarone-loaded liposomes were found to reduce the hemodynamic effects of the drug while enhancing its antiarrhythmic effects thereby decreasing the chances of death due to lethal arrhythmia (Takahama et al., 2013). In another study, Guan and colleagues developed a liposomal gel as a topical delivery method for propranolol hydrochloride (Guan et al., 2015). The propranolol content in skin seems to have increased as a result of the liposomal gel. Furthermore, propranolol concentrations reached a value of 1.17 mg/mL and 0.41 mg/mL, respectively, when administered orally and topically.

In terms of niosomal nanocarriers' formulations, Arzani and colleagues prepared and characterized various carvedilol-loaded niosomal formulations, an antihypertensive that function to block both alpha and beta receptors hence utilized in arrhythmia treatment (Arzani, Haeri, Daeihamed, Bakhtiari-Kaboutaraki, & Dadashzadeh, 2015). They discovered that the bilosomes they created improved the bioavailability of the encapsulated carvedilol. Ammar and colleagues developed diltiazem-loaded niosomes to achieve nasal drug delivery (Ammar, Haider, Ibrahim, & El Hoffy, 2017), since diltiazem has a low oral bioavailability and a large first-pass effect. The hydrophilic drug was protected within the niosomes' cores in combination with the nasal route offers both swift delivery and avoidance of first-pass metabolism. In Wistar rats, the pharmacokinetic efficiency of the niosomal formulation was assessed. They observed an obvious increment in the plasma half-life (t1/2) of the drug. An increment in the area under the drug concentration versus time curve (AUC) and the mean residence time (MRT), as well as a decrease in the elimination rate constant (Ke), was observed in in-vivo studies. These favorable pharmacokinetic attributes indicated that nasal niosomal diltiazem may be a viable substitute to the traditional oral formulation of the drug. Oral administration of atenolol, a beta-blocker, can result in a number of unwanted side effects, such as diarrhea, nausea, ischemic colitis, mesenteric arterial thrombosis. Atenolol-loaded poly(lactic-*co*-glycolic acid) nanoparticles were formulated by Chourasiya, Bohrey, and Pandey (2016). The formulation was designed through optimizing formulation variables using 3^3 factorial design. The final formulation showed spherical-shaped particles and loaded 71% of the drug with slow and sustained release drug profile. Mouez and colleagues employed transnasal permeation after encapsulating verapamil in composite chitosan-transfersomal vesicles to resolve the drug's poor bioavailability (around 10%–20%) due to hepatic first-pass clearance (Mouez, Nasr, Abdel-Mottaleb, Geneidi, & Mansour, 2016).

Al-Kassas and colleagues enhanced the systemic bioavailability of propranolol hydrochloride by developing a drug delivery method focused on the dissemination of chitosan nanoparticles in a carbopol and poloxamer gel (Al-Kassas et al., 2016). Owing to the hydrophobic nature of interactions between propranolol hydrochloride and chitosan, the nanoparticles-gel delivery system manifested a prolonged release of the encapsulated drug as demonstrated by permeation experiments performed on the skin of pig ears, with only 11% of propranolol being released from the gel formulation after 24 h. Üstündag-Okur and colleagues used non-toxic Solid Lipid Nanoparticles (SLNs) made of compritol, lecithin, and poloxamer to increase the oral bioavailability of nebivolol, a beta blocker (Üstündağ-Okur, Yurdasiper, Gündoğdu, & Homan Gökçe, 2015). In another study, Shah used a hot homogenization method to prepare an enteric compression coated formulation of carvedilol-incorporated SLNs to prevent first-pass metabolism, causing an increased drug bioavailability (Shah, Madan, & Lin, 2015). An intranasal drug delivery system comprising SLNs was proposed by Aboud and co-workers to improve the bioavailability of carvedilol (Aboud, El Komy, Ali, El Menshawe, & Abd Elbary, 2016). Within the intranasal SLNs, the absolute bioavailability of carvedilol was significantly higher than that of oral carvedilol, as demonstrated by the in vivo pharmacokinetic experiments carried out on rabbits. Poly(ethylene-*co*-vinyl acetate) based polymeric nanoparticles coated with chitosan were developed by Varshosaz et al. for delivering carvedilol to the lungs (Varshosaz, Taymouri, & Hamishehkar, 2014). The formulation existed as a dry inhalable mucoadhesive powder and their study results showed that mannitol spray-dried, mucoadhesive nanoparticles were suitable for pulmonary delivery of carvedilol through inhalation.

It was hypothesized that cardiac macrophages could be an efficient vehicle for delivering amiodarone to the heart, which is a particularly important idea in view of lowering the overall dose and off-site drug accumulation. To test this hypothesis, Ahmed et al. created a supramolecular nanocarrier made of L-lysine cross-linked succinyl-cyclodextrin that binds amiodarone via supramolecular host-guest interplay and has a large affinity for macrophages (Ahmed et al., 2019). Rapid absorption of nanoparticles by cardiac macrophages results in accumulation of nanoparticles in the heart, according to biodistribution studies at the single-cell and organ stage. The use of nanoparticles to aid amiodarone delivery led to a 250% increase in selectively delivering the drug to cardiac tissue. Elgart's team developed a self nano-emulsifying drug delivery system (SNEDDs) to improve amiodarone oral bioavailability (Elgart, Cherniakov, Aldouby, Domb, & Hoffman, 2013). Oral administration of amiodarone-SNEDDs and talinolol-SNEDDs showed higher Cmax and AUC with less variability. In vitro lipolysis of amiodarone-SNEDDs showed high drug concentration in the medium available for absorption. Talinolol-SNEDDs showed Pgp inhibition. LNCs, liposomes, niosomes, SLNs, and PNPs are all common nano-carriers used for specific antiarrhythmic drug delivery. Each technique has its own set of features that can be applied to a variety of situations. Each of these strategies is chosen based on the agent's hydrophilicity or hydrophobicity, target tissue, and administration route.

Nanocarriers in thrombosis

The formation of a life-threatening obstructive blood clot, also known as a thrombus, within a blood vessel is referred to as thrombosis. This clot has the potential to impede or block blood flow in the affected region as shown in Fig. 4.3. Ischemic heart disease (acute coronary syndrome), venous thromboembolism (VTE), and stroke are all caused by thrombosis, which occurs as the most frequent underlying pathology. Heparin, streptokinase (SK) tissue

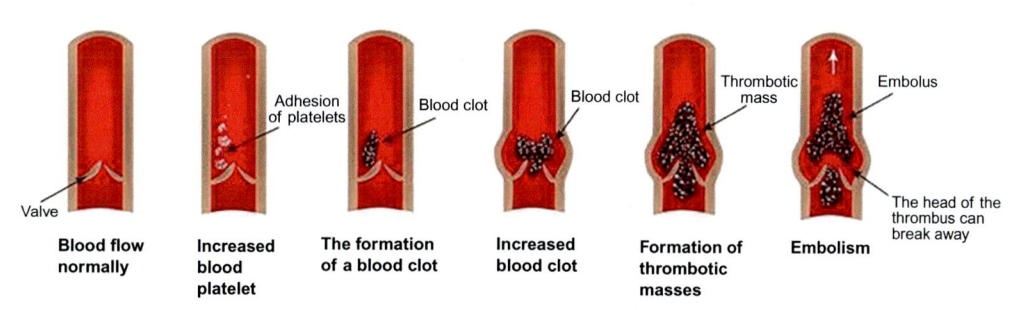

FIG. 4.3

Formation of thrombosis and emboli in veins. *No Permission Required.*

plasminogen activator (tPA), urokinase plasminogen activator (uPA), and recombinant plasminogen activator (rtPA) are currently available fibrinolytic or thrombolytic agents (Jinatongthai et al., 2017; Onishi, St Ange, Dordick, & Linhardt, 2016). Thrombus can be solubilized or the development of thrombus can be greatly hampered using these agents. These protein-based clinical thrombolytic agents, on the other hand, may cause allergic reactions, have a shorter half-life, or be inactivated. Besides, these agents have little therapeutic effect on account of their poor targeting performance and low accumulation efficiencies. Under such circumstances, nanomedicines have been extensively studied for the diagnosis and treatment of thrombosis because fast recanalization of the arteries occluded by thrombus is needed for improved outcomes and for minimizing the risk of mortality in stroke or acute myocardial infarction.

Hirulog, a synthetic precursor of hirudin, a natural inhibitor of thrombin, was administered to the thrombus by employing lipid nanoparticles bearing a fibrin-binding peptide. Antithrombin activity was significantly higher within the aortic tree of ApoE-deficient mice following administration of fibrin-targeting hirulog-carrying particles compared to non-targeted particles (Peters et al., 2009). Intravenous tPA infusion has a number of disadvantages, including poor effectiveness and an increased risk of bleeding complications. Liu et al. looked into the prospects of magnetic-targeting tPA for localized thrombolysis in a rat embolic model and developed PEGylated thermosensitive magnetoliposomes (Liu, Hsu, Chen, Wu, & Ma, 2019). Under the direction of an external magnet traveling along the iliac artery, tPA-loaded PEGylated thermosensitive magnetoliposomes (tPA equivalent of 0.2 mg/kg) were intra-arterially administered. With just 20% of a daily dose of free tPA, magnetic tPA thermosensitive magnetoliposomes gathered in the thrombus-affected area and achieved successful target thrombolysis and significantly restored iliac blood flow after clot lodging. Kawata et al. tested a similar tPA delivery nanosystem made of basic gelatin and zinc acetate in a swine acute myocardial infarction model (Kawata et al., 2012). In vitro, tPA activity was decreased to about half that of free tPA inside this nanosystem, and it was completely recoverable using transthoracic ultrasound.

D-Phenylalanyl-L-prolyl-L-arginyl-chloromethyl ketone (PPACK) bound to perfluorocarbon nanoparticles outperformed both heparin and uncomplexed PPACK in impeding thrombosis, and created a localized clotting barrier that kept operating successfully even after the rapid disappearance of systemic effects (Myerson, He, Lanza, Tollefsen, & Wickline, 2011). In another study, PPACK-liposomes provided before an arterial injury substantially sped up the time to arterial occlusion compared to free PPACK. McCarthy et al. created a multimodal nano-agent using magnetofluorescent cross-linked dextran-coated iron oxide nanoparticles conjugated to tPA in an attempt to build a theranostic construct with fibrinolytic operations (McCarthy et al., 2012). To target thrombus, nanoparticles were functionalized with an activated FXIIIa-sensitive peptide. The FXIIIa-targeted fibrinolytic nano-agent was assessed via intravital fluorescence microscopy and effectively bound to the margins of intravascular thrombi. The pulmonary emboli were lysed by FXIII1-targeted agent with similar efficacy as free tPA in in-vivo fibrinolysis studies.

Korin et al. explained how they used universal hemodynamic phenomena to deliver targeted tPA to stenotic arteries (Korin et al., 2012). The authors created micro-aggregates of poly-lacticglycolic acid nanoparticles coated with tPA because occlusions in blood vessels cause local increases in shear stress. These micro-aggregates remained stable under physiologic flow conditions with shear stress values up to 70 dyn/cm^2, but exposure to abnormally high shear stress in the regions of vascular occlusion/stenosis resulted in their rapid breakup and local drug release. Shear-activated tPA coated nanoparticles quickly dissolved ferric chloride-induced arterial thrombi in mouse mesenteric arteries, with full clearance of occluding thrombi within 5 min of application, compared to free drug. Urokinase-coated, self-assembled chitosan and tripolyphosphate nanoparticles were created by Jin, Zhang, Sun, Zhang, and Zhang (2013). When Urokinase-Type Plasminogen Activator (uPA)-carrying nanoparticles were administered to a rabbit model of thrombosis, the thrombolytic effect was significantly improved compared to free uPA. These studies suggest that the nanocarriers-based drug delivery systems have great potential to be translated to clinical practice for treatment and diagnosis of cardiovascular ailments including thrombosis.

Nanocarriers in atherosclerosis

Atherosclerosis (AS) is characterized by fatty buildup or plaque accumulation due to the chronic inflammation of the arteries and on the blood vessel wall (Fig. 4.4). It is a major contributor to a multitude of other CVDs including

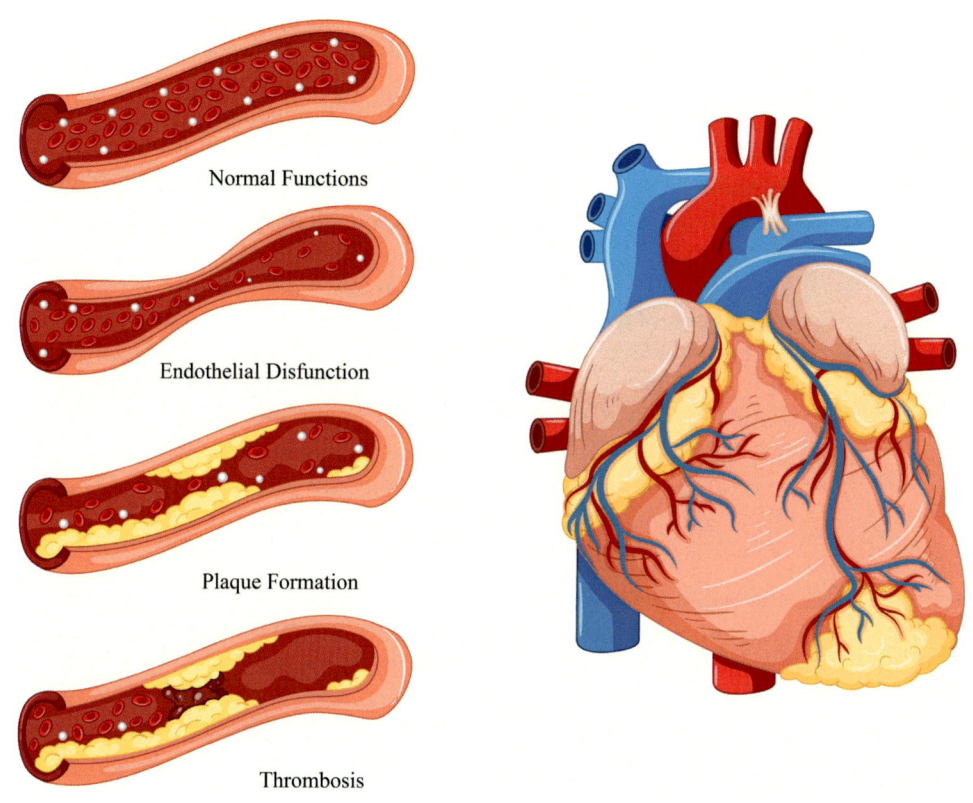

Normal Functions

Endothelial Disfunction

Plaque Formation

Thrombosis

FIG. 4.4

Illustration of the formation of atherosclerosis. *No Permission Required.*

peripheral arterial diseases (PADs) or peripheral vascular diseases (PVD), ischemic heart disease or thrombosis. A dysregulated cholesterol metabolism, an impaired endothelial function, and simultaneously a faulty immune response are at the root of the pathogenesis of atherosclerosis. Plaque, which is the build-up of fatty deposits, waste, and cellular components on the inner wall of the blood vessel, is the signature characteristic of atherosclerosis. Coronary angioplasty and stent implantation are standard surgical procedures, up to 50% of CVD patients who undergo angioplasty experience re-stenosis at the intervention site within 6 months of the procedure. Depending upon the site of atherosclerotic plague, it can give rise to ischemic heart disease, PAD, aortic diseases on several other complications. AS treatment approaches can be sub-divided into five classes: (1) controlling cholesterol metabolism, (2) anti-inflammatory and (3) anti-oxidation techniques, (4) inhibiting foam cell formation, and (5) reducing activity of metalloproteinases. Established therapeutic strategies against atherosclerosis include the treatment of hypertension and hyperlipidemia, as well as the regulation of hemostasis to prevent thrombosis. Statins (e.g., atorvastatin and simvastatin) are commonly used first-line medications for lowering low-density lipoprotein (LDL) cholesterol and preventing atherosclerotic plaque formation. Other dyslipidemia and inflammation medications, such as proprotein convertase subtilisin kexin type 9 (PCSK9) inhibitors and cholesteryl ester transfer protein inhibitors, are still in the preclinical stages. Antiplatelet medications (such as aspirin and clopidogrel), anticoagulants (such as warfarin and heparin), beta-blockers, angiotensin-converting enzyme inhibitors, and other pharmacological strategies are often commonly used. However, because of their low aqueous solubility and substantial first-pass effect, most of these drugs have undesirable systemic bioavailability and serious adverse drug effects. Some statins, for example, have a poor oral bioavailability of up to 30% due to their low water solubility and large molecular weight. Statins can cause side effects such as muscle pain, fatigue, and weakness when taken orally. Nanotechnology-based strategies could be the mainstay for prevention of the side effects of such drugs and improving their therapeutic efficacies. Furthermore, different peptides and proteins can be attached to NPs in accordance with the disease's microenvironment to target lesions.

A possible treatment mechanism for atherosclerosis is cholesterol metabolism regulation. High-density lipoprotein (HDL), also known as "healthy cholesterol," transports cholesterol in the form of bile acids from atherosclerotic lesions to the liver, where it is recycled or excreted, reversing the cholesterol transport pathway. HDL extracts cholesterol from walls of the arteries and prevents the formation of new plaques by regulating the reverse transport of cholesterol. Banik et al. created a polymer-lipid hybrid NP made of synthetic biodegradable materials that can act same as HDL (Banik et al., 2018). Cholesterol removal ability was expressed by the NPs with anti-oxidative and lipid reduction properties. In vivo therapeutic assay of the targeted-NP formulations in mice showed good anti-inflammatory and lipid reduction properties when compared with non-targeted NPs. The newly constructed synthetic targeted NP could potentiate therapeutic interventions for heart diseases. Marrache et al. developed a fluorescent HDL-mimicking polymer lipid NP platform, comprising a PLGA hydrophobic core with an HDL-like hydrophobic core and cholesteryl oleate (CO) (Marrache & Dhar, 2013). Apart from the targeted and therapeutic effects of reconstituted HDL, Jiang et al. loaded lovastatin into NPs to improve their efficacy (Jiang et al., 2019).

Statins are a group of lipid-lowering drugs that inhibits the 3-hydroxy-3-methyl-glutarylcoenzyme A (HMGCoA) reductase and are the most widely used cholesterol-lowering drugs. Most of the drugs in this class belong to Class II of the biopharmaceutical classification system (BCS) thus show low bioavailability due to their intrinsic low water-solubility property. In the work of Duivenvoorden et al., an injectable reconstituted HDL nanoparticle was used to deliver statins to atherosclerotic plaques, resulting in increased statin systemic bioavailability and anti-inflammatory impact (Duivenvoorden et al., 2014). Andalib et al. found the anti-inflammatory potential of atorvastatin, and that atorvastatin encapsulated micelles had an enhanced anti-inflammatory potential (Andalib, Molhemazar, & Danafar, 2018). In another study, Yu et al. created a lipid-polymer hybrid nanoparticles with a Col IV targeting peptide for the delivery of GW3965, a Liver X Receptor antagonist and core regulator of lipid metabolism and transport that suppresses inflammatory signaling in macrophages. The concentration of GW3965 at atherosclerotic lesion sites was significantly increased, resulting in significantly increased curative effects (Yu et al., 2017).

Similarly, anti-inflammatory medicines also play vital role in the treatment of AS. Interleukin 10, methotrexate, andrographolide, rapamycin, betamethasone, and hyaluronan are all anti-inflammatory drugs that are mostly encapsulated in target-specific NPs for selective drug release, which can reduce systemic side effects. Tsoref et al. came up with a new E-selectin binding copolymer based on N-(2-hydroxypropyl) methacrylamide (HPMA) and demonstrated its role in the regression and stabilization of atherosclerotic plaques (Tsoref et al., 2018). McCarthy et al. carried out the modification of particles with light-activated therapeutic moieties and near-infra-red (NIR) fluorophores in addition to the anti-inflammatory influence of peptides or drugs (McCarthy, Korngold, Weissleder, & Jaffer, 2010). The particles can eliminate inflammatory macrophages, lower inflammation, and stabilize lesions due to their photoactivity. Flores et al. recently reported that antibody-mediated CD47 blockade can also reduce inflammation, which in fact is a novel approach (Flores et al., 2020).

Regulation of Oxidative stress and inflammatory responses feed off each other, hastening the progression of AS. Anti-oxidation can minimize inflammation as well as the development of ox-LDL by removing oxygen-free radicals. Chmielowski et al. employed NPs made up of an amphiphilic shell and a degradable ferulic acid-based polymer linked to diglycolic acid (1cM-PFAG) to prevent the formation of foam cells by lowering ox-LDL uptake by down-regulating Scavenger Receptors (Chmielowski et al., 2017). Selenium, which functions as a component of enzymes acting against free radicals within cells was employed in another study by Zhu et al. for preparing selenium quantum dots and investigated as a new strategy for AS treatment (Zhu et al., 2019). In an attempt to create a reactive oxygen species (ROS)-eliminating NP, Wang et al. covalently conjugated a hydrogen-peroxide-eliminating compound of phenylboronic acid pinacol ester onto a cyclic polysaccharide beta-cyclodextrin (TPCD) as a superoxide dismutase mimetic agent (Wang et al., 2018). The TPCD NPs reduced systemic and local oxidative stress and inflammation as well as decreased inflammatory cell infiltration in atherosclerotic plaques and the results suggest that TPCD NPs could be a potential strategy for antiatherosclerotic nanotherapy.

Excessive macrophage or smooth muscle cells (SMC) phagocytosis of oxidatively modified LDL (ox-LDL) is a crucial cause for AS, causing foam cell formation. Ox-LDL is an entry point for the treatment of AS, in addition to being an appealing biological target. Inspired by the negative charge on ox-LDL, Moretti et al. created negatively charged NPs to achieve competitive combination with scavenger receptors (SRs). They discovered that certain NPs inhibited human-derived macrophages from absorbing ox-LDL. Furthermore, plaque stabilization is a novel treatment concept in which MMPs degrade the extracellular matrix and reduce plaque stability (Moretti et al., 2018). Kheirolomoom et al. developed multifunctional NPs to facilitate molecular targeting and miRNA delivery. They used anti-miRNA therapeutics to minimize miR-712 expression, inhibit metalloproteinase activity, and successfully stabilize plaques (Kheirolomoom et al., 2015).

Nanoburrs are small particles that pass across the bloodstream and bind to affected arteries, delivering medicine to damaged tissue directly (Rajiv, Hardik, Vaibhav, & Vimal, 2011). Nanoburrs have tiny protein fragments on their surfaces that enable them to adhere to damaged arterial walls. They will release drugs once stuck (such as paclitaxel, which inhibits cell division and helps to prevent the growth of scar tissue that can clog arteries). The nanoburrs are aimed at the basement membrane, which lines the arterial walls and is only revealed when those walls are weakened. As a result, the nanoburrs may be used to administer medicines to patients with restricted arteries or to treat atherosclerosis and other inflammatory cardiovascular diseases. Since, targeted peptides are bound to an outer shell rather than directly to the drug-carrying nucleus, which would require a more complicated chemical reaction, the nanoburrs' structure could make it easier to produce surface-modified systems. This nature also lowers the risk of nanoparticles bursting and aids in the controlled release of drugs.

Nanocarriers in aortic aneurysm and other aortic diseases

Since the aorta is the body's largest artery, any diseases or illnesses that could harm it will place patients at serious and life-threatening risk. They can impact the aorta in virtually any area of the body, including the chest (thoracic aorta) or the abdomen (abdominal aorta). Aortic aneurysm (AAA) is a permanent protrusion in the artery that will gradually rupture the sidewall. Aneurysms of the aorta are among the main causes of deaths and morbidities among the elderly (Gillum, 1995).

Extracellular matrix (ECM) proteins degraded by the matrix metalloproteinases (MMPs), inflammatory cell invasion of vessel walls, reactive oxygen species (ROS) overproduction, elasticity lamellar aortic degeneration, and vascular smooth muscle cell (VSMC) degradation have all been linked to aneurysm formation (Joviliano, Ribeiro, & Tenorio, 2017; Quintana & Taylor, 2019). Invasive therapies for aortic aneurysms are currently the standard of care, for instance, open-chest surgery for replacing the affected portion with vascular prostheses and endovascular stent-graft implantation. Nevertheless, clinically only a few medications can be used to prevent aortic aneurysms. A general characteristic of the pathology of aortic disorders is vascular calcification. Certain medications that target aneurysm dilation or rupture have recently been studied in animal models and have furnished outstanding findings on inhibiting aneurysms (Habashi et al., 2011; Steinmetz et al., 2005). However, further research is needed to improve these promising drugs' effectiveness in clinical trials, presumably because oral, intra-arterial or parenteral administrations will subsequently lead to unintended side effects and only small drug doses finally reach the wall because of the rapid blood flow associated with high pressure (Karlsson, Gnarpe, Bergqvist, Lindbäck, & Pärsson, 2009; Kurosawa, Matsumura, & Yamanouchi, 2013).

In connection with nanocarriers-based approaches for the treatment of aneurysms, Nosoudi et al. developed an elastin-antibody binding PLA NPs loaded with the hydroxamic acid matrix metalloproteinase inhibitor (MMP inhibitor), Batimastat (BB-94), which could be used to prevent abdominal aortic aneurysm (AAA) in systemic delivery (Nosoudi et al., 2015). MMP activity was 56% lower at the injury site relative to the thoracic aorta, and NPs substantially inhibited the growth of aorta diameter. Based on these findings, Lei et al. developed bovine serum albumin (BSA) NPs loaded with ethylenediaminetetraacetic acid (EDTA) and delivered them to aneurysm sites to eliminate calcification (Lei, Nosoudi, & Vyavahare, 2014). The targeted chelation therapy with EDTA-loaded albumin nanoparticles regressed arterial calcification without causing systemic side effects.

The pathology of aortic aneurysm is known to involve elastic degeneration and Sinha et al. loaded 1,1-dioctadecyl-3,3,3,3-tetramethylindotricarbocyanine iodide (DIR) dyes and surface maleimide groups conjugated to thiolate elastin/IgG antibodies onto PLA NPs to attack the degraded elastin layer (Sinha et al., 2014). To aid in the visualization and monitoring of NPs, DIR dyes were permanently attached to them. To begin, isolated rat aorta was treated with elastase to simulate elastin degradation in vitro, and it was discovered that the number of NPs and the efficiency of elastase attachment increased as the amount of elastic damage increased. They also used a popular aortic aneurysm model ($CaCl_2$-induced rat model) as well as other two vascular disease models to evaluate targeting effectiveness (atherosclerosis and vascular medial calcification). Their results strongly suggest that NPs were vulnerable to elastic injury and precise spatial accumulation occurred even under high-shear hemodynamic conditions, as $CaCl_2$ can mimic the deterioration of the elastic lamina in the abdominal aorta of rats.

Despite the fact that both approaches have some effects on animal models, the common problem is that they are only effective in the early stages of aortic aneurysms. Nosoudi et al. used NPs to extract mineral deposits and restore the elastic layer to reverse the production of moderate aneurysms in $CaCl_2$-induced AAA rat models, simulating the clinical situation and treating moderate-sized aneurysms (Nosoudi et al., 2016). EDTA was used to dissolve calcified deposits in the arteries, and then pentagalloyl glucose (PGG) was used to release the polyphenol, which stabilizes elastin and improves elastic fiber deposition. The chelators EDTA and PGG were used to minimize macrophage recruitment, MMP function, calcification, and elastin degradation in the aorta, resulting in an increase in vascular elastance. Dhital and Vyavahare also developed PGG-loaded nanoparticles that were tested in AAA mice induced by porcine pancreatic elastase (PPE) (Dhital & Vyavahare, 2020). Their results supported that site-specific delivery of PGG with targeted NPs could be used for treatment of already developed AAA and suggested that such therapy can restore arterial homeostasis and reverse inflammatory markers.

Cheng et al. developed a ROS-responsive nano-therapy capable of releasing Rapamycin (RAP; a candidate that has been shown to prevent aneurysm formation) (Cheng et al., 2018). Meanwhile, they created a multifunctional nano-therapy for targeted treatment based on the pathophysiology of aortic aneurysm, which is related to ROS. They targeted aneurysmal sites with a nano-platform made up of ROS-responsive materials that can release therapeutic molecules when activated by ROS. Shirasu et al. developed a unique nanosystem that relies on the penetration of NPs to meet the aneurysm wall defect (Shirasu et al., 2016). To inject the NPs containing rapamycin, they used an elastase-induced AAA rat model. The rapamycin NPs successfully attached to the damaged wall structures as revealed by the microscopic examination.

Nanocarriers in ischemic heart disease and myocardial infarction

When the blood flow to the heart muscle is decreased due to a partial or total blockage of the arteries supplying it with blood, it is referred to as Ischemic Heart Disease or Myocardial Ischemia. Coronary heart disease or acute coronary syndrome are the two other terminologies used for referring to the same condition. The most common symptom of advanced ischemic heart disease is an infarction (heart attack), which is the permanent death (necrosis) of heart muscle caused by a chronic lack of oxygen supply (ischemia; Fig. 4.5). Ischemic cardiomyopathy is a generalized weakening of the heart muscle caused by decreased blood flow to the heart.

Traditional clinical approaches for ischemia and myocardial infarction rely on surgical revascularization procedures, such as coronary stenting or coronary

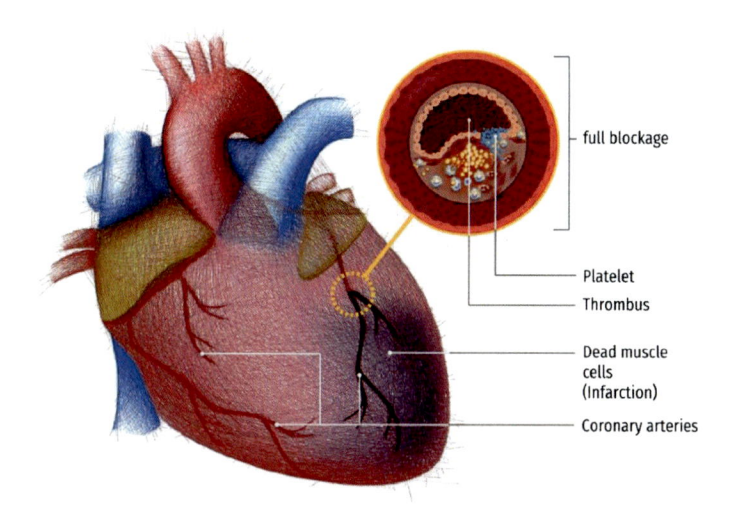

full blockage

Platelet

Thrombus

Dead muscle
cells
(Infarction)

Coronary arteries

FIG. 4.5

Illustration of myocardial infarction. *Reproduced with permission from Acharya, U. R., Fujita, H., Oh, S. L., Hagiwara, Y., Tan, J. H., & Adam, M. (2017). Application of deep convolutional neural network for automated detection of myocardial infarction using ECG signals. Information Sciences, 415, 190–198. https://doi.org/10.1016/j.ins.2017.06.027.*

artery bypass grafts (CABG). Although the novel therapeutics using cells (especially stem cells) (Gao et al., 2013; Madigan & Atoui, 2018; Terashvili & Bosnjak, 2019), exosomes (Davidson & Yellon, 2018), growth factors (Crafts, Jensen, Blocher-Smith, & Markel, 2015; Rebouças, Santos-Magalhães, & Formiga, 2016), and genes (Oggu et al., 2017) treatment strategies are emerging and have shown significant research outcomes, numerous challenges still exist in translating those technologies into clinical practice.

Nanoparticles of various types (inorganic, organic, and hybrid) engineered to target ischemic cardiac cells are promising candidates for myocardial infarction treatment. Lipid NPs are widely regarded as promising therapeutic delivery candidates to the infarcted area of heart. They have a similar morphology to cell membranes (Gao et al., 2013) and can hold both lipophilic and hydrophilic substances (Saludas, Pascual-Gil, Roli, Garbayo, & Blanco-Prieto, 2018). They have successfully demonstrated the ability to deliver a variety of biomaterials, including low molecular weight drugs, imaging agents, peptides, proteins, and nucleic acids, into the target tissues (Cheraghi, Negahdari, Daraee, & Eatemadi, 2017). Micelles have also been reported to be promising vehicles for the delivery of cardioprotective drugs, which are required during the acute stage of MI, as well as drugs that control infarct healing during the chronic stage of MI. Liposomes, on the other hand, are better at delivering pro-angiogenic drugs to the infarct microvasculature (Paulis et al., 2012). However, some therapeutic cargoes have been needed to achieve cardiac safety after a myocardial infarction. Numerous agents have been loaded into nanoparticles for various therapeutic strategies of MI. Recent research has shown that encapsulating ROS, Puerarin, or Baicalin into micelles or lipids will reduce the size of infarcts in ischemic hearts in animals (Dong et al., 2017; Vong et al., 2018; Zhang, Wang, & Pan, 2016). For cardiomyocyte apoptosis prevention, IGF1, liraglutide, Nitroxyl radical, Cyclosporine A, Pitavastatin, or 2,2,6,6-tetramethyl piperidine-1-oxyl was embarked on lipid-based NPs and sent to the animals' ischemic heart (Asanuma et al., 2017; Qi et al., 2017; Tokutome et al., 2019). Intravenous injection of collapsin response mediator protein-2 (CRMP2) lipid with the size of 50 nm was shown fibrosis reducing in the mice chronic MI heart (Zhou, Zhao, Liu, et al., 2015). For vasculogenesis enhancement, VEGF, FGF1, Ang-1, stromal cell-derived factor-1 (SDF-1), or CCR2 was loaded in NPs and delivered to the ischemic myocardial tissue to stimulate angiogenesis (Ding et al., 2020; Fan et al., 2020; Lu et al., 2015; Oduk et al., 2018). A couple of studies have also validated the promising potentials of carbon nanotubes (CNTs) in cardiac tissue engineering, such as in the support of cardiomyocyte function and growth and acceleration of the gap junction formation (Martinelli et al., 2013; Sun et al., 2017). Table 4.1 enlists various approaches that have been used for the delivery of therapeutics to treat cardiac MI.

Conclusion

A number of parameters are essential for an effective design of nanoscale drug delivery systems for the treatment of cardiovascular diseases. It is important to understand the basics of various heart diseases, such as their etiology, pathological manifestations, clinical features, and desired therapies to better design a suitable nanocarrier for the disease under investigation. Second, appropriate heart-targeted drug carriers must be chosen, and their active targeting properties must be controlled in order to improve the safety of loaded drugs. Finally, researchers should select the right targeting ligands to ensure that the nanocarriers bind specifically with complementary molecules on the targeted cells' surfaces. Keeping in view the above, effective nano-drug delivery systems are being developed for many drugs and nutraceuticals that can potentially treat CVDs. Targeted delivery of nanocarriers to the heart is not only beneficial in reducing toxicity but also significantly increases drug availability. While heart-targeted nanocarriers delivery holds promise for transporting drugs to affected cardiac tissue and regulating drug release in the body, these systems have drawbacks, including toxicity, immunogenicity, and undesirable pharmacokinetic conduct. The desired nanocarriers' fabrication techniques aim to ensure better absorption in target cells, the appropriate determination of a nanomaterial's half-life in a biological cell, the biological protection of NPs at the cellular level more favorable drug accumulation in affected tissue than in unaffected areas,

Table 4.1 List of selective studies employing organic/inorganic/hybrid nanoparticles for the delivery of therapeutics to repair infarcted myocardial tissue.

Nanoparticles	Therapeutic agents	MI model	Dose/ administration route	Results	References
Micellar	Rapamycin	Diabetic mouse, I/R	0.75 mg/kg/day, p.o. 10 weeks before I/R	Improvement in cardiac activity; reduction in infarct size	Samidurai et al. (2017)
Micelles	ROS	Mouse, acute	3 mg, intramyocardial	Reduction in infarct size and a boost in heart activity	Vong et al. (2018)
PLGA	VEGF	Mouse, chronic	0.06, 2.6, or 0.6 pg, intramyocardial	An improvement in heart function, increment in wall thickness, vasculogenesis, and reduction of infarct size	Oduk et al. (2018)
PK3	Nox2-miRNA	Mouse, chronic	5 µg/kg, intramyocardial	Reduction in infarct size, enhanced fractional shortening and ejection fraction	Yang et al. (2017)
Micelles	Nitroxyl radical	Dog, I/R	3 mg/kg, intravenous	Reduction in infarct size and decrease in myocardial apoptosis	Asanuma et al. (2017)
Lipid	siRNA CRMP2	Mouse, chronic	70 µg/kg, intravenous	Decreased heart failure post-MI, fibrosis, and mortality	Zhou et al. (2018)
Silicon	ERK1/2 inhibitor	Rat, acute	33 µg, intravenous	Reduced hypertrophy	Ferreira et al. (2017)
PLGA	Irbesartan	Mouse, I/R	3 mg/kg, intravenous	Decrease in infarct size and LV remodeling	Nakano et al. (2016)
PLGA	Pitavastatin	Pig, I/R	8–32 mg/kg, intravenous	Reduced infarct size, improved LV remodeling and decreased cardiomyocyte apoptosis	Ichimura et al. (2016)
Lipid	Puerarin	Rat, chronic	50 mg/kg, intravenous	Decrease in oxidative stress and infarct size	Dong et al. (2017)
Lipid	Baicalin	Rat, chronic	10 mg/kg, intravenous	Reduced oxidative stress and infarct size	Dong et al. (2017)
Dendrimer	MicroRNA-1 inhibitor	Mouse, acute	15 µg, intravenous	Reduced infarct size and cardiomyocyte apoptosis	Xue et al. (2018)
PEG-polyoxymethylestyrene	2,2,6,6-Tetramethylpiperidine-1-oxyl	Dod, I/R	3 mg/kg, intravenous	Decreased infarct size, apoptosis, and reduced ventricular fibrillation	Asanuma et al. (2017)

Continued

Table 4.1 List of selective studies employing organic/inorganic/hybrid nanoparticles for the delivery of therapeutics to repair infarcted myocardial tissue—cont'd

Nanoparticles	Therapeutic agents	MI model	Dose/administration route	Results	References
Graphene oxide gold nanosheets (GO-Au)	Chitosan-GO-Au scaffold	Rat, acute	5 × 2 mm scaffold	Improved the cardiac contractility and restored ventricular functions	Park et al. (2015)
OPF/graphene oxide hydrogel	–	Rat, acute	100 µL, intramyocardial	Improved load-dependent ejection fraction/fractional shortening of heart function	Saravanan et al. (2018)
Graphene oxide complex	IL-4 pDNA	Mouse, acute	50 µL, intramyocardial	Lowering of inflammation, improved heart function, and mitigated fibrosis	Zhou et al. (2018)
Graphene oxide	Mesenchymal stem cell	Rat, I/R	One million MSCs	Improved the engraftment and therapeutic efficacy of MSCs, which promoted cardiac tissue repair and cardiac function	Park et al. (2015)
Gold/10	PEG coated	Mouse, acute	100 µL/day, intravenous, 7 days	Decreased infarct size, improved systolic functions, inhibited cardiac fibrosis, no effect on apoptosis and hypertrophy	Han et al. (2018)
DNAzyme-conjugated AuNPs	Silence TNF-α	Rat, acute	100 µL, intramyocardial	Substantial anti-inflammatory advantages and enhanced cardiac function	Tian et al. (2018)
Organic-inorganic hybrid hollow mesoporous organosilica nanoparticles (HMONs)	Hepatocyte growth factor (HGF) gene-transfected BMMSCs	Rat, acute	2×10^6 HGF gene-transfected BMMSCs, intramyocardial	Decrease in number of apoptotic cardiomyocytes, reduction in infarct scar size, relief in interstitial fibrosis, a rise in angiogenesis, and improved cardiac functions	Zhu et al. (2016)
recombinant baculovirus complexed with Tat/DNA nanoparticles	Angiopoietin-1 gene	Rat, acute	300 µL, intramyocardial	Increase in capillary density, reduction in infarct size, and enhancement of cardiac functions	Paul et al. (2011)
PEG-PLGA	Liraglutide	Rat, chronic	380 µg, intramyocardial	Reduction in the size of infarct, preservation of wall thickness, stimulation of angiogenesis, prevention of apoptosis of cardiomyocyte, and improved heart activity	Qi et al. (2017)

and more efficient avoidance of drug toxicity. The nanocarriers-based drug delivery to cardiovascular system is still in its infancy and continuously emerging. The research conducted so far is mainly based on in-vitro and animal studies, however, in future, new revolutionized nanocarriers could play a key role in designing innovative strategies for the treatment of CVDs.

References

Aboud, H. M., El Komy, M. H., Ali, A. A., El Menshawe, S. F., & Abd Elbary, A. (2016). Development, optimization, and evaluation of carvedilol-loaded solid lipid nanoparticles for intranasal drug delivery. *AAPS PharmSciTech*, *17*(6), 1353–1365. https://doi.org/10.1208/s12249-015-0440-8.

Ahmed, M. S., Rodell, C. B., Hulsmans, M., Kohler, R. H., Aguirre, A. D., Nahrendorf, M., et al. (2019). A supramolecular nanocarrier for delivery of amiodarone anti-arrhythmic therapy to the heart. *Bioconjugate Chemistry*, *30*(3), 733–740. https://doi.org/10.1021/acs.bioconjchem.8b00882.

Al-Kassas, R., Wen, J., Cheng, A. E. M., Kim, A. M. J., Liu, S. S. M., & Yu, J. (2016). Transdermal delivery of propranolol hydrochloride through chitosan nanoparticles dispersed in mucoadhesive gel. *Carbohydrate Polymers*, *153*, 176–186. https://doi.org/10.1016/j.carbpol.2016.06.096.

Ammar, H. O., Haider, M., Ibrahim, M., & El Hoffy, N. M. (2017). In vitro and in vivo investigation for optimization of niosomal ability for sustainment and bioavailability enhancement of diltiazem after nasal administration. *Drug Delivery*, *24*(1), 414–421. https://doi.org/10.1080/10717544.2016.1259371.

Andalib, S., Molhemazar, P., & Danafar, H. (2018). In vitro and in vivo delivery of atorvastatin: A comparative study of anti-inflammatory activity of atorvastatin loaded copolymeric micelles. *Journal of Biomaterials Applications*, *32*(8), 1127–1138. https://doi.org/10.1177/0885328217750821.

Arzani, G., Haeri, A., Daeihamed, M., Bakhtiari-Kaboutaraki, H., & Dadashzadeh, S. (2015). Niosomal carriers enhance oral bioavailability of carvedilol: Effects of bile salt-enriched vesicles and carrier surface charge. *International Journal of Nanomedicine*, *10*, 4797–4813. https://doi.org/10.2147/IJN.S84703.

Asanuma, H., Sanada, S., Yoshitomi, T., Sasaki, H., Takahama, H., Ihara, M., et al. (2017). Novel synthesized radical-containing nanoparticles limit infarct size following ischemia and reperfusion in canine hearts. *Cardiovascular Drugs and Therapy*, *31*(5–6), 501–510. https://doi.org/10.1007/s10557-017-6758-6.

Banik, B., Wen, R., Marrache, S., Kumar, A., Kolishetti, N., Howerth, E. W., et al. (2018). Core hydrophobicity tuning of a self-assembled particle results in efficient lipid reduction and favorable organ distribution. *Nanoscale*, *10*(1), 366–377. https://doi.org/10.1039/c7nr06295h.

Bobo, D., Robinson, K. J., Islam, J., Thurecht, K. J., & Corrie, S. R. (2016). Nanoparticle-based medicines: A review of FDA-approved materials and clinical trials to date. *Pharmaceutical Research*, 2373–2387. https://doi.org/10.1007/s11095-016-1958-5.

Cheng, J., Zhang, R., Li, C., Tao, H., Dou, Y., Wang, Y., et al. (2018). A targeting nanotherapy for abdominal aortic aneurysms. *Journal of the American College of Cardiology*, *72*(21), 2591–2605. https://www.jacc.org/doi/full/10.1016/j.jacc.2018.08.2188.

Cheraghi, M., Negahdari, B., Daraee, H., & Eatemadi, A. (2017). Heart targeted nanoliposomal/nanoparticles drug delivery: An updated review. *Biomedicine and Pharmacotherapy*, *86*, 316–323. https://doi.org/10.1016/j.biopha.2016.12.009.

Chmielowski, R. A., Abdelhamid, D. S., Faig, J. J., Petersen, L. K., Gardner, C. R., Uhrich, K. E., et al. (2017). Athero-inflammatory nanotherapeutics: Ferulic acid-based poly(anhydride-ester) nanoparticles attenuate foam cell formation by regulating macrophage lipogenesis and reactive oxygen species generation. *Acta Biomaterialia, 57,* 85–94. https://doi.org/10.1016/j.actbio.2017.05.029.

Chourasiya, V., Bohrey, S., & Pandey, A. (2016). Formulation, optimization, characterization and in-vitro drug release kinetics of atenolol loaded PLGA nanoparticles using 33 factorial design for oral delivery. *Materials Discovery, 5,* 1–13. https://doi.org/10.1016/j.md.2016.12.002.

Crafts, T. D., Jensen, A. R., Blocher-Smith, E. C., & Markel, T. A. (2015). Vascular endothelial growth factor: Therapeutic possibilities and challenges for the treatment of ischemia. *Cytokine, 71*(2), 385–393. https://doi.org/10.1016/j.cyto.2014.08.005.

Davidson, S. M., & Yellon, D. M. (2018). Exosomes and cardioprotection—A critical analysis. *Molecular Aspects of Medicine, 60,* 104–114. https://doi.org/10.1016/j.mam.2017.11.004.

Dhital, S., & Vyavahare, N. R. (2020). Nanoparticle-based targeted delivery of pentagalloyl glucose reverses elastase-induced abdominal aortic aneurysm and restores aorta to the healthy state in mice. *PLoS One, 15*(3). https://doi.org/10.1371/journal.pone.0227165.

Ding, Y., Zhao, A., Liu, T., Wang, Y., Gao, Y., Li, J., et al. (2020). An injectable nanocomposite hydrogel for potential application of vascularization and tissue repair. *Annals of Biomedical Engineering, 48*(5), 1511–1523. https://doi.org/10.1007/s10439-020-02471-7.

Dong, Z., Guo, J., Xing, X., Zhang, X., Du, Y., & Lu, Q. (2017). RGD modified and PEGylated lipid nanoparticles loaded with puerarin: Formulation, characterization and protective effects on acute myocardial ischemia model. *Biomedicine and Pharmacotherapy, 89,* 297–304. https://doi.org/10.1016/j.biopha.2017.02.029.

Duivenvoorden, R., Tang, J., Cormode, D. P., Mieszawska, A. J., Izquierdo-Garcia, D., Ozcan, C., et al. (2014). A statin-loaded reconstituted high-density lipoprotein nanoparticle inhibits atherosclerotic plaque inflammation. *Nature Communications, 5.* https://doi.org/10.1038/ncomms4065.

Elgart, A., Cherniakov, I., Aldouby, Y., Domb, A. J., & Hoffman, A. (2013). Improved oral bioavailability of BCS class 2 compounds by self nano-emulsifying drug delivery systems (SNEDDS): The underlying mechanisms for amiodarone and talinolol. *Pharmaceutical Research, 30*(12), 3029–3044. https://doi.org/10.1007/s11095-013-1063-y.

Fan, C., Tang, Y., Zhao, M., Lou, X., Pretorius, D., Menasche, P., et al. (2020). CHIR99021 and fibroblast growth factor 1 enhance the regenerative potency of human cardiac muscle patch after myocardial infarction in mice. *Journal of Molecular and Cellular Cardiology, 141,* 1–10. https://doi.org/10.1016/j.yjmcc.2020.03.003.

Ferreira, M. P. A., Ranjan, S., Kinnunen, S., Correia, A., Talman, V., Mäkilä, E., et al. (2017). Drug-loaded multifunctional nanoparticles targeted to the endocardial layer of the injured heart modulate hypertrophic signaling. *Small, 13*(33), 1701276. https://doi.org/10.1002/smll.201701276.

Flores, A. M., Hosseini-Nassab, N., Jarr, K. U., Ye, J., Zhu, X., Wirka, R., et al. (2020). Pro-efferocytic nanoparticles are specifically taken up by lesional macrophages and prevent atherosclerosis. *Nature Nanotechnology, 15*(2), 154–161. https://doi.org/10.1038/s41565-019-0619-3.

Gao, L. R., Pei, X. T., Ding, Q. A., Chen, Y., Zhang, N. K., Chen, H. Y., et al. (2013). A critical challenge: Dosage-related efficacy and acute complication intracoronary injection of autologous bone marrow mesenchymal stem cells in acute myocardial infarction. *International Journal of Cardiology, 168*(4), 3191–3199. https://doi.org/10.1016/j.ijcard.2013.04.112.

Gillum, R. F. (1995). Epidemiology of aortic aneurysm in the United States. *Journal of Clinical Epidemiology, 48*(11), 1289–1298. https://doi.org/10.1016/0895-4356(95)00045-3.

Guan, Y., Zuo, T., Chang, M., Zhang, F., Wei, T., Shao, W., et al. (2015). Propranolol hydrochloride-loaded liposomal gel for transdermal delivery: Characterization and in vivo evaluation.

International Journal of Pharmaceutics, *487*(1–2), 135–141. https://doi.org/10.1016/j.ijpharm.2015.04.023.

Habashi, J. P., Doyle, J. J., Holm, T. M., Aziz, H., Schoenhoff, F., Bedja, D., et al. (2011). Angiotensin II type 2 receptor signaling attenuates aortic aneurysm in mice through ERK antagonism. *Science*, *332*(6027), 361–365. https://doi.org/10.1126/science.1192152.

Han, J., Kim, Y. S., Lim, M. Y., Kim, H. Y., Kong, S., Kang, M., et al. (2018). Dual roles of graphene oxide to attenuate inflammation and elicit timely polarization of macrophage phenotypes for cardiac repair. *ACS Nano*, *12*(2), 1959–1977. https://doi.org/10.1021/acsnano.7b09107.

Heidenreich, P. (2015). Heart failure prevention and team-based interventions. *Heart Failure Clinics*, *11*(3), 349–358. https://doi.org/10.1016/j.hfc.2015.03.001.

Ichimura, K., Matoba, T., Nakano, K., Tokutome, M., Honda, K., Koga, J. I., et al. (2016). A translational study of a new therapeutic approach for acute myocardial infarction: Nanoparticle-mediated delivery of pitavastatin into reperfused myocardium reduces ischemia-reperfusion injury in a preclinical porcine model. *PLoS One*, *11*(9). https://doi.org/10.1371/journal.pone.0162425.

Jiang, C., Qi, Z., Tang, Y., Jia, H., Li, Z., Zhang, W., et al. (2019). Rational design of lovastatin-loaded spherical reconstituted high density lipoprotein for efficient and safe anti-atherosclerotic therapy. *Molecular Pharmaceutics*, 3284–3291. https://doi.org/10.1021/acs.molpharmaceut.9b00445.

Jin, H. J., Zhang, H., Sun, M. L., Zhang, B. G., & Zhang, J. W. (2013). Urokinase-coated chitosan nanoparticles for thrombolytic therapy: Preparation and pharmacodynamics in vivo. *Journal of Thrombosis and Thrombolysis*, *36*(4), 458–468. https://doi.org/10.1007/s11239-013-0951-7.

Jinatongthai, P., Kongwatcharapong, J., Foo, C. Y., Phrommintikul, A., Nathisuwan, S., Thakkinstian, A., et al. (2017). Comparative efficacy and safety of reperfusion therapy with fibrinolytic agents in patients with ST-segment elevation myocardial infarction: A systematic review and network meta-analysis. *The Lancet*, *390*(10096), 747–759. https://doi.org/10.1016/S0140-6736(17)31441-1.

Joviliano, E. E., Ribeiro, M. S., & Tenorio, E. J. R. (2017). MicroRNAs and current concepts on the pathogenesis of abdominal aortic aneurysm. *Brazilian Journal of Cardiovascular Surgery*, *23*(3), 215–224. https://doi.org/10.21470/1678-9741-2016-0050.

Karlsson, L., Gnarpe, J., Bergqvist, D., Lindbäck, J., & Pärsson, H. (2009). The effect of azithromycin and Chlamydophilia pneumonia infection on expansion of small abdominal aortic aneurysms—A prospective randomized double-blind trial. *Journal of Vascular Surgery*, *50*(1), 23–29. https://doi.org/10.1016/j.jvs.2008.12.048.

Kawata, H., Uesugi, Y., Soeda, T., Takemoto, Y., Sung, J. H., Umaki, K., et al. (2012). A new drug delivery system for intravenous coronary thrombolysis with thrombus targeting and stealth activity recoverable by ultrasound. *Journal of the American College of Cardiology*, *60*(24), 2550–2557. https://doi.org/10.1016/j.jacc.2012.08.1008.

Kheirolomoom, A., Kim, C. W., Seo, J. W., Kumar, S., Son, D. J., Gagnon, M. K. J., et al. (2015). Multifunctional nanoparticles facilitate molecular targeting and miRNA delivery to inhibit atherosclerosis in ApoE-/- mice. *ACS Nano*, *9*(9), 8885–8897. https://doi.org/10.1021/acsnano.5b02611.

Korin, N., Kanapathipillai, M., Matthews, B. D., Crescente, M., Brill, A., Mammoto, T., et al. (2012). Shear-activated nanotherapeutics for drug targeting to obstructed blood vessels. *Science*, *337*(6095), 738–742. https://doi.org/10.1126/science.1217815.

Kurosawa, K., Matsumura, J. S., & Yamanouchi, D. (2013). Current status of medical treatment for abdominal aortic aneurysm. *Circulation Journal*, *77*(12), 2860–2866. https://doi.org/10.1253/circj.CJ-13-1252.

Lamprecht, A., Bouligand, Y., & Benoit, J. P. (2002). New lipid nanocapsules exhibit sustained release properties for amiodarone. *Journal of Controlled Release, 84*(1–2), 59–68. https://doi.org/10.1016/S0168-3659(02)00258-4.

Lei, Y., Nosoudi, N., & Vyavahare, N. (2014). Targeted chelation therapy with EDTA-loaded albumin nanoparticles regresses arterial calcification without causing systemic side effects. *Journal of Controlled Release, 196*, 79–86. https://doi.org/10.1016/j.jconrel.2014.09.029.

Liu, C. H., Hsu, H. L., Chen, J. P., Wu, T., & Ma, Y. H. (2019). Thrombolysis induced by intravenous administration of plasminogen activator in magnetoliposomes: Dual targeting by magnetic and thermal manipulation. *Nanomedicine: Nanotechnology, Biology and Medicine, 20*, 101992. https://doi.org/10.1016/j.nano.2019.03.014.

Lu, W., Xie, Z. Y., Tang, Y., Yao, L. B., Fu, C., & Ma, G. (2015). Photoluminescent mesoporous silicon nanoparticles with siCCR2 improve the effects of mesenchymal stromal cell transplantation after acute myocardial infarction. *Theranostics, 5*(10), 1068–1082. https://doi.org/10.7150/thno.11517.

Madigan, M., & Atoui, R. (2018). Therapeutic use of stem cells for myocardial infarction. *Bioengineering, 5*(2). https://doi.org/10.3390/bioengineering5020028.

Marrache, S., & Dhar, S. (2013). Biodegradable synthetic high-density lipoprotein nanoparticles for atherosclerosis. *Proceedings of the National Academy of Sciences, 110*(23), 9445–9450. https://doi.org/10.1073/pnas.1301929110.

Martinelli, V., Cellot, G., Fabbro, A., Bosi, S., Mestroni, L., & Ballerini, L. (2013). Improving cardiac myocytes performance by carbon nanotubes platforms. *Frontiers in Physiology, 4*. https://doi.org/10.3389/fphys.2013.00239.

McCarthy, J. R., Korngold, E., Weissleder, R., & Jaffer, F. A. (2010). A light-activated theranostic nanoagent for targeted macrophage ablation in inflammatory atherosclerosis. *Small, 6*(18), 2041–2049. https://doi.org/10.1002/smll.201000596.

McCarthy, J. R., Sazonova, I. Y., Erdem, S. S., Hara, T., Thompson, B. D., Patel, P., et al. (2012). Multifunctional nanoagent for thrombus-targeted fibrinolytic therapy. *Nanomedicine, 7*(7), 1017–1028. https://doi.org/10.2217/nnm.11.179.

Moretti, A., Li, Q., Chmielowski, R., Joseph, L. B., Moghe, P. V., & Uhrich, K. E. (2018). Nanotherapeutics containing lithocholic acid-based amphiphilic scorpion-like macromolecules reduce in vitro inflammation in macrophages: Implications for atherosclerosis. *Nanomaterials, 8*(2), 84. https://doi.org/10.3390/nano8020084.

Mouez, M. A., Nasr, M., Abdel-Mottaleb, M., Geneidi, A. S., & Mansour, S. (2016). Composite chitosan-transfersomal vesicles for improved transnasal permeation and bioavailability of verapamil. *International Journal of Biological Macromolecules, 93*, 591–599. https://doi.org/10.1016/j.ijbiomac.2016.09.027.

Myerson, J., He, L., Lanza, G., Tollefsen, D., & Wickline, S. A. (2011). Thrombin-inhibiting perfluorocarbon nanoparticles provide a novel strategy for the treatment and magnetic resonance imaging of acute thrombosis. *Journal of Thrombosis and Haemostasis, 9*(7), 1292–1300. https://doi.org/10.1111/j.1538-7836.2011.04339.x.

Nakano, Y., Matoba, T., Tokutome, M., Funamoto, D., Katsuki, S., Ikeda, G., et al. (2016). Nanoparticle-mediated delivery of irbesartan induces cardioprotection from myocardial ischemia-reperfusion injury by antagonizing monocyte-mediated inflammation. *Scientific Reports, 6*. https://doi.org/10.1038/srep29601.

Nosoudi, N., Chowdhury, A., Siclari, S., Karamched, S., Parasaram, V., Parrish, J., et al. (2016). Reversal of vascular calcification and aneurysms in a rat model using dual targeted therapy with EDTA-and PGG-loaded nanoparticles. *Theranostics, 6*(11), 1975–1987. https://doi.org/10.7150/thno.16547.

Nosoudi, N., Nahar-Gohad, P., Sinha, A., Chowdhury, A., Gerard, P., Carsten, C. G., et al. (2015). Prevention of abdominal aortic aneurysm progression by targeted inhibition of matrix metalloproteinase activity with batimastat-loaded nanoparticles. *Circulation Research, 117*(11), e80–e89. https://doi.org/10.1161/CIRCRESAHA.115.307207.

Oduk, Y., Zhu, W., Kannappan, R., Zhao, M., Borovjagin, A. V., Oparil, S., et al. (2018). VEGF nanoparticles repair the heart after myocardial infarction. *American Journal of Physiology—Heart and Circulatory Physiology, 314*(2), H278–H284. https://doi.org/10.1152/ajpheart.00471.2017.

Oggu, G. S., Sasikumar, S., Reddy, N., Ella, K. K. R., Rao, C. M., & Bokara, K. K. (2017). Gene delivery approaches for mesenchymal stem cell therapy: Strategies to increase efficiency and specificity. *Stem Cell Reviews and Reports, 13*(6), 725–740. https://doi.org/10.1007/s12015-017-9760-2.

Onishi, A., St Ange, K., Dordick, J. S., & Linhardt, R. J. (2016). Heparin and anticoagulation. *Frontiers in Bioscience—Landmark, 21*(7), 1372–1392. https://doi.org/10.2741/4462.

Park, J., Kim, B., Han, J., Oh, J., Park, S., Ryu, S., et al. (2015). Graphene oxide flakes as a cellular adhesive: Prevention of reactive oxygen species mediated death of implanted cells for cardiac repair. *ACS Nano, 9*(5), 4987–4999. https://doi.org/10.1021/nn507149w.

Paul, A., Binsalamah, Z. M., Khan, A. A., Abbasia, S., Elias, C. B., Shum-Tim, D., et al. (2011). A nanobiohybrid complex of recombinant baculovirus and Tat/DNA nanoparticles for delivery of Ang-1 transgene in myocardial infarction therapy. *Biomaterials, 32*(32), 8304–8318. https://doi.org/10.1016/j.biomaterials.2011.07.042.

Paulis, L. E., Geelen, T., Kuhlmann, M. T., Coolen, B. F., Schäfers, M., Nicolay, K., et al. (2012). Distribution of lipid-based nanoparticles to infarcted myocardium with potential application for MRI-monitored drug delivery. *Journal of Controlled Release, 162*(2), 276–285. https://doi.org/10.1016/j.jconrel.2012.06.035.

Peters, D., Kastantin, M., Kotamraju, V. R., Karmali, P. P., Gujraty, K., Tirrell, M., et al. (2009). Targeting atherosclerosis by using modular, multifunctional micelles. *Proceedings of the National Academy of Sciences, 106*(24), 9815–9819. https://doi.org/10.1073/pnas.0903369106.

Qi, Q., Lu, L., Li, H., Yuan, Z., Chen, G., Lin, M., et al. (2017). Spatiotemporal delivery of nanoformulated liraglutide for cardiac regeneration after myocardial infarction. *International Journal of Nanomedicine, 12*, 4835–4848. https://doi.org/10.2147/IJN.S132064.

Quintana, R. A., & Taylor, W. R. (2019). Cellular mechanisms of aortic aneurysm formation. *Circulation Research, 124*(4), 607–618. https://doi.org/10.1161/CIRCRESAHA.118.313187.

Rajiv, J., Hardik, J., Vaibhav, S., & Vimal, A. (2011). Nanoburrs: A novel approach in the treatment of cardiovascular disease. *International Research Journal of Pharmacy, 2*(5), 91–92.

Rebouças, J. D. S., Santos-Magalhães, N. S., & Formiga, F. R. (2016). Cardiac regeneration using growth factors: Advances and challenges. *Arquivos Brasileiros de Cardiologia, 107*(3), 271–275. https://doi.org/10.5935/abc.20160097.

Saludas, L., Pascual-Gil, S., Roli, F., Garbayo, E., & Blanco-Prieto, M. J. (2018). Heart tissue repair and cardioprotection using drug delivery systems. *Maturitas, 110*, 1–9. https://doi.org/10.1016/j.maturitas.2018.01.011.

Samidurai, A., Salloum, F. N., Durrant, D., Chernova, O. B., Kukreja, R. C., & Das, A. (2017). Chronic treatment with novel nanoformulated micelles of rapamycin, Rapatar, protects diabetic heart against ischaemia/reperfusion injury. *British Journal of Pharmacology, 174*(24), 4771–4784. https://doi.org/10.1111/bph.14059.

Saravanan, S., Sareen, N., Abu-El-Rub, E., Ashour, H., Sequiera, G. L., Ammar, H. I., et al. (2018). Graphene oxide-gold nanosheets containing chitosan scaffold improves ventricular contractility and function after implantation into infarcted heart. *Scientific Reports.* https://doi.org/10.1038/s41598-018-33144-0.

Shah, M. K., Madan, P., & Lin, S. (2015). Elucidation of intestinal absorption mechanism of carvedilol-loaded solid lipid nanoparticles using Caco-2 cell line as an in-vitro model. *Pharmaceutical Development and Technology, 20*(7), 877–885. https://doi.org/10.3109/10837450.2014.938857.

Shirasu, T., Koyama, H., Miura, Y., Hoshina, K., Kataoka, K., Watanabe, T., et al. (2016). Nanoparticles effectively target rapamycin delivery to sites of experimental aortic aneurysm in rats. *PLoS One*, *11*(6), e0157813. https://doi.org/10.1371/journal.pone.0157813.

Sinha, A., Shaporev, A., Nosoudi, N., Lei, Y., Vertegel, A., Lessner, S., et al. (2014). Nanoparticle targeting to diseased vasculature for imaging and therapy. *Nanomedicine: Nanotechnology, Biology, and Medicine*, *10*(5), e1003–e1012. https://doi.org/10.1016/j.nano.2014.02.002.

Steinmetz, E. F., Buckley, C., Shames, M. L., Ennis, T. L., Vanvickle-Chavez, S. J., Mao, D., et al. (2005). Treatment with simvastatin suppresses the development of experimental abdominal aortic aneurysms in normal and hypercholesterolemic mice. *Annals of Surgery*, *241*(1), 92–101. https://doi.org/10.1097/01.sla.0000150258.36236.e0.

Sun, H., Tang, J., Mou, Y., Zhou, J., Qu, L., Duval, K., et al. (2017). Carbon nanotube-composite hydrogels promote intercalated disc assembly in engineered cardiac tissues through β1-integrin mediated FAK and RhoA pathway. *Acta Biomaterialia*, *48*, 88–99. https://doi.org/10.1016/j.actbio.2016.10.025.

Takahama, H., Shigematsu, H., Asai, T., Matsuzaki, T., Sanada, S., Fu, H. Y., et al. (2013). Liposomal Amiodarone augments anti-arrhythmic effects and reduces Hemodynamic adverse effects in an ischemia/reperfusion rat model. *Cardiovascular Drugs and Therapy*, *27*(2), 125–132. https://doi.org/10.1007/s10557-012-6437-6.

Terashvili, M., & Bosnjak, Z. J. (2019). Stem cell therapies in cardiovascular disease. *Journal of Cardiothoracic and Vascular Anesthesia*, *33*(1), 209–222. https://doi.org/10.1053/j.jvca.2018.04.048.

Tian, A., Yang, C., Zhu, B., Wang, W., Liu, K., Jiang, Y., et al. (2018). Polyethylene-glycol-coated gold nanoparticles improve cardiac function after myocardial infarction in mice. *Canadian Journal of Physiology and Pharmacology*, *96*(12), 1318–1327. https://doi.org/10.1139/cjpp-2018-0227.

Timmis, A., Townsend, N., Gale, C., Grobbee, R., Maniadakis, N., Flather, M., et al. (2018). European society of cardiology: Cardiovascular disease statistics 2017. *European Heart Journal*, 508–579. https://doi.org/10.1093/eurheartj/ehx628.

Tokutome, M., Matoba, T., Nakano, Y., Okahara, A., Fujiwara, M., Koga, J.-I., et al. (2019). Peroxisome proliferator-activated receptor-gamma targeting nanomedicine promotes cardiac healing after acute myocardial infarction by skewing monocyte/macrophage polarization in preclinical animal models. *Cardiovascular Research*, *115*(2), 419–431. https://doi.org/10.1093/cvr/cvy200.

Tsoref, O., Tyomkin, D., Amit, U., Landa, N., Cohen-Rosenboim, O., Kain, D., et al. (2018). E-selectin-targeted copolymer reduces atherosclerotic lesions, adverse cardiac remodeling, and dysfunction. *Journal of Controlled Release*, *288*, 136–147. https://doi.org/10.1016/j.jconrel.2018.08.029.

Üstündağ-Okur, N., Yurdasiper, A., Gündoğdu, E., & Homan Gökçe, E. (2015). Modification of solid lipid nanoparticles loaded with nebivolol hydrochloride for improvement of oral bioavailability in treatment of hypertension: Polyethylene glycol versus chitosan oligosaccharide lactate. *Journal of Microencapsulation*, *33*(1), 30–42. https://doi.org/10.3109/02652048.2015.1094532.

Varshosaz, J., Taymouri, S., & Hamishehkar, H. (2014). Fabrication of polymeric nanoparticles of poly(ethylene-co-vinyl acetate) coated with chitosan for pulmonary delivery of carvedilol. *Journal of Applied Polymer Science*, *131*(1). https://doi.org/10.1002/app.39694.

Vong, L. B., Bui, T. Q., Tomita, T., Sakamoto, H., Hiramatsu, Y., & Nagasaki, Y. (2018). Novel angiogenesis therapeutics by redox injectable hydrogel—Regulation of local nitric oxide generation for effective cardiovascular therapy. *Biomaterials*, *167*, 143–152. https://doi.org/10.1016/j.biomaterials.2018.03.023.

Wang, Y., Li, L., Zhao, W., Dou, Y., An, H., Tao, H., et al. (2018). Targeted therapy of atherosclerosis by a broad-spectrum reactive oxygen species scavenging nanoparticle with intrinsic anti-inflammatory activity. *ACS Nano, 12*(9), 8943–8960. https://doi.org/10.1021/acsnano.8b02037.

Wang, Y., Su, W., Li, Q., Li, C., Wang, H., Li, Y., et al. (2013). Preparation and evaluation of lidocaine hydrochloride-loaded TAT-conjugated polymeric liposomes for transdermal delivery. *International Journal of Pharmaceutics, 441*(1–2), 748–756. https://doi.org/10.1016/j.ijpharm.2012.10.019.

World Health Organization (WHO). (2017). *Cardiovascular diseases (CVDs) fact sheet.* https://www.who.int/news-room/fact-sheets/detail/cardiovascular-diseases-(cvds).

Xue, X., Shi, X., Dong, H., You, S., Cao, H., Wang, K., et al. (2018). Delivery of microRNA-1 inhibitor by dendrimer-based nanovector: An early targeting therapy for myocardial infarction in mice. *Nanomedicine: Nanotechnology, Biology, and Medicine, 14*(2), 619–631. https://doi.org/10.1016/j.nano.2017.12.004.

Yang, J., Brown, M. E., Zhang, H., Martinez, M., Zhao, Z., Bhutani, S., et al. (2017). High-throughput screening identifies microRNAs that target Nox2 and improve function after acute myocardial infarction. *American Journal of Physiology—Heart and Circulatory Physiology, 312*(5), H1002–H1012. https://doi.org/10.1152/ajpheart.00685.2016.

Yu, M., Amengual, J., Menon, A., Kamaly, N., Zhou, F., Xu, X., et al. (2017). Targeted nanotherapeutics encapsulating liver X receptor agonist GW3965 enhance antiatherogenic effects without adverse effects on hepatic lipid metabolism in Ldlr $-$/$-$ mice. *Advanced Healthcare Materials, 6*(20). https://doi.org/10.1002/adhm.201700313.

Zhang, S., Wang, J., & Pan, J. (2016). Baicalin-loaded PEGylated lipid nanoparticles: Characterization, pharmacokinetics, and protective effects on acute myocardial ischemia in rats. *Drug Delivery, 23*(9), 3696–3703. https://doi.org/10.1080/10717544.2016.1223218.

Zhou, J., Yang, X., Liu, W., Wang, C., Shen, Y., Zhang, F., et al. (2018). Injectable OPF/graphene oxide hydrogels provide mechanical support and enhance cell electrical signaling after implantation into myocardial infarct. *Theranostics, 8*(12), 3317–3330. https://doi.org/10.7150/thno.25504.

Zhou, L. S., Zhao, G. L., Liu, Q., et al. (2015). Silencing collapsin response mediator protein-2 reprograms macrophage phenotype and improves infarct healing in experimental myocardial infarction model. *Journal of Inflammation, 12*, 11. https://doi.org/10.1186/s12950-015-0053-8.

Zhu, M. L., Wang, G., Wang, H., Guo, Y. M., Song, P., Xu, J., et al. (2019). Amorphous nano-selenium quantum dots improve endothelial dysfunction in rats and prevent atherosclerosis in mice through Na+/H+ exchanger 1 inhibition. *Vascular Pharmacology, 115*, 26–32. https://doi.org/10.1016/j.vph.2019.01.005.

Zhu, K., Wu, M., Lai, H., Guo, C., Li, J., Wang, Y., et al. (2016). Nanoparticle-enhanced generation of gene-transfected mesenchymal stem cells for in vivo cardiac repair. *Biomaterials, 74*, 188–199. https://doi.org/10.1016/j.biomaterials.2015.10.010.

Zhuge, Y., Xie, M.-Q., Li, L., Wang, F., Gao, F., & Zheng, Z.-F. (2016). Preparation of liposomal amiodarone and investigation of its cardiomyocyte-targeting ability in cardiac radiofrequency ablation rat model. *International Journal of Nanomedicine, 11*, 2359. https://doi.org/10.2147/ijn.s98815.

CHAPTER 5

Improved pulmonary drug delivery through nanocarriers

Introduction

Since the introduction of the very first nano-sized drug carrier (nanocarrier) in late 1970s; research in the field of pharmaceutical sciences has got great momentum (Kreuter, Täuber, & Illi, 1979; Rosen & Abribat, 2005). Currently, continuous efforts are made toward optimizing the activity of already marketed drugs as well as enhancing their therapeutic efficacy and administration modes in pharmaceutical research (Rosen & Abribat, 2005). For years, pulmonary route was only used to treat asthma locally among the so many routes of drug administration, e.g., oral, topical, transdermal, subcutaneous, intravenous, and pulmonary. Pulmonary route of drug administration however has recently got wider attention in the field of drug delivery (Grenha, Seijo, & Remuñán-López, 2005) because it possesses various distinctive features that are ideal for drug transport including larger alveolar surface-areas of nearly $100\,m^2$ (Groneberg, Witt, Wagner, Chung, & Fischer, 2003; Zhang, Shen, & Nagai, 2001) augmenting the absorption of drug, minimal epithelial barriers enhancing drug absorption, excessive vascularization promoting systemic drug delivery (Rosen & Abribat, 2005) and above all; avoidance of first-pass metabolism of drug (Steimer, Haltner, & Lehr, 2005; Vanbever, Ben-Jebria, Mintzes, Langer, & Edwards, 1999). All of these characteristics may serve to enhance bioavailability and therapeutic efficacy of the therapeutic agent administered via this route.

Pulmonary route of drug administration is getting wider scientific attention in the emerging field of nano-medicine complementing its recognition in drug delivery (Oberdörster, Oberdörster, & Oberdörster, 2005; Sham, Zhang, Finlay, Roa, & Löbenberg, 2004). The applications of nano-medicine are widespread however their therapeutic and theranostic ones are worth mentioning. As nanocarriers can carry therapeutic agents to specific sites of the body for targeted drug delivery (Kagan, Bayir, & Shvedova, 2005; LaVan, McGuire, & Langer, 2003), they and can also be used as diagnostic nano-scale agents/devices to repair tissue damage and monitor health (Hood, 2004; LaVan et al., 2003).

Nanocarriers for Organ-Specific and Localized Drug Delivery. https://doi.org/10.1016/B978-0-12-821093-2.00008-6

Drug delivery to specific body sites via nanocarriers having mean diameter less than 1000 nm is one coinciding area in nano-medicine (Farrugia & Groves, 1999; Pison, Welte, Giersig, & Groneberg, 2006). The concentration of drug at specific site of action is further enhanced when it is loaded and delivered via nanocarriers. This is a particular benefit of employing nanocarriers for drug delivery and this feature of nanocarriers based drugs is of prime importance as most of the currently administered drugs must first cross different biological impediments before reaching their ultimate site of action but cannot pass easily through membranes (Tamai & Tsuji, 2000). Thus a decrease in the drug concentration is observed because of non-specific distribution to other undesired sites of the body or its pre-mature inactivation (Balthasar et al., 2005; Torchilin, 2000). Therefore nanomedicines are widely proposed to be employed for targeted delivery of drugs to specific body sites and subsequently to reduce most of the unwanted adverse effects resulting from the use of non-specific administration of drugs (Balthasar et al., 2005; Fiorito, Serafino, Andreola, Togna, & Togna, 2006; Torchilin, 2000).

Distribution of nanocarrier-based drugs within the body can be controlled through intelligent manipulation of particles properties, i.e., surface properties and size which is an additional benefit of using nanocarriers based drugs (Araujo, Löbenberg, & Kreuter, 1999). Coating the drug-loaded nanocarriers' surfaces with different ligands results in altered distribution of the loaded drug in the body (Löbenberg, Araujo, Von Briesen, Rodgers, & Kreuter, 1998). Moreover, covalent attachment of specific ligands and modification can promote targeted delivery of drugs to specific population of cells in the body (Araujo et al., 1999; Balthasar et al., 2005).

Nanocarriers are constantly being investigated as suitable carriers for delivery of drugs to specific cell populations within the respiratory tract (Groneberg et al., 2003). As a result, nanocarriers-based inhalational drugs may play a vital role in near future for the treatment of various pulmonary ailments, e.g., lung cancer, cystic fibrosis, asthma, and chronic pulmonary infections like Tuberculosis (Courrier, Butz, & Vandamme, 2002; Gelperina, Kisich, Iseman, & Heifets, 2005). Ernie Hood has stated in his published review that nearly half of all drug delivery and design strategies might use nanotechnological approaches within 10 years (Hood, 2004). This chapter describes the current application of nanocarriers as effective drug carrier systems in the lungs with an emphasis on their mechanism of action and toxicology.

Pulmonary delivery of drugs and lung physiology

Respiratory tract may serve as a suitable way for provision of local and systemic medication because of its big alveolar surface area (Grenha et al., 2005; Zhang et al., 2001), extensive pulmonary vascularization (Groneberg et al., 2003), and escaping of drugs from first-pass metabolism (Gelperina et al., 2005; Steimer et al., 2005) as mentioned earlier. Pulmonary route of drug delivery may serve as a non-invasive way of drug administration through aerosolized particles inhalation (LaVan et al., 2003). The efficiency of aerosolized particles application is based on and related to their density and size. This route of drug administration permits delivery of drugs to deeper regions in respiratory tract as discussed later (LaVan et al., 2003; Sham et al., 2004).

Pulmonary surfactant forms an essential film coating the surface of the alveoli to inhibit the alveolar region collapse (Grenache & Gronowski, 2006; Schief, Antia, Discher, Hall, & Vogel, 2003; Wüstneck, Wüstneck, Fainerman, Miller, & Pison, 2001). It primarily reduces alveolar surface tension during breathing and stabilizing the terminal airways and encourages exchange of gases at lung air interface (Piknova, Schram, & Hall, 2002). It is produced in type II pneumocytes (Grenache & Gronowski, 2006; Rosen & Abribat, 2005) where it is stored within lamellar bodies of type II pneumocytes (storage granules) before being secreted via exocytosis to the air-water interface as shown in Fig. 5.1 (Grenache & Gronowski, 2006).

It is necessary for the pulmonary surfactant to spread and compress during breathing as well as it should be fluid and quite rigid to stabilize the alveolar region. The composition of surfactant layer is almost 10% protein and 90% lipid (Jon, 1998). SP-A, SP-B, SP-C, and SP-D are the main four proteins

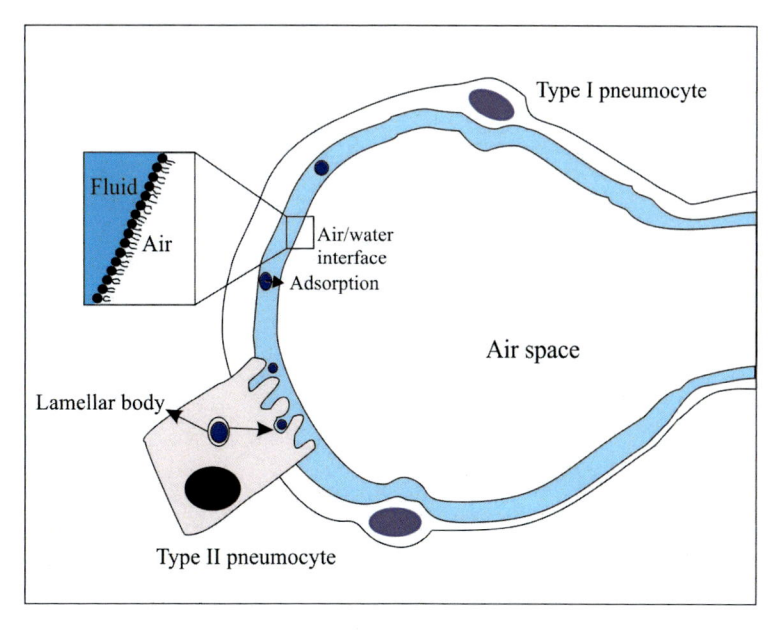

FIG. 5.1

Synthesis and secretion of lung surfactant that coats the alveolar surface of the air-water interface. *No Permission Required.*

present in surfactant film. These proteins have different hydrophobicities where SP-B and SP-C are lipophilic and SP-A and SP-D are hydrophilic in nature (Jon, 1998) and show prime roles in attachment, transport, and exchange processes in the air lung interface (Pérez-Gil & Keough, 1998). Zwitterionic phosphatidylcholines are major lipid contributors present in 80% of the surfactant followed by negatively charged phosphatidylglycerols (Yu & Possmayer, 2003) with approximately 15% of total lipids contents.

It has been shown in the literature that samples of surfactants from patients with Acute Respiratory Distress Syndrome and asthma possess an associated surface potential. It has also been shown that the associated surface potential of surfactant model systems composed of solutions of lipid can be due to the presence of SP-C. Such information about human surfactant films' surface potential may permit the surface charge manipulation of drug delivery systems or drug particles for optimum interactions of surfactant film and inhaled particles (Gill et al., 2007).

Types and classification of nanocarriers used in drug delivery

In a broader sense, nano-carrier can be classified as organic-based, inorganic-based, or a hybrid combination of both as shown in Fig. 5.2. Organic nano-carriers include but are not limited to lipid-based systems, i.e., liposomes and nanoemulsions (NE), and polymer based systems, i.e., polymeric conjugates, micelles, and dendrimers while inorganic types of nano-carriers consist of metal-based nanostructures, silica-based nanoparticles (NPs), and quantum dots (QD) (Jabr-Milane et al., 2008).

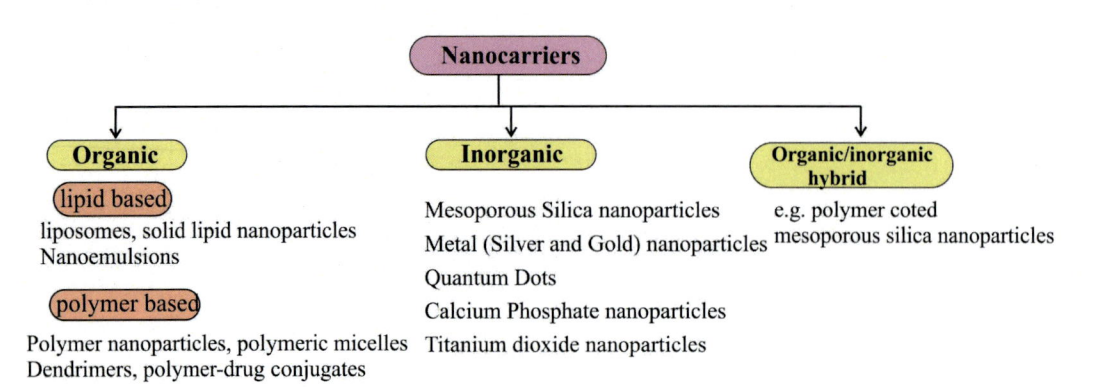

FIG. 5.2

General classification of nano-carriers employed as drug delivery systems. *No Permission Required.*

Organic-based drugs carriers

Solid lipid nanoparticles (SLNs)

Solid lipid nanoparticles are nano-sized colloidal drug carriers ranging in size from 50 to 1000 nm and were initially developed in early 1990s (Müller, Mäder, & Gohla, 2000). SLNs are constructed by dispersion of solid melted lipid(s) in water and stabilized with emulsifiers. They provide an extensive lipid matrix for dissolution or dispersion of lipophilic drugs (Malam, Loizidou, & Seifalian, 2009). For the preparation of SLNs, a number of lipids that are solid at room temperature have been used including free fatty acids, mono-, di-, or triglycerides, waxes, free fatty alcohols, and steroids. As nano drug carriers, they offer numerous advantages over their other counterparts such as controlled drug delivery, biocompatibility, high drug loading, enhanced bioavailability of lipophilic drugs, greater stability as well as economical and easy and economical production in large-scale (Mehnert & Mäder, 2012; Zeb et al., 2017). Drugs can be loaded in different parts of SLNs and for its loading in lipid matrix, or in shell surrounding the lipid core or in the core surrounded by a lipid shell, different models have been proposed, i.e., homogeneous matrix/solid solution, drug-enriched shell and drug-enriched core models (Müller, Radtke, & Wissing, 2002).

SLNs have been widely employed as drug carriers for a number of drug molecules since their development. The favorable and versatile characteristics of these nanocarriers make them excellent drug delivery systems to minimize and overcome the drawbacks of using conventional approaches. In addition, ionic and hydrophilic drugs can also be incorporated into this delivery system and it is made possible due to recent advancements in SLNs systems, i.e., conjugation of drug with lipids of SLNs and polymer-lipid hybrid SLNs preparations. However, SLNs are removed quickly from the blood circulation by reticule endothelial system (RES). Other obstacles that restrain SLNs from becoming excellent drug delivery nanocarriers are relatively harder to encapsulate ionic or hydrophilic drugs and control the rate and extent of drug release from SLNs (Din et al., 2017).

Liposomes

Liposomes got greater research attention in the last few decades in nanotechnology especially as drug delivery system for antitumor drugs. Liposomes are spherical vesicles based drug delivery systems having a central aqueous core surrounded by lipid bilayers as shown in Fig. 5.3. They are single or multiple bilayered membrane structures formed from the assembly of natural or synthetic lipids. Liposomes that are made of a single membrane bilayer are known as unilamellar and may be small or large depending on their sizes whereas the term multilamellar is used for those liposomes that have more than one bilayer (Torchilin, 2005). They owe a number of advantages over conventional drug

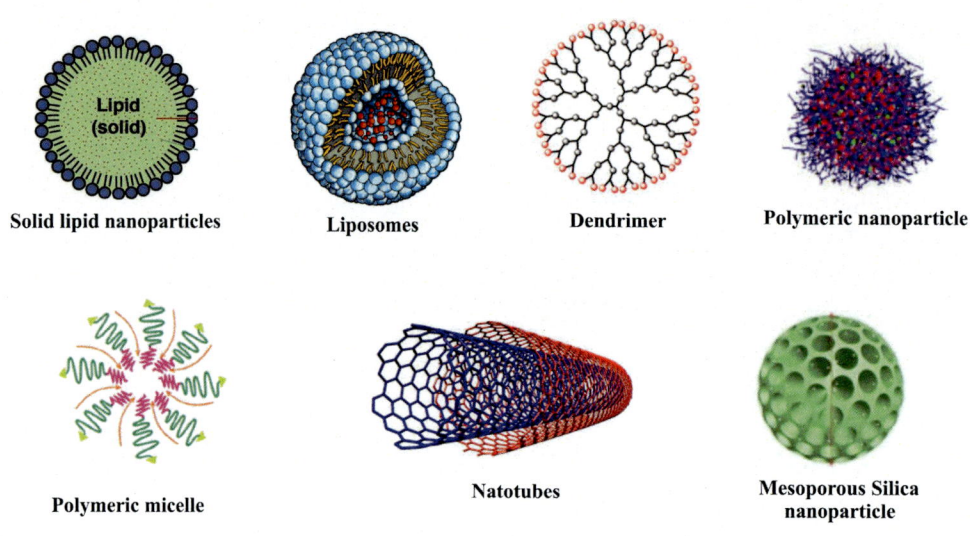

FIG. 5.3

Representative structures of different nanocarriers used for drug delivery. *No Permission Required.*

delivery systems including but not limited to protection of loaded drug from environmental factors, enhanced drug delivery properties, enhanced product performance features, prevention of early degradation of the loaded drug, cost-effective formulations of expensive drugs, and efficient treatment with little systemic toxicities. In addition, drugs loaded in liposomes have markedly altered pharmacokinetic profiles as compared to their free drug solutions (He et al., 2019). Moreover, they can be coated with polymers, e.g., polyethylene glycol (PEG) and poly(lactic-*co*-glycolic acid) (PLGA) to produce PEGylated or stealth liposomes to extend their half-life in blood circulation. Hence, liposomes are widely employed as model carriers for different bioactive agents including vaccines, drugs, nutraceutical and cosmetic products. Liposomes have distinctive quality of accommodating both lipophilic and hydrophilic drugs in their lipids bilayers and aqueous cores, respectively. Moreover, their biocompatible and biodegradable nature makes them excellent therapeutic carriers (Deshpande, Biswas, & Torchilin, 2013).

Dendrimers

Dendrimers are branched macromolecular drug carriers made of natural or synthetic macromolecules including amino acids, nucleotides, and sugars where multiple arms originate from a central core. The extensive regular branching pattern, stepwise synthesis, multi-valency, mono-dispersed phase, spherical shape, and characteristic molecular weight properties enable them to accommodate wide varieties of drugs. Dendrimers that are synthesized in a stepwise synthetic procedure possess distinctive properties because of their excellently arranged patterns in comparison to those that are produced by normal polymerization procedures that have irregular branching patterns (Aoki & Ichimura, 2001). A classical dendrimer molecule is made of extremely branched layers of repeating units, various active terminal groups, and an initiator core. The architectural design of dendrimers allows an excellent control over its size, shape, surface functionality, and branching length (Basu, Sandanaraj, & Thayumanavan, 2002).

Drugs can be loaded to the cavities in the dendrimers cores through hydrogen bonding, hydrophobic interaction, or true chemical bonds. The extensive branched structures of dendrimers result in exterior bulky groups to which drugs and or targeting ligands can be attached and functionalized for obtaining site-specific delivery purposes. Preclinical development of dendrimers based drug delivery systems has been largely focused on dendrimer-drug conjugates. These conjugates are formed by covalently attaching drugs to core or terminal groups of the dendrimers. Dendrimers have been widely used for gene delivery, immunology, magnetic resonance imaging, vaccines, and various drug delivery applications (Pedziwiatr-Werbicka et al., 2019). Being structurally controlled and monodispersed system with well-defined molecular weight and size, dendrimers-drug conjugates are considered the carriers of choice over other polymer-based drug delivery systems. The drug-dendrimer linkage type is of great importance because the drug is needed to be released at the site of action in active form. Drugs are attached to dendrimer surface through disulfide (susceptible to reduction) or acid labile bonds to normally take benefits from acidic or reducing environments near the cancerous cells for their release. Thus, every drug molecule is specifically detached and released from dendrimers surface over an extended period of time (Gothwal, Malik, Gupta, & Jain, 2020).

Polymeric nanoparticles (PNPs)

Polymers have got great deal of attention in drug delivery in the last few decades because of a number of their attractive features. Polymeric nanoparticles are nano-sized (10–1000 nm) solid colloidal particles made up of biodegradable polymers (Chan, Valencia, Zhang, Langer, & Farokhzad, 2010). They are categorized as nanospheres (matrix type) or nanocapsules (reservoir type) depending on their structural organization where drug is entrapped in the polymer matrix in case of nanospheres, while the drug is dissolved in liquid core of oil or water and then encapsulated by a solid polymeric membrane in nanocapsules. The chemical conjugation or adsorption of drugs on the surface of PNPs is possible (Prabhu, Patravale, & Joshi, 2015). Based on the composition and desired features of PNPs, various methods of their preparations have been developed that may be classified broadly into two categories, i.e., dispersion of pre-formed polymers and direct polymerization of monomers methods. Salting out, solvent evaporation, dialysis, nano-precipitation, and supercritical fluid technology are the methods to prepare PNPs under the dispersion of pre-formed polymers method. On the other hand, emulsification polymerization, mini-emulsion polymerization, micro-emulsion polymerization, interfacial polymerization, and radical polymerization are the techniques for PNPs production through direct polymerization of monomers method (Rao & Geckeler, 2011).

Various biodegradable and biocompatible natural or synthetic polymers have been used for the construction of PNPs. These polymers are degraded to their respective individual monomers in the body and being biodegradable, they are removed through normal metabolic processes from the body (Mishra, Patel, & Tiwari, 2010). Polyglycolic acid (PGA), Polylactic acid (PLA), PLGA, PEG, *N*-(2-hydroxypropyl)methacrylamide (HPMA) copolymer, polycaprolactone (PCL), polyglutamic acid, and polyaspartic acid are some of the synthetic polymers mostly used for PNPs synthesis. On the other hand, Chitosan, albumin, collagen, alginate, dextran, heparin, and gelatin are the examples of most widely used natural polymers for the construction of PNPs (Wang, Wang, Chen, & Shin, 2009). Good storage and in-vivo stability properties, higher drug loading, homogeneous size distribution, controllable and better physicochemical properties, extended drug circulation times, and controlled drug release pattern are some of the salient features that distinguish PNPs from their colloidal counterparts, e.g., liposomes and polymeric micelles (El-Say & El-Sawy, 2017).

Polymeric micelles (PMs)

Polymeric micelles are colloidal particles having size range from 10 to 100 nm resulting from self-assembly of synthetic amphiphilic di- or tri-block copolymers in aqueous environments (Musacchio & Torchilin, 2019). Such copolymers, being amphiphilic in nature (containing both hydrophilic and lipophilic portions) when exposed to an aqueous milieu, they form micelles above certain concentration known as critical miceller concentration (CMC). The lipophilic portion of block copolymer makes the core of micelle, while hydrophilic portion forms its shell thus PMs possess a core/shell structure with a hydrophilic shell and a hydrophobic core (Amin, Butt, Amjad,& Kesharwani, 2017; Biswas, Kumari, Lakhani, & Ghosh, 2016). The lipophilic core of micelles allows entrapment of lipophilic drugs and controls the drug release properties of PMs whereas; the hydrophilic shell stabilizes the core, ensuring water solubility of PMs and controls in-vivo pharmacokinetics. Drugs are loaded into the PMs through either chemical attachment or via physical entrapment (Kim, Shi, Kim, Park, & Cheng, 2010). Dialysis, oil-in-water emulsion, solvent evaporation, co-solvent evaporation, and freeze-drying are some of the most widely used methods for micelles preparation (Ahmad, Shah, Siddiq, & Kraatz, 2014).

Inorganic nanocarriers

Carbon nanotubes (CNTs)

Carbon nanotubes (CNTs) are well-known nano-sized cylindrical nanocarriers offering the potentials for the treatment and diagnosis of diseases. They are divided into two main classes namely single-wall carbon nanotubes (SWCNTs) and multi-wall carbon nanotube (MWCNTs) according to their structures. SWCNTs are made of single tube-shaped carbon benzene rings whereas MWCNTs have multiple concentric layers of such benzene rings (Kavosi et al., 2018). An attractive quality of CNTs is that their cylindrical or fiber-like structure allows for higher amounts of cargo loading and extensive functionalization thus increasingly used for delivery of drugs. CNTs that are used as drug delivery carriers ranging in diameter from 0.4 to 100 nm and can be functionalized with different biocompatible and biodegradable materials, e.g., lipids, carbohydrates, and proteins (Hartgerink, Clark, & Ghadiri, 1998) thus hold greater potential for effective delivery of drugs. They were first discovered in 1991 by Iijima and belong to the fullerenes family of carbon (Iijima, 1991). Laser ablation, arc discharge, and thermal or plasma-enhanced chemical vapor deposition are some of the well-known techniques for the preparation of CNTs (Prasek et al., 2011).

Nano-needle shape, hollow monolithic architecture, ultrahigh surface area, higher length to diameter ratio (>200:1), high thermal and electrical conductivity, ultralightweight, high mechanical strength and their surface modification ability make them attractive nanocarriers for drug delivery (Madani, Naderi, Dissanayake, Tan, & Seifalian, 2011; Yan, Chan-Park, & Zhang, 2007). Needle-like shape of these nanocarriers makes them penetrate cell membranes easily via endocytosis or "needle-like penetration" (Yaron et al., 2011). Toxic nature and poor water solubility are the main drawbacks of CNTs however, their surface functionalization property can render them biocompatible, water-soluble, serum stable, and less/nontoxic (Vardharajula et al., 2012).

Mesoporous silica nanoparticles (MSNs)

Silica (SiO_2) based nanomaterials have got greater applications in biomedicine due to their simple synthesis methods and availability for large-scale production. Among such materials, mesoporous silica-based nanocarriers are of prime importance in drug delivery because they have honeycomb-like structures and thus are able to carry higher amounts of drugs (Slowing, Vivero-Escoto, Wu, & Lin, 2008). They possess various attractive properties like excellent biocompatibility, high drug loading capacity, large surface area and pore volume, controllable pore diameters (2–50 nm) with narrow pore size distribution, loading of both lipophilic and hydrophilic drugs, good thermal and chemical stability thus make them promising drug nanocarriers (Li et al., 2017). Moreover, easy surface functionalization quality enables them for targeted and controlled drug delivery applications and thus can enhance therapeutic efficacy and reduce toxicity of loaded drugs (Wang et al., 2015).

Organic/inorganic hybrid nanocarriers

To combine the advantageous properties of organic and inorganic materials in a single carrier Organic/inorganic hybrid nano-frameworks have been established. Such approach has been employed by functionalizing organic materials on the surface of inorganic NPs to enhance the efficiency and selectivity of anticancer drugs (Du & Chen, 2004). For example, MSNs surface functionalized with polyethyleneimine (PEI) not only enhanced their cellular uptake but also produced cationic surface for effective delivery of nucleic acid (Xia et al., 2009). In another study, MSNs integrated with hyperbranched PEI resulted in a high drug loading and sustained intracellular delivery of short interfering RNA (siRNA). These PEI/MSNs hybrid nanocarriers successfully crossed endosomes and effectively reached tumor environment (Prabhakar et al., 2016).

General characteristics of nanocarriers

In a broad definition nanocarriers are particles having one dimension of size smaller than 1000 nm, however, the latest definitions impeded the term of the nanocarriers to particles below 100 nm size (Hood, 2004; Kayser, Lemke, & Hernández-Trejo, 2005). Literature reflects that nanoparticles generally used as drug carriers are between 50 and 500 nm in size (Bosquillon, Lombry, Préat, & Vanbever, 2001; Yokoyama, 2005). Nanocarriers possess various unique and interesting characteristics that make them very suitable and reliable for drug delivery. One of the most critical characteristics of nano-sized drug carriers is their large surface area that contributes to better biological activities (Oberdörster, Celein, Ferin, & Weiss, 1995).

Nanocarriers with a certain size and large surface area boost up the potential for chemical reactions to take place on the particles' surface. Moreover, the number of intermediate compounds formed during reactions that may be more toxic than the original compound can be reduced by modifying the nanocarriers' surfaces with specific materials such as antibodies or ligands (Hood, 2004). However, the size of the drug carrier in the nanometer range is advantageous as it enables higher uptake of them into cells and cellular structures as compared to larger sized particles (Labhasetwar, 2005; Pison et al., 2006). Generally, drugs administered without carriers, i.e., dissolved molecules are susceptible to not being proficient to go across the membrane as the hydrophilic structure of m-embrane bilayers contrasts with the hydrophobic nature of drug molecules (Tamai & Tsuji, 2000). Thus, the modification of nanocarrier's surface for the development of an effective carrier has wide potential for exceptional site-specific and targeted drug delivery (Yokoyama, 2005). Moreover, with medication of the nanocarriers' surface, they acquire the ability to target the drug molecules to particular organs or even to distinct intracellular regions such as the cytoplasm, endo-lysosomes, mitochondria, and nucleus (Labhasetwar, 2005).

An important constraint for the matrix of nanocarrier is that it should be biocompatible and biodegradable. The applications of non-biodegradable matrixes are best for mechanistic studies in animal models but not preferable for therapeutic use. Currently, several different biodegradable and biocompatible materials are available to develop nanocarriers. Different techniques and methods have been established to develop and modify biodegradable nanocarriers. Biodegradable nano-sized particles can be produced by employing synthetic or natural macromolecules such as polylactic-*co*-glycolic acid, polycyano-acrylates, gelatin, chitosan, or serum albumin. Azarmi et al. have reported several types of nanoparticles made of synthetic or natural materials. In one example, they developed poly-butyl-cyano-acrylate nanoparticles for doxorubicin delivery and studied the use of gelatin as biomaterial to synthesize nanoparticles to treat H549 and A460 lung cancer cells in vitro (Azarmi et al., 2006).

Drug loading and site-specific targeting

There are several approaches such as covalent attachment, encapsulation, or adsorption that are employed to load drug onto nanocarriers' surface or matrix (Kreuter, 1991). It is most preferred that a system having long systemic circulation should be developed for parenteral administration so as there will be more time available for the drug to reach its targets. Targeted delivery to various regions can be achieved by integrating the surface of the particles with antibodies or ligands (Popielarski, Pun, & Davis, 2005). Surface modifications of nanocarriers can also be achieved by attaching oligosaccharides, polymers, peptides, proteins, and other targeting agents to the surface of the particles for specific targets (Torchilin, 2005). In addition to target specificity, Illum et al. reported that surface coating of nanocarriers enables them to escape from macrophages uptake (Illum, Davis, Müller, Mak, & West, 1987) and Araujo et al. have concluded that specific amounts of coating material are required to avoid the uptake of particles by macrophages (Araujo et al., 1999).

Drug targeting is achieved either with active or passive targeting methods. It is imperative to know the difference between active and passive targeting in drug delivery. Active targeting is achieved by the process of modifying the nanocarriers' surface with targeting ligands for targeting purposes to specific organs or regions within the body (Moffatt & Cristiano, 2006). On the other hand, passive targeting gets the benefit of the body's defense mechanisms; as such, nanosized drug particles used for passive targeting primarily interact with phagocytic immune cells and macrophages (Gu & Roy, 2004). However, at the tumor site, passive targeting can be attributed to the leaky nature of tumor blood vessels and reduced lymphatic drainage which allows nanoparticles to gather in tumor tissue (known as enhanced permeability and retention (EPR) effect) (Brigger, Dubernet, & Couvreur, 2012; Nakamura, Mochida, Choyke, & Kobayashi, 2016).

Nanocarrier's mechanism and criteria for successful deposition within the respiratory tract upon inspiration

The main mechanism by which nanocarriers deposit after inhalation in different regions of the respiratory tract is the Brownian diffusion (Oberdörster et al., 2005). Large-sized particles' deposition in the lungs is however mainly based on inertial impaction and gravitational settling. Inertial impaction is a process where particles from the gas streamlines are deviated upon interaction with objects and deposited on the walls of the surrounding environment as shown in Fig. 5.4 (Coates & Ho, 1998).

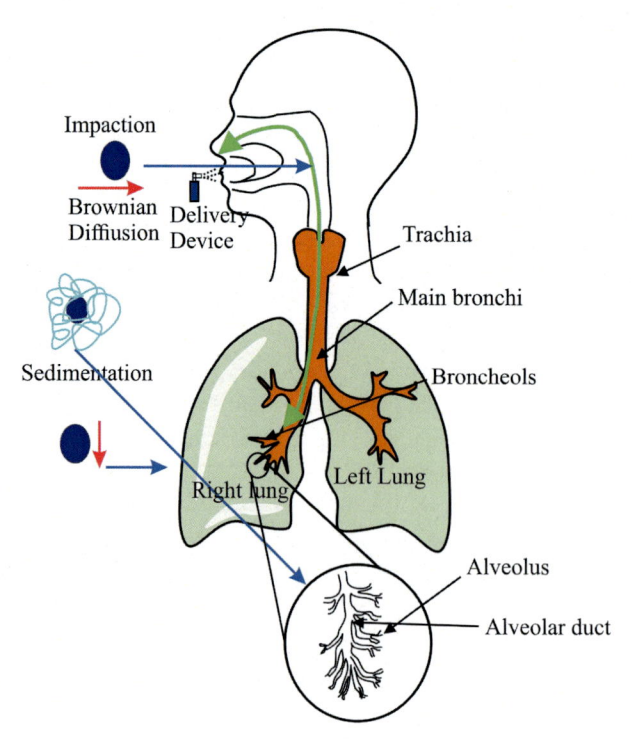

FIG. 5.4

Deposition mechanisms of nanocarriers within the lungs. *No Permission Required.*

The deposition of nanocarriers in pulmonary tract upon inhalation depends on the particles' surface properties, density, and size and all these features of nanocarriers also influence the interaction of nanocarriers with biological systems (Lam, James, McCluskey, & Hunter, 2004). Nanocarriers' size especially is a key contributing factor that determines whether the nanocarrier will be effectively deposited deep into the lungs or if they will merely be breathed out. The particles' size which is used for inhalation therapy is normally stated in terms of the mass median aerodynamic diameter (MMAD). The standard size for effective deposition is that the particles should be in nanometer size range to evade deposition in the upper respiratory tract and permit them to cross the mouth, larynx, pharynx, and lower airways while at the same time being large enough to avoid breathed out (Tsapis, Bennett, Jackson, Weitz, & Edwards, 2002). Thus, the particle's size and density indicated in the MMAD of particles are essential physical characteristics for pulmonary drug delivery. Several studies have depicted that particles of 100–500 nm in diameter (Bair, 1995; Rabinow, 2004) in suitable vehicles (e.g., dry powders or aerosols) can be effectively deposited in different regions of the respiratory tract (Sham et al., 2004; Tsapis et al., 2002).

Nanocarriers' mechanism of action and deposition in the respiratory tract upon inspiration

Diffusion is the main mechanism responsible for the deposition of nanocarriers in the respiratory tract, however depending on the vehicle used for their delivery, inertial compaction and sedimentation are also observed for their deposition (Groneberg et al., 2003; Sham et al., 2004). The size of nanocarriers plays a vital role in determining the area of deposition in the respiratory tract. Deposition of particles having size of about 1 nm will mainly take place in the upper regions of the respiratory tract, e.g., larynx, pharynx, and nose. Particles having size range of about 5 nm will deposit in the bronchi and tracheal regions while for adequate deposition in the deeper alveolar regions, the particles size should be in 20 nm range (Groneberg et al., 2003). Particles having size greater than 5 µm are deposited mainly via inertial impaction while gravitational sedimentation is the main mechanism responsible for the deposition of those particles that have size in the range of 0.5–3 µm.

Nanocarriers interaction with lungs' epithelia and macrophages

Nanocarriers will interact with specific cell populations within the respiratory tract depending on the location where they are deposited. The primary cells that will interact in the deeper regions of the respiratory tract with nanocarriers are type I and II pneumocytes as well as ciliated epithelial cells (Groneberg et al., 2003) however, very little is known about the basic mechanisms of their interactions. Intracellular uptake of the nanocarriers is mainly mediated through receptor-mediated endocytosis where different opsonin molecules (e.g., glycolipid, proteins, and glycerolipids) attach to the nanocarrier's surface forming a recognizable complex. The macrophages recognize this complex formed on the particles' surface and bind them with the complex on their own cells ultimately allowing particles uptake via pseudopod extensions process (Fig. 5.5). The degradation of the pulmonary drug delivery vehicles is accomplished via acid hydrolysis upon fusion of phagosomes (containing engulfed particles) with lysosomes (having acid hydrolases enzymes). The drug molecule itself as well as the nanocarriers are destroyed by enzymes' action in this process (Hoet, Brüske-Hohlfeld, & Salata, 2004) and thus they must have the ability to escape the acid environment of lysosomes so as to retain their full activity.

Delivery of the nanocarriers to deeper alveolar region in the respiratory tract allows alveolar macrophage targeting and thus such drug delivery systems may be effectively used for treating respiratory diseases involving or caused

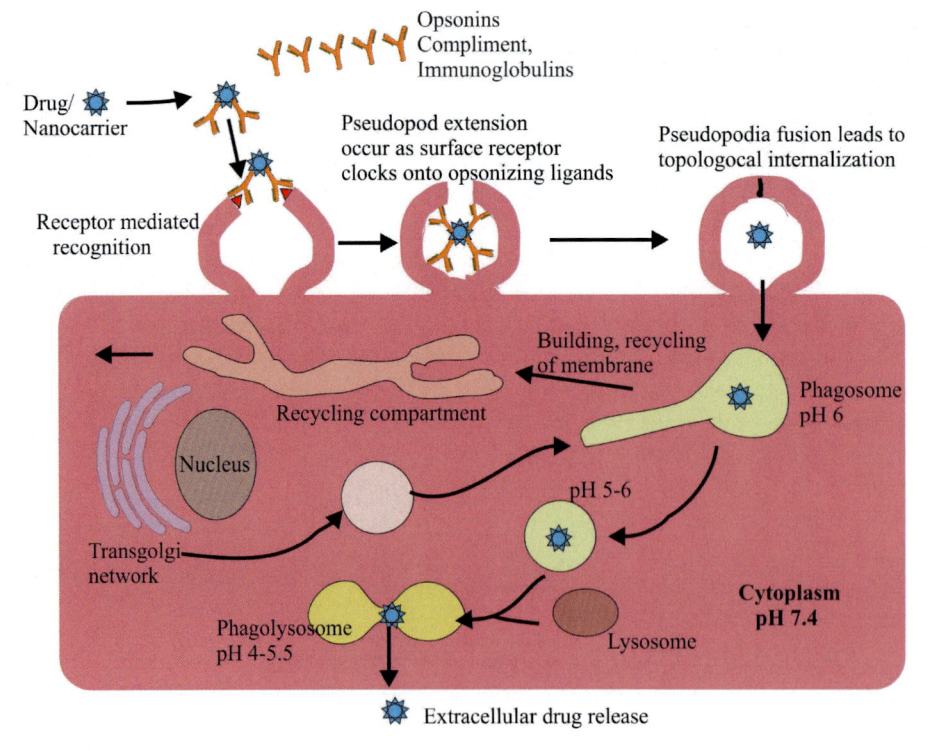

FIG. 5.5

Proposed mechanisms of uptakes and cellular trafficking of drug/nanocarriers. *No Permission Required.*

by these cells (Rabinow, 2004). As an example, the causative agent of Tuberculosis (Mycobacterium) resides within the alveolar macrophages and such alveolar macrophages targeted drug delivery will show improved and enhanced therapeutic results for treating tuberculosis (Gelperina et al., 2005). Moreover, nanocarriers have the ability to carry DNA, genes and antigens thus have promising potentials to be used for pulmonary delivery of vaccines (Kohli & Alpar, 2004).

Nanocarriers' interaction with lung surfactant

The study of nanocarriers' interaction with pulmonary surfactant is of great significance as the results will be terrible if such drug delivery systems disrupt the surfactant film covering the alveolar surface. Therefore using Langmuir Blodgett trough model, many surfactant film interaction studies have been accomplished as this model allows one to simulate the physiological state existing within the respiratory tract (Yu & Possmayer, 2003).

In a study, Stuart et al. investigated the effect of various nanoparticles on lung surfactant film. Similarly, di-palmitoyl phosphatidylcholine (DPPC; the main component of native pulmonary surfactant) interaction with different nanoparticles has been investigated in another study. In this study, lipid monolayer of DPPC was used for simulation of the lung surfactant film (Discher, Schief, Vogel, & Hall, 1999). The objective of this study was to explore whether the integrity of the surfactant layer will be compromised on the alveolar region deposition of nanoparticles. These disruptive interactions may lead to dosage-form-related incompatibilities and are vital to be investigated in preclinical studies for the design and construction of pulmonary delivery of nanoparticles or nano-medical devices. Size of the nano-drug delivery system is also of great importance as it has measurable impact on the lipid film surface tension. The study of Stuart et al. has revealed that size-dependent impact on surface tension and incorporation of nanoparticles occurs in lipid (DPPC) layer. However, their study also showed that the surfactant film was not significantly destabilized by the nanoparticles (Diana et al., 2006). The process shown in this study can serve as a reliable test for setting limits for deposition and nanotoxicological properties evaluation of pulmonary dosage-forms and nano-drug delivery systems.

An analogous protocol was adopted by Mu and Seow (2006) for evaluation of nanoparticles' effect on surface tension. They studied the effect of nanoparticles made from D-α-Tocopheryl polyethylene glycol succinate (α-TPGS) and various other biodegradable carrier matrix substances. α-TPGS is believed to have stabilizing effect on pulmonary surfactant hence used in pulmonary drug delivery mostly. The comparison of maximum surface pressure derived from compression isotherms for DPPC with different amounts of nanoparticles revealed slight decrease when compared with pure DPPC maximum surface tension. This study also showed that the model surfactant film integrity was least disrupted with α-TPGS coated particles hence have a significant potential to be used as a carrier in pulmonary drug delivery systems. However, true and natural lung surfactant has a complex nature comprising of proteins and lipids so only those experiments that are conducted on natural lung surfactant will enable one to deeply understand the use and relationship of nanoparticles in humans for pulmonary drug delivery systems (Fleming & Keough, 1988).

Nanocarriers' retention and clearance mechanisms in lungs

For successful deposition of nanocarriers in lungs, they must have the ability to overcome the defense mechanism of pulmonary route. In order to reach epithelial cells, pulmonary nano-drug delivery system should possess the ability to bypass mucociliary clearance of the upper bronchi and tracheal region (Grenache & Gronowski, 2006). The upper parts of the airway (tracheal and bronchi regions) possess ciliated epithelial cells whose major function is to trap inhaled dust particles and get rid of them via mucociliary action. Each ciliated epithelial cell has nearly 200 cilia on their surface and makes 30%–65% of airway epithelium that's why the chances of particles elimination are enormously high (Groneberg et al., 2003). Periciliary fluid is a mucus layer on these ciliated epithelial cells that effectively capture particles and boost their clearance from airway by mucociliary mechanism. Mucociliary clearance pushes the mucus toward the glottis and the entrapped particles are either eliminated through gastrointestinal route or excreted through mouth (Barlow, Clouter-Baker, Donaldson, MacCallum, & Stone, 2005).

Alveolar macrophages will clear the particles whenever they enter into lower regions (e.g., alveoli) of airways. The alveolar macrophages reside within surfactant monolayer (see Fig. 5.1) and as previously discussed, nanoparticles are interacting with them (Groneberg et al., 2003). The activated alveolar macrophages release mediators like chemokines and cytokines (immunological response mediators) that allow for subsequent phagocytosis and particles clearance (Hoet et al., 2004). The disadvantage of using particulate drug delivery for pulmonary application is that it provokes pulmonary inflammation leading to toxicity and adverse health effects. In addition, oxidative stress is produced in alveolar cells because alveolar antioxidants are depleted in pulmonary inflammation.

Experiments have shown that the lung surfactant plays a significant role in protecting, eliminating, and clearance of inhaled particles through expiration

(Barlow et al., 2005). The surface tension in trachea and bronchi regions is high as compared to alveolar region and a pressure gradient is established in expiration process which forces the particles toward lung surfactant and helps in their elimination. Alveolar macrophages rest in the aqueous region of surfactant film and alveolar clearance is activated as soon as the particles are deposited in the aqueous region of the surfactant film. Thus, various factors including surface properties and size will determine the clearance mechanisms upon successful deposition of inhaled nanocarriers within the respiratory tract.

Toxicological aspects of inhaled nanocarriers

Toxicologists are mainly concerned with the mechanism of lung injury and inflammation caused by inhalation of smaller particle size nanoparticles in the pulmonary drug delivery system (Nel, Xia, Mädler, & Li, 2006). When the surface area of larger sized particles is compared to that of the smaller sized particle, it is observed that the surface area of the smaller sized particle promotes interactions with the biological system and causes reasonable adverse effects. In particular, compared with the large particles in the lungs of rats, it has been shown that nanoparticles have more significant inflammatory potential per unit mass (Steimer et al., 2005). However, the experiments related to the toxicity evaluation of nanoparticles are mainly carried out in-vitro, and very few in-vivo studies have been cited in the literature (Oberdörster et al., 2005).

Production of reactive oxygen species (ROS) followed by oxidative stress is the main mechanism by which nanocarriers induce toxic effects in biological system (Nel et al., 2006). A hierarchical oxidative stress model has been proposed to fully elucidate the role of oxidative stress in pulmonary toxicity caused by nanocarriers. Reactive oxygen species are generated as by-products in normal cellular respiration according to this model. These ROS are neutralized by antioxidants (e.g., glutathione) and antioxidant enzymes normally present in sufficient amounts in the cells. The amount of ROS is sufficiently increased when there is severe lung injury as in the case of inhalation of nano-sized particles, so the amount of antioxidants in the cells is not well enough to neutralize these ROS. This leads to the build-up of reactive molecules such as oxidized glutathione in the cells (Nel et al., 2006). It is also known that ROS production within the mitochondria increases in the presence of nanocarriers. The accumulation of oxide ions (O^{-2}) is considered to be the result of the stagnation of particles in mitochondria, which hampers the electron transport chain and follows energy production. In such scenario, cells can identify a high ratio of oxidized glutathione to non-oxidized glutathione, ultimately leading to inflammation induction. The interaction between nanoparticles and biological systems can lead to toxicity in several other ways (Azarmi et al., 2006).

Oberdörster and his co-workers carried out experiments to show the effects and significance of nanocarriers' surface chemistry on their toxicity. They performed a study in which rats were exposed to hot fumes (480°C) of polytetrafluoroethylene (PTFE) in-vivo. Nanoparticles in the size range of 18 nm were produced upon application of heat and it was observed that inhalation of these fumes caused severe acute lung injury in the rats leading to higher mortality rates 4 h after inhalation (Oberdörster et al., 1995).

A follow-up study by Oberdörster explored the effects of inhalation of fine and ultrafine titanium dioxide (TiO_2) particles on the lungs of mice and rats. Excessive inflammatory response was observed on the inhalation of ultra-fine (20 nm) particles which was determined by an increase in the number of neutrophils in the lungs after 24 h. On the other hand, exposure of animals to the same amount of fine (250 nm) titanium dioxide particles did not lead to excessive immune responses as detected with ultra-fine particles, thus highlighting the importance of particle size for toxicity. Further detailed study showed that when the surface area of ultra-fine and fine titanium dioxide particles were compared to the percentage of neutrophils in the lung, their induced responses were similar (Oberdörster et al., 2005). These results show that the surface area of nano-carriers is directly related to their toxicity and can be used to predict the lung toxicity (Nel et al., 2006).

Pulmonary drug delivery applications of nanocarriers

Drug-loaded nanocarriers have the ability to treat various pulmonary as well as other systemic diseases (Sham et al., 2004). Targeted delivery potentials of these nanosystems empower them to deliver therapeutic drugs to specific site of action thus may be used efficiently for the treatment of lung cancer, cystic fibrosis, TB as well as various other respiratory tract ailments (Gelperina et al., 2005). Particularly, the aerosolized nano-drug delivery systems for pulmonary applications may enhance patient compliance as it is a non-invasive mode of delivering drugs (Pandey & Khuller, 2005a, 2005b). In addition, particular areas within the respiratory tract may be targeted through intelligent manipulation of the nanocarriers' or droplet size because the size of pulmonary nano-drug delivery system is a key determinant for effective deposition in the lungs (Groneberg et al., 2003). Various studies have been conducted on different delivery systems comprised of biodegradable and non-biodegradable carriers to evaluate them to be used for efficient pulmonary drug delivery and discussed under relevant headings as:

Polymeric nanoparticles (PNs) in pulmonary drug delivery

Polymeric nanoparticles are extensively studied as drug delivery systems for parenteral administration however their application in pulmonary route has also been widely recognized (Heidi & Xiao, 2009). Dailey and his team conducted a study to check the inflammatory response of biodegradable co-polymeric nanoparticles composed of di-ethylaminopropylamine polyvinyl alcohol-grafted poly(lactic-*co*-glycolic acid) co-polymer and poly(lactic-*co*-glycolic acid) and non-biodegradable polystyrene-based nanospheres having 220 nm and 75 nm sizes. They intratracheally instilled these polymeric nanoparticles in murine lung and studied the inflammatory responses in terms of Lactate dehydrogenase (LDH) release, MIP-2 mRNA induction, protein concentration, and polymorphonucleocyte (PMN) concentration in the bronchial alveolar lavage fluid (BALF). It was concluded from their study that the designed biodegradable polymeric-based nanoparticles for pulmonary application were quite less active in provoking the inflammatory responses as compared with polystyrene (non-biodegradable) based particles having nearly similar size (Dailey et al., 2006).

As polymers and their degradation products can interact with lung surfactant and can adversely affect its normal function therefore, proper in-vitro lung surfactant models and in vivo studies are essential to ascertain the pulmonary acceptability of these nanocarrier systems. Though, cationic lipids-based gene delivery systems are extensively evaluated clinically as compared to polymer-based gene carriers (De Fougerolles, 2008), however, cationic polymers are considered one of the most popular gene delivery vectors for lungs (De Smedt, Demeester, & Hennink, 2000). Low transfection efficiency and cytotoxicity problems of PNs have to be properly addressed for successful DNA delivery to lungs (Chollet, Favrot, Hurbin, & Coll, 2002), although polyethyleneimine (PEI) and polyamino acids based nanoparticles have shown to be excellent carriers for DNA delivery both in vitro and in vivo (Kichler, Leborgne, Coeytaux, & Danos, 2001). Different modification strategies, e.g., conjugation of ligands like transferrin on PEI surface and modification with PEGs/liposomes have been applied to address such problems associated with PNs (Ko, Kale, Hartner, Papahadjopoulos-Sternberg, & Torchilin, 2009; Rudolph et al., 2002).

Liposomes in pulmonary drug delivery

Liposomes are among the most widely investigated nanocarriers for controlled drug delivery to the lung (Zeng, Martin, & Marriott, 1995). Since they are made of lipids resembling compounds endogenous to lung surfactant, they seem particularly suitable for pulmonary drug delivery (Justo & Moraes, 2003). The first liposomal product launched in market was Alveofact containing synthetic lung surfactant and used for respiratory distress syndrome (RDS) via pulmonary instillation (Proquitté et al., 2007). Liposomal formulations have typically been delivered to the lungs via nebulizers in the liquid state as aerosols (Schreier, Gonzalez-Rothi, & Stecenko, 1993). However, drug stability, leakage, and short half-life are the main concerns associated with liquid state formulation (Taylor, Taylor, Kellaway, & Stevens, 1990). In order to address such issues, liposomal formulations in dry form (proliposomes) have been intensively developed in recent years (Chennakesavulu et al., 2018; Joshi & Misra, 2001; Wang et al., 2009). Such formulations in dry form have proved to be very promising for the pulmonary delivery of various drugs and a number of such formulations are in clinical trials.

As discussed earlier, drugs can also be administered via pulmonary route for their systemic action. In a study, Bi et al. verified the hypoglycemic activity of spray-freeze-dried insulin-loaded liposomes in rats. Insulin-loaded liposomes were prepared with reverse-phase evaporation method, spray-freeze-dried and delivered by dry powder inhaler (DPI). These liposomes lowered the blood glucose level in rats upon their administration via intratracheal instillation and the hypoglycemic effect lasted obviously longer as compared to insulin solution (Bi, Shao, Wang, & Zhang, 2008). In another study, Shahiwala et al. checked the effectiveness of intratracheally instilled levonorgestrel liposomes and compared them with oral administration of the same. Drug-loaded liposomes were prepared using reversed-phase evaporation method and then freeze-dried. When intratracheally administered, they observed a decrease in bioavailability of the drug for liposomal formulation however this formulation maintained effective plasma concentration of the drug for up to 60 h (Shahiwala & Misra, 2004). In a recent study, Hamedinasab et al. developed chitosan-coated liposomes for pulmonary delivery of N-acetylcysteine (NAC) to lung by inhalation. Results of their study showed that chitosan coating of liposomes resulted in prolonged release of NAC from CH-coated liposomes. The flow cytometry results of this study indicated that CH-coated liposomes were efficiently deposited in epithelial cells as compared with the CH-uncoated liposomes. Moreover, the in-vivo study result of this formulation showed good deposition and retention in lung as compared with CH-uncoated liposomes (Hamedinasab, Rezayan, Mellat, Mashreghi, & Jaafari, 2019).

Solid lipid nanoparticles in pulmonary delivery

Although SLNs is not fully cherished for pulmonary delivery due to their toxicological concerns however, when physiological lipids that have little or no toxicity issuer are used for their preparation, they are expected to be better carriers than polymer-based systems (Müller, Rühl, Runge, Schulze-Forster, & Mehnert, 1997; Paliwal, Paliwal, Kenwat, Kurmi, & Sahu, 2020). It is possible that SLNs may be successful drug carriers for pulmonary delivery if they are formulated in aqueous suspensions or dry powder formulations for their aerosol administration using nebulizers or DPIs (Lingayat, Zarekar, & Shendge, 2017). Various studies have been reported on pulmonary applications of SLNs as systemic delivery carriers for macromolecules (Liu et al., 2008) or as local delivery carriers for small molecules (Pandey & Khuller, 2005a). Pandey and Khuller studied the chemotherapeutic effects of SLNs loaded with rifampicin, isoniazid, and pyrazinamide against experimental TB and observed the sustained and slow release of drugs from SLNs in-vitro and in-vivo (Pandey & Khuller, 2005b). Liu et al. studied the pulmonary delivery application of novel nebulizer-compatible SLNs containing insulin for systemic effects. They concluded from their study that their formulation could be successfully used as pulmonary carrier for insulin delivery and can provide novel solutions for systemic delivery of proteins (Liu et al., 2008).

In a study, Videira and his co-workers studied the deposition and clearance of SLNs after inhalation as aerosolized insoluble particles with gamma-scintigraphy imaging. It was found that inhaled particles began to translocate to regional lymph nodes a few minutes after deposition indicating that inhalation can serve as an effective route for delivering drugs to lymphatic system and lipid particles can be used as potential drug carriers for lung cancer therapy (Videira et al., 2002; Videira, Gano, Santos, Neves, & Almeida, 2006).

Dendrimer-based nanocarriers in pulmonary delivery

Research in dendrimers-based drug delivery has been mainly centered on the delivery of DNA and genes to cells for gene/antisense therapy and much research has been focused on their use as non-viral gene vectors . Regarding their use as systemic delivery agents for drugs and macromolecules through pulmonary route, several studies have been published in the literature. In a study, Bai et al., prepared low molecular weight heparin (LMWH)-dendrimer complex for prevention of deep vein thrombosis. The LMWH-dendrimer complex was prepared using electrostatic interactions with different PAMAM dendrimers and screened both the efficacy and safety drug-dendrimer formulations for deep vein thrombosis prevention in-vitro and in-vivo. They concluded from their study that positively charged dendrimers can serve as pulmonary delivery agents for relatively large molecular weight negatively charged drugs. These carriers bind oppositely charged drugs most likely through electrostatic interactions and increased amounts of drug molecules are absorbed via charge neutralization. Moreover, initial safety evaluation through bronchoalveolar lavage analysis and frog palate model study indicates that dendrimers can be effectively used for LMWH delivery through pulmonary route (Bai, Thomas, & Ahsan, 2007). In an additional study, this research group studied pegylated dendrimers (mPEG-dendrimer) in order to enhance circulation time and lung absorption of the drug. They observed and concluded that absorption and half-life LMWH administered through pulmonary route can be enhanced by encapsulating the drug in dendrimeric micelles. Further, their study suggests that LMWH loaded mPEG-dendrimer could potentially be used as noninvasive delivery system for treating thromboembolic disorder (Bai & Ahsan, 2009).

Nasr et al. formulated polyamidoamine (PAMAM) dendrimers and evaluated them as nanocarriers for pulmonary delivery of poorly water-soluble anti-asthma drug beclometasone dipropionate (BDP) using different generations of dendrimers. They evaluated dendrimers for drug solubility, in-vitro drug release, and aerosolization properties using three nebulizers, i.e., air-jet, actively vibrating-mesh, and passively vibrating-mesh nebulizers. For all dendrimer formulations, drug release was less than 35% after 8 h in in-vitro studies. The solubility of drug using dendrimers was increased by increasing the dendrimer generation and by using higher pH media. Their study demonstrated that BDP-dendrimers have potential for pulmonary application via inhalation using air-jet and vibrating-mesh nebulizers. In addition, the aerosol characteristics were influenced by nebulizer design rather than dendrimer generation (Nasr, Najlah, D'Emanuele, & Elhissi, 2014). Several other studies have also been reported regarding the use of dendrimers in pulmonary drug delivery however, their potential use in this route of drug administration still remains a challenge and needs further research to achieve higher biocompatibility and lower cytotoxicity.

Carbon nanotubes (CNTs) for pulmonary drug delivery

Carbon nanotubes may prove advantageous for pulmonary delivery of drugs. Though, the literature about CNTs and pulmonary drug delivery is very scarce, some studies show their use in lungs targeted drug delivery to specific cell populations using animal model systems (Fiorito et al., 2006). Nevertheless, the toxicity of CNTs must first be properly evaluated and addressed before their effective application in pulmonary drug delivery. Lam et al. performed studies on evaluation of toxicological effects of CNTs in mice lungs upon their intratracheal instillation after 7 and 90 days post exposure. Their study results concluded that all single-walled CNTs products produced dose-dependent epithelioid granulomas and interstitial inflammation in some cases in 7 days group animals. These inflammatory lesions persisted and were more pronounced in the 90 days group of animals (Lam et al., 2004). Similar results have been reported by Warheit et al.; for CNTs upon their intratracheal instillation into rat lungs where they caused pulmonary inflammation and toxicity post exposure (Warheit et al., 2004). Thus, more research is needed in-depth in this particular field for the construction of biocompatible CNTs with least un-desirable effects for pulmonary drug delivery.

Conclusion

Drug delivery through pulmonary route has numerous advantages such as extensive vascularization, large alveolar surface area for drug absorption, non-invasive nature, presence of very thin epithelial barrier, and escape of the drug from first-pass metabolism as compared to other routes for drug administration presence of a thin epithelial barrier. Pharmaceutical industries' research is focusing on pulmonary drug delivery and this field is gaining great research attention persistently for targeted and site-specific drug delivery applications of inhaled nanoparticles. Moreover, this route of drug administration can also be exploited for systemic drug delivery applications.

In the field of drug delivery, nanometer-size-ranged medicines are gaining greater scientific interest. It is hoped that there will be reduction in the unwanted side effects of drugs seen with nonspecific drug carriers and improved site-specific/targeted delivery of drugs will be achieved through the use of nanocarriers (having diameter less than 300 nm). Nanocarriers have already revolutionized targeted delivery in intravenous route by intelligent manipulation of their certain properties like surface properties, surface charge, size, and surface functionalization. Appropriate sized nanocarriers can be deposited in specific sites within the lungs and thus may prove advantageous for the treatment of various pulmonary diseases like cystic fibrosis, cancer, and tuberculosis via such targeted drug delivery strategies.

The environmental and health concerns should also be considered when developing nanocarriers system that exists currently and may become more common in the near future. For construction of nanocarriers, better personal protection as well as clearly defined handling measures and their application tools needs to be discussed properly, understood and such regulation should be developed in industries. More detailed and extensive studies are required for safe administration of nanocarriers to properly understand the exact nature of the interactions between particles and biological systems. Similarly, more information about cellular uptake and trafficking of nanoparticles within the cells is needed as such information will be helpful in formulating biocompatible and low toxicity pulmonary drug delivery system. Ultimately, this will lead to further improvements in site-specific and targeted delivery of drugs to desired organs and cell populations.

References

Ahmad, Z., Shah, A., Siddiq, M., & Kraatz, H. B. (2014). Polymeric micelles as drug delivery vehicles. *RSC Advances, 4*(33), 17028–17038. https://doi.org/10.1039/c3ra47370h.

Amin, M. C. I. M., Butt, A. M., Amjad, M. W., & Kesharwani, P. (2017). Polymeric micelles for drug targeting and delivery. In *Nanotechnology-based approaches for targeting and delivery of drugs and genes* (pp. 167–202). Elsevier Inc. https://doi.org/10.1016/B978-0-12-809717-5.00006-3.

Aoki, K. I., & Ichimura, K. (2001). Dendrimers and dendrons, concepts, synthesis, applications dendrimers and dendrons, concepts, synthesis, applications. *Chemistry Letters, 38*, 990–991.

Araujo, L., Löbenberg, R., & Kreuter, J. (1999). Influence of the surfactant concentration on the body distribution of nanoparticles. *Journal of Drug Targeting, 6*(5), 373–385. https://doi.org/10.3109/10611869908996844.

Azarmi, S., Tao, X., Chen, H., Wang, Z., Finlay, W. H., Löbenberg, R., et al. (2006). Formulation and cytotoxicity of doxorubicin nanoparticles carried by dry powder aerosol particles. *International Journal of Pharmaceutics, 319*(1–2), 155–161. https://doi.org/10.1016/j.ijpharm.2006.03.052.

Bai, S., & Ahsan, F. (2009). Synthesis and evaluation of pegylated dendrimeric nanocarrier for pulmonary delivery of low molecular weight heparin. *Pharmaceutical Research, 26*(3), 539–548. https://doi.org/10.1007/s11095-008-9769-y.

Bai, S., Thomas, C., & Ahsan, F. (2007). Dendrimers as a carrier for pulmonary delivery of enoxaparin, a low-molecular weight heparin. *Journal of Pharmaceutical Sciences, 96*(8), 2090–2106. https://doi.org/10.1002/jps.20849.

Bair, J. W. (1995). The ICRP human respiratory tract model for radiological protection. In *Vol. 60, Issue 4. Radiation protection dosimetry* (pp. 307–310). Oxford University Press. https://doi.org/10.1093/oxfordjournals.rpd.a082732.

Balthasar, S., Michaelis, K., Dinauer, N., Von Briesen, H., Kreuter, J., & Langer, K. (2005). Preparation and characterisation of antibody modified gelatin nanoparticles as drug carrier system for uptake in lymphocytes. *Biomaterials, 26*(15), 2723–2732. https://doi.org/10.1016/j.biomaterials.2004.07.047.

Barlow, P. G., Clouter-Baker, A., Donaldson, K., MacCallum, J., & Stone, V. (2005). Carbon black nanoparticles induce type II epithelial cells to release chemotaxins for alveolar macrophages. *Particle and Fibre Toxicology, 2*. https://doi.org/10.1186/1743-8977-2-11.

Basu, S., Sandanaraj, B. S., & Thayumanavan, S. (2002). *Molecular recognition in dendrimers*. Polymer Science and Technology.

Bi, R., Shao, W., Wang, Q., & Zhang, N. (2008). Spray-freeze-dried dry powder inhalation of insulin-loaded liposomes for enhanced pulmonary delivery. *Journal of Drug Targeting, 16*(9), 639–648. https://doi.org/10.1080/10611860802201134.

Biswas, S., Kumari, P., Lakhani, P. M., & Ghosh, B. (2016). Recent advances in polymeric micelles for anti-cancer drug delivery. *European Journal of Pharmaceutical Sciences, 83*, 184–202. https://doi.org/10.1016/j.ejps.2015.12.031.

Bosquillon, C., Lombry, C., Préat, V., & Vanbever, R. (2001). Influence of formulation excipients and physical characteristics of inhalation dry powders on their aerosolization performance. *Journal of Controlled Release, 70*(3), 329–339. https://doi.org/10.1016/S0168-3659(00)00362-X.

Brigger, I., Dubernet, C., & Couvreur, P. (2012). Nanoparticles in cancer therapy and diagnosis. *Advanced Drug Delivery Reviews, 64*, 24–36. https://doi.org/10.1016/j.addr.2012.09.006.

Chan, J. M., Valencia, P. M., Zhang, L., Langer, R., & Farokhzad, O. C. (2010). Polymeric nanoparticles for drug delivery. *Methods in Molecular Biology (Clifton, N.J.), 624*, 163–175. https://doi.org/10.1007/978-1-60761-609-2_11.

Chennakesavulu, S., Mishra, A., Sudheer, A., Sowmya, C., Reddy, C. S., & Bhargav, E. (2018). Pulmonary delivery of liposomal dry powder inhaler formulation for effective treatment of idiopathic pulmonary fibrosis. *Asian Journal of Pharmaceutical Sciences*, 91–100. https://doi.org/10.1016/j.ajps.2017.08.005.

Chollet, P., Favrot, M. C., Hurbin, A., & Coll, J. L. (2002). Side-effects of a systemic injection of linear polyethylenimine-DNA complexes. *The Journal of Gene Medicine, 4*(1), 84–91. https://doi.org/10.1002/jgm.237.

Coates, A. L., & Ho, S. L. (1998). Drug administration by jet nebulization. *Pediatric Pulmonology, 26*(6), 412–423. https://doi.org/10.1002/(SICI)1099-0496(199812)26:6<412::AID-PPUL6>3.0.CO;2-O.

Courrier, H. M., Butz, N., & Vandamme, T. F. (2002). Pulmonary drug delivery systems: Recent developments and prospects. *Critical Reviews in Therapeutic Drug Carrier Systems, 19*(4–5), 425–498. https://doi.org/10.1615/CritRevTherDrugCarrierSyst.v19.i45.40.

Dailey, L. A., Jekel, N., Fink, L., Gessler, T., Schmehl, T., Wittmar, M., et al. (2006). Investigation of the proinflammatory potential of biodegradable nanoparticle drug delivery systems in the lung. *Toxicology and Applied Pharmacology*, *215*(1), 100–108. https://doi.org/10.1016/j.taap.2006.01.016.

De Fougerolles, A. R. (2008). Delivery vehicles for small interfering RNA in vivo. *Human Gene Therapy*, *19*(2), 125–132. https://doi.org/10.1089/hum.2008.928.

De Smedt, S. C., Demeester, J., & Hennink, W. E. (2000). Cationic polymer based gene delivery systems. *Pharmaceutical Research*, *17*(2), 113–126. https://doi.org/10.1023/A:1007548826495.

Deshpande, P. P., Biswas, S., & Torchilin, V. P. (2013). Current trends in the use of liposomes for tumor targeting. *Nanomedicine*, *8*(9), 1509–1528. https://doi.org/10.2217/nnm.13.118.

Diana, S., Raimar, L., Tabitha, K., Shirzad, A., Leticia, E., Wilson, R., et al. (2006). Biophysical investigation of nanoparticle interactions with lung surfactant model systems. *Journal of Biomedical Nanotechnology*, 245–252. https://doi.org/10.1166/jbn.2006.031.

Din, F. U., Aman, W., Ullah, I., Qureshi, O. S., Mustapha, O., Shafique, S., et al. (2017). Effective use of nanocarriers as drug delivery systems for the treatment of selected tumors. *International Journal of Nanomedicine*, *12*, 7291–7309. https://doi.org/10.2147/IJN.S146315.

Discher, B. M., Schief, W. R., Vogel, V., & Hall, S. B. (1999). Phase separation in monolayers of pulmonary surfactant phospholipids at the air-water interface: Composition and structure. *Biophysical Journal*, *77*(4), 2051–2061. https://doi.org/10.1016/S0006-3495(99)77046-3.

Du, J., & Chen, Y. (2004). Organic-inorganic hybrid nanoparticles with a complex hollow structure. *Angewandte Chemie, International Edition*, *43*(38), 5084–5087. https://doi.org/10.1002/anie.200454244.

El-Say, K. M., & El-Sawy, H. S. (2017). Polymeric nanoparticles: Promising platform for drug delivery. *International Journal of Pharmaceutics*, 675–691. https://doi.org/10.1016/j.ijpharm.2017.06.052.

Farrugia, C. A., & Groves, M. J. (1999). Gelatin behaviour in dilute aqueous solution: Designing a nanoparticulate formulation. *Journal of Pharmacy and Pharmacology*, *51*(6), 643–649. https://doi.org/10.1211/0022357991772925.

Fiorito, S., Serafino, A., Andreola, F., Togna, A., & Togna, G. (2006). Toxicity and biocompatibility of carbon nanoparticles. *Journal of Nanoscience and Nanotechnology*, *6*(3), 591–599. https://doi.org/10.1166/jnn.2006.125.

Fleming, B. D., & Keough, K. M. W. (1988). Surface respreading after collapse of monolayers containing major lipids of pulmonary surfactant. *Chemistry and Physics of Lipids*, *49*(1–2), 81–86. https://doi.org/10.1016/0009-3084(88)90067-9.

Gelperina, S., Kisich, K., Iseman, M. D., & Heifets, L. (2005). The potential advantages of nanoparticle drug delivery systems in chemotherapy of tuberculosis. *American Journal of Respiratory and Critical Care Medicine*, *172*(12), 1487–1490. https://doi.org/10.1164/rccm.200504-613PP.

Gill, S., Löbenberg, R., Ku, T., Azarmi, S., Roa, W., & Prenner, E. J. (2007). Nanoparticles: Characteristics, mechanisms of action, and toxicity in pulmonary drug delivery—A review. *Journal of Biomedical Nanotechnology*, *3*(2), 107–119. https://doi.org/10.1166/jbn.2007.015.

Gothwal, A., Malik, S., Gupta, U., & Jain, N. K. (2020). Toxicity and biocompatibility aspects of dendrimers. In *Pharmaceutical applications of dendrimers* (pp. 251–274). Elsevier.

Grenache, D. G., & Gronowski, A. M. (2006). Fetal lung maturity. *Clinical Biochemistry*, *39*(1), 1–10. https://doi.org/10.1016/j.clinbiochem.2005.10.008.

Grenha, A., Seijo, B., & Remuñán-López, C. (2005). Microencapsulated chitosan nanoparticles for lung protein delivery. *European Journal of Pharmaceutical Sciences*, *25*(4–5), 427–437. https://doi.org/10.1016/j.ejps.2005.04.009.

Groneberg, D. A., Witt, C., Wagner, U., Chung, K. F., & Fischer, A. (2003). Fundamentals of pulmonary drug delivery. *Respiratory Medicine*, *97*(4), 382–387. https://doi.org/10.1053/rmed.2002.1457.

Gu, H., & Roy, K. (2004). Topical permeation enhancers efficiently deliver polymer micro and nanoparticles to epidermal Langerhans' cells. *Journal of Drug Delivery Science and Technology*, *14*(4), 265–273. https://doi.org/10.1016/s1773-2247(04)50047-3.

Hamedinasab, H., Rezayan, A. H., Mellat, M., Mashreghi, M., & Jaafari, M. R. (2019). Development of chitosan-coated liposome for pulmonary delivery of N-acetylcysteine. *International Journal of Biological Macromolecules*. https://doi.org/10.1016/j.ijbiomac.2019.11.190.

Hartgerink, J. D., Clark, T. D., & Ghadiri, M. R. (1998). Peptide nanotubes and beyond. *Chemistry—A European Journal*, *4*(8), 1367–1372. https://doi.org/10.1002/(SICI)1521-3765(19980807)4:8<1367::AID-CHEM1367>3.0.CO;2-B.

He, H., Lu, Y., Qi, J., Zhu, Q., Chen, Z., & Wu, W. (2019). Adapting liposomes for oral drug delivery. *Acta Pharmaceutica Sinica B*, *9*(1), 36–48. https://doi.org/10.1016/j.apsb.2018.06.005.

Heidi, M., & Xiao, W. (2009). Nanomedicine in pulmonary delivery. *International Journal of Nanomedicine*, 299. https://doi.org/10.2147/ijn.s4937.

Hoet, P. H., Brüske-Hohlfeld, I., & Salata, O. V. (2004). Nanoparticles–known and unknown health risks. *Journal of Nanobiotechnology*, *2*.

Hood, E. (2004). Nanotechnology: Looking as we leap. *Environmental Health Perspectives*, *112*(13), A740–A749.

Iijima, S. (1991). Helical microtubules of graphitic carbon. *Nature*, *354*(6348), 56–58. https://doi.org/10.1038/354056a0.

Illum, L., Davis, S. S., Müller, R. H., Mak, E., & West, P. (1987). The organ distribution and circulation time of intravenously injected colloidal carriers sterically stabilized with a blockcopolymer—Poloxamine 908. *Life Sciences*, *40*(4), 367–374. https://doi.org/10.1016/0024-3205(87)90138-X.

Jabr-Milane, L., van Vlerken, L., Devalapally, H., Shenoy, D., Komareddy, S., Bhavsar, M., et al. (2008). Multi-functional nanocarriers for targeted delivery of drugs and genes. *Journal of Controlled Release*, *130*(2), 121–128. https://doi.org/10.1016/j.jconrel.2008.04.016.

Jon, G. (1998). Pulmonary surfactant: Functions and molecular composition. *Biochimica et Biophysica Acta (BBA)—Molecular Basis of Disease*, 79–89. https://doi.org/10.1016/s0925-4439(98)00060-x.

Joshi, M., & Misra, A. N. (2001). Pulmonary disposition of budesonide from liposomal dry powder inhaler. *Methods and Findings in Experimental and Clinical Pharmacology*, *23*(10), 531–536. https://doi.org/10.1358/mf.2001.23.10.677118.

Justo, O. R., & Moraes, A. M. (2003). Incorporation of antibiotics in liposomes designed for tuberculosis therapy by inhalation. *Drug Delivery: Journal of Delivery and Targeting of Therapeutic Agents*, *10*(3), 201–207. https://doi.org/10.1080/713840401.

Kagan, V. E., Bayir, H., & Shvedova, A. A. (2005). Nanomedicine and nanotoxicology: Two sides of the same coin. *Nanomedicine: Nanotechnology, Biology, and Medicine*, *1*(4), 313–316. https://doi.org/10.1016/j.nano.2005.10.003.

Kavosi, A., Noei, S. H. G., Madani, S., Khalighfard, S., Khodayari, S., Khodayari, H., et al. (2018). The toxicity and therapeutic effects of single-and multi-wall carbon nanotubes on mice breast cancer. *Scientific Reports*, *8*, 1–12.

Kayser, O., Lemke, A., & Hernández-Trejo, N. (2005). The impact of nanobiotechnology on the development of new drug delivery systems. *Current Pharmaceutical Biotechnology*, 3–5. https://doi.org/10.2174/1389201053167158.

Kichler, A., Leborgne, C., Coeytaux, E., & Danos, O. (2001). Polyethylenimine-mediated gene delivery: A mechanistic study. *The Journal of Gene Medicine*, *3*(2), 135–144. https://doi.org/10.1002/jgm.173.

Kim, S., Shi, Y., Kim, J. Y., Park, K., & Cheng, J. X. (2010). Overcoming the barriers in micellar drug delivery: Loading efficiency, in vivo stability, and micelle-cell interaction. *Expert Opinion on Drug Delivery, 7*(1), 49–62. https://doi.org/10.1517/17425240903380446.

Ko, Y. T., Kale, A., Hartner, W. C., Papahadjopoulos-Sternberg, B., & Torchilin, V. P. (2009). Self-assembling micelle-like nanoparticles based on phospholipid-polyethyleneimine conjugates for systemic gene delivery. *Journal of Controlled Release, 133*(2), 132–138. https://doi.org/10.1016/j.jconrel.2008.09.079.

Kohli, A. K., & Alpar, H. O. (2004). Potential use of nanoparticles for transcutaneous vaccine delivery: Effect of particle size and charge. *International Journal of Pharmaceutics, 275*(1–2), 13–17. https://doi.org/10.1016/j.ijpharm.2003.10.038.

Kreuter, J. (1991). Nanoparticle-based dmg delivery systems. *Journal of Controlled Release, 16*(1–2), 169–176. https://doi.org/10.1016/0168-3659(91)90040-K.

Kreuter, J., Täuber, U., & Illi, V. (1979). Distribution and elimination of poly (methyl-2-14C-methacrylate) nanoparticle radioactivity after injection in rats and mice. *Journal of Pharmaceutical Sciences, 68*(11), 1443–1447. https://doi.org/10.1002/jps.2600681129.

Labhasetwar, V. (2005). Nanotechnology for drug and gene therapy: The importance of understanding molecular mechanisms of delivery. *Current Opinion in Biotechnology, 16*(6), 674–680. https://doi.org/10.1016/j.copbio.2005.10.009.

Lam, C. W., James, J. T., McCluskey, R., & Hunter, R. L. (2004). Pulmonary toxicity of single-wall carbon nanotubes in mice 7 and 90 days after intratracheal instillation. *Toxicological Sciences, 77*(1), 126–134. https://doi.org/10.1093/toxsci/kfg243.

LaVan, D. A., McGuire, T., & Langer, R. (2003). Small-scale systems for in vivo drug delivery. *Nature Biotechnology, 21*(10), 1184–1191. https://doi.org/10.1038/nbt876.

Li, Y., Li, N., Pan, W., Yu, Z., Yang, L., & Tang, B. (2017). Hollow mesoporous silica nanoparticles with tunable structures for controlled drug delivery. *ACS Applied Materials and Interfaces, 9*(3), 2123–2129. https://doi.org/10.1021/acsami.6b13876.

Lingayat, V. J., Zarekar, N. S., & Shendge, R. S. (2017). Solid lipid nanoparticles: A review. *2. Nanoscience and nanotechnology research* (pp. 67–72). Science and Education Publishing.

Liu, J., Gong, T., Fu, H., Wang, C., Wang, X., Chen, Q., et al. (2008). Solid lipid nanoparticles for pulmonary delivery of insulin. *International Journal of Pharmaceutics, 356*(1–2), 333–344. https://doi.org/10.1016/j.ijpharm.2008.01.008.

Löbenberg, R., Araujo, L., Von Briesen, H., Rodgers, E., & Kreuter, J. (1998). Body distribution of azidothymidine bound to hexyl-cyanoacrylate nanoparticles after i.v. injection to rats. *Journal of Controlled Release, 50*(1–3), 21–30. https://doi.org/10.1016/S0168-3659(97)00105-3.

Madani, S. Y., Naderi, N., Dissanayake, O., Tan, A., & Seifalian, A. M. (2011). A new era of cancer treatment: Carbon nanotubes as drug delivery tools. *International Journal of Nanomedicine, 6*.

Malam, Y., Loizidou, M., & Seifalian, A. M. (2009). Liposomes and nanoparticles: Nanosized vehicles for drug delivery in cancer. *Trends in Pharmacological Sciences, 30*(11), 592–599. https://doi.org/10.1016/j.tips.2009.08.004.

Mehnert, W., & Mäder, K. (2012). Solid lipid nanoparticles: Production, characterization and applications. *Advanced Drug Delivery Reviews, 64*, 83–101. https://doi.org/10.1016/j.addr.2012.09.021.

Mishra, B., Patel, B. B., & Tiwari, S. (2010). Colloidal nanocarriers: A review on formulation technology, types and applications toward targeted drug delivery. *Nanomedicine: Nanotechnology, Biology, and Medicine, 6*(1), 9–24. https://doi.org/10.1016/j.nano.2009.04.008.

Moffatt, S., & Cristiano, R. J. (2006). Uptake characteristics of NGR-coupled stealth PEI/pDNA nanoparticles loaded with PLGA-PEG-PLGA tri-block copolymer for targeted delivery to human monocyte-derived dendritic cells. *International Journal of Pharmaceutics, 321*(1–2), 143–154. https://doi.org/10.1016/j.ijpharm.2006.05.007.

Mu, L., & Seow, P. H. (2006). Application of TPGS in polymeric nanoparticulate drug delivery system. *Colloids and Surfaces B: Biointerfaces*, *47*(1), 90–97. https://doi.org/10.1016/j.colsurfb.2005.08.016.

Müller, R. H., Mäder, K., & Gohla, S. (2000). Solid lipid nanoparticles (SLN) for controlled drug delivery—A review of the state of the art. *European Journal of Pharmaceutics and Biopharmaceutics*, *50*(1), 161–177. https://doi.org/10.1016/S0939-6411(00)00087-4.

Müller, R. H., Radtke, M., & Wissing, S. A. (2002). Solid lipid nanoparticles (SLN) and nanostructured lipid carriers (NLC) in cosmetic and dermatological preparations. In. *Advanced Drug Delivery Reviews*, *54*, S131–S155. https://doi.org/10.1016/S0169-409X(02)00118-7.

Müller, R. H., Rühl, D., Runge, S., Schulze-Forster, K., & Mehnert, W. (1997). Cytotoxicity of solid lipid nanoparticles as a function of the lipid matrix and the surfactant. *Pharmaceutical Research*, *14*(4), 458–462. https://doi.org/10.1023/A:1012043315093.

Musacchio, T., & Torchilin, V. P. (2019). Advances in polymeric and lipid-core micelles as drug delivery systems. In *Polymeric biomaterials* Taylor & Francis Group.

Nakamura, Y., Mochida, A., Choyke, P. L., & Kobayashi, H. (2016). Nanodrug delivery: Is the enhanced permeability and retention effect sufficient for curing cancer? *Bioconjugate Chemistry*, *27*(10), 2225–2238. https://doi.org/10.1021/acs.bioconjchem.6b00437.

Nasr, M., Najlah, M., D'Emanuele, A., & Elhissi, A. (2014). PAMAM dendrimers as aerosol drug nanocarriers for pulmonary delivery via nebulization. *International Journal of Pharmaceutics*, *461*(1–2), 242–250. https://doi.org/10.1016/j.ijpharm.2013.11.023.

Nel, A., Xia, T., Mädler, L., & Li, N. (2006). Toxic potential of materials at the nanolevel. *Science*, *311*(5761), 622–627. https://doi.org/10.1126/science.1114397.

Oberdörster, G., Celein, R. M., Ferin, J., & Weiss, B. (1995). Association of particulate air pollution and acute mortality: Involvement of ultrafine particles? *Inhalation Toxicology*, *7*(1), 111–124. https://doi.org/10.3109/08958379509014275.

Oberdörster, G., Oberdörster, E., & Oberdörster, J. (2005). Nanotoxicology: An emerging discipline evolving from studies of ultrafine particles. *Environmental Health Perspectives*, *113*(7), 823–839. https://doi.org/10.1289/ehp.7339.

Paliwal, R., Paliwal, S. R., Kenwat, R., Kurmi, B. D., & Sahu, M. K. (2020). Solid lipid nanoparticles: A review on recent perspectives and patents. *Expert Opinion on Therapeutic Patents*, *30*(3), 179–194. https://doi.org/10.1080/13543776.2020.1720649.

Pandey, R., & Khuller, G. K. (2005a). Antitubercular inhaled therapy: Opportunities, progress and challenges. *Journal of Antimicrobial Chemotherapy*, *55*(4), 430–435. https://doi.org/10.1093/jac/dki027.

Pandey, R., & Khuller, G. K. (2005b). Solid lipid particle-based inhalable sustained drug delivery system against experimental tuberculosis. *Tuberculosis*, *85*(4), 227–234. https://doi.org/10.1016/j.tube.2004.11.003.

Pedziwiatr-Werbicka, E., Milowska, K., Dzmitruk, V., Ionov, M., Shcharbin, D., & Bryszewska, M. (2019). Dendrimers and hyperbranched structures for biomedical applications. *European Polymer Journal*, *119*, 61–73. https://doi.org/10.1016/j.eurpolymj.2019.07.013.

Pérez-Gil, J., & Keough, K. M. W. (1998). Interfacial properties of surfactant proteins. *Biochimica et Biophysica Acta-Molecular Basis of Disease*, *1408*(2–3), 203–217. https://doi.org/10.1016/S0925-4439(98)00068-4.

Piknova, B., Schram, V., & Hall, S. B. (2002). Pulmonary surfactant: Phase behavior and function. *Current Opinion in Structural Biology*, *12*(4), 487–494. https://doi.org/10.1016/S0959-440X(02)00352-4.

Pison, U., Welte, T., Giersig, M., & Groneberg, D. A. (2006). Nanomedicine for respiratory diseases. *European Journal of Pharmacology*, *533*(1–3), 341–350. https://doi.org/10.1016/j.ejphar.2005.12.068.

Popielarski, S. R., Pun, S. H., & Davis, M. E. (2005). A nanoparticle-based model delivery system to guide the rational design of gene delivery to the liver. 1. Synthesis and characterization. *Bioconjugate Chemistry, 16*(5), 1063–1070. https://doi.org/10.1021/bc050113d.

Prabhakar, N., Zhang, J., Desai, D., Casals, E., Gulin-Sarfraz, T., Näreoja, T., et al. (2016). Stimuli-responsive hybrid nanocarriers developed by controllable integration of hyperbranched PEI with mesoporous silica nanoparticles for sustained intracellular siRNA delivery. *International Journal of Nanomedicine, 11*, 6591–6608. https://doi.org/10.2147/IJN.S120611.

Prabhu, R. H., Patravale, V. B., & Joshi, M. D. (2015). Polymeric nanoparticles for targeted treatment in oncology: Current insights. *International Journal of Nanomedicine, 10*, 1001–1018. https://doi.org/10.2147/IJN.S56932.

Prasek, J., Drbohlavova, J., Chomoucka, J., Hubalek, J., Jasek, O., Adam, V., et al. (2011). Methods for carbon nanotubes synthesis—Review. *Journal of Materials Chemistry, 21*(40), 15872–15884. https://doi.org/10.1039/c1jm12254a.

Proquitté, H., Dushe, T., Hammer, H., Rüdiger, M., Schmalisch, G., & Wauer, R. R. (2007). Observational study to compare the clinical efficacy of the natural surfactants Alveofact and Curosurf in the treatment of respiratory distress syndrome in premature infants. *Respiratory Medicine, 101*(1), 169–176. https://doi.org/10.1016/j.rmed.2006.03.033.

Rabinow, B. E. (2004). Nanosuspensions in drug delivery. *Nature Reviews Drug Discovery, 3*(9), 785–796. https://doi.org/10.1038/nrd1494.

Rao, J. P., & Geckeler, K. E. (2011). Polymer nanoparticles: Preparation techniques and size-control parameters. *Progress in Polymer Science (Oxford), 36*(7), 887–913. https://doi.org/10.1016/j.progpolymsci.2011.01.001.

Rosen, H., & Abribat, T. (2005). The rise and rise of drug delivery. *Nature Reviews Drug Discovery, 4*(5), 381–385. https://doi.org/10.1038/nrd1721.

Rudolph, C., Schillinger, U., Plank, C., Gessner, A., Nicklaus, P., Müller, R. H., et al. (2002). Nonviral gene delivery to the lung with copolymer-protected and transferrin-modified polyethylenimine. *Biochimica et Biophysica Acta-General Subjects, 1573*(1), 75–83. https://doi.org/10.1016/S0304-4165(02)00334-3.

Schief, W. R., Antia, M., Discher, B. M., Hall, S. B., & Vogel, V. (2003). Liquid-crystalline collapse of pulmonary surfactant monolayers. *Biophysical Journal, 84*(6), 3792–3806. https://doi.org/10.1016/S0006-3495(03)75107-8.

Schreier, H., Gonzalez-Rothi, R. J., & Stecenko, A. A. (1993). Pulmonary delivery of liposomes. *Journal of Controlled Release, 24*(1–3), 209–223. https://doi.org/10.1016/0168-3659(93)90180-D.

Shahiwala, A., & Misra, A. (2004). Pulmonary absorption of liposomal levonorgestrel. *AAPS PharmSciTech, 5*(1). https://doi.org/10.1208/pt050113.

Sham, J. O. H., Zhang, Y., Finlay, W. H., Roa, W. H., & Löbenberg, R. (2004). Formulation and characterization of spray-dried powders containing nanoparticles for aerosol delivery to the lung. *International Journal of Pharmaceutics, 269*(2), 457–467. https://doi.org/10.1016/j.ijpharm.2003.09.041.

Slowing, I. I., Vivero-Escoto, J. L., Wu, C. W., & Lin, V. S. Y. (2008). Mesoporous silica nanoparticles as controlled release drug delivery and gene transfection carriers. *Advanced Drug Delivery Reviews, 60*(11), 1278–1288. https://doi.org/10.1016/j.addr.2008.03.012.

Steimer, A., Haltner, E., & Lehr, C.-M. (2005). Cell culture models of the respiratory tract relevant to pulmonary drug delivery. *Journal of Aerosol Medicine,* 137–182. https://doi.org/10.1089/jam.2005.18.137.

Tamai, I., & Tsuji, A. (2000). Transporter-mediated permeation of drugs across the blood-brain barrier. *Journal of Pharmaceutical Sciences, 89*(11), 1371–1388. https://doi.org/10.1002/1520-6017(200011)89:11<1371::AID-JPS1>3.0.CO;2-D.

Taylor, K. M. G., Taylor, G., Kellaway, I. W., & Stevens, J. (1990). The stability of liposomes to nebulisation. *International Journal of Pharmaceutics, 58*(1), 57–61. https://doi.org/10.1016/0378-5173(90)90287-E.

Torchilin, V. P. (2000). Drug targeting. *European Journal of Pharmaceutical Sciences, 11*(2), S81–S91. https://doi.org/10.1016/S0928-0987(00)00166-4.

Torchilin, V. P. (2005). Recent advances with liposomes as pharmaceutical carriers. *Nature Reviews Drug Discovery, 4*(2), 145–160. https://doi.org/10.1038/nrd1632.

Tsapis, N., Bennett, D., Jackson, B., Weitz, D. A., & Edwards, D. A. (2002). Trojan particles: Large porous carriers of nanoparticles for drug delivery. *Proceedings of the National Academy of Sciences,* 12001–12005. https://doi.org/10.1073/pnas.182233999.

Vanbever, R., Ben-Jebria, A., Mintzes, J. D., Langer, R., & Edwards, D. A. (1999). Sustained release of insulin from insoluble inhaled particles. *Drug Development Research, 48*(4), 178–185. https://doi.org/10.1002/(SICI)1098-2299(199912)48:4<178::AID-DDR5>3.0.CO;2-I.

Vardharajula, S., Ali, S. Z., Tiwari, P. M., Eroğlu, E., Vig, K., Dennis, V. A., et al. (2012). Functionalized carbon nanotubes: Biomedical applications. *International Journal of Nanomedicine, 7,* 5361–5374. https://doi.org/10.2147/IJN.S35832.

Videira, M. A., Botelho, M. F., Santos, A. C., Gouveia, L. F., Pedroso De Lima, J. J., & Almeida, A. J. (2002). Lymphatic uptake of pulmonary delivered radiolabelled solid lipid nanoparticles. *Journal of Drug Targeting, 10*(8), 607–613. https://doi.org/10.1080/1061186021000054933.

Videira, M. A., Gano, L., Santos, C., Neves, M., & Almeida, A. J. (2006). Lymphatic uptake of lipid nanoparticles following endotracheal administration. *Journal of Microencapsulation, 23*(8), 855–862. https://doi.org/10.1080/02652040600788221.

Wang, X., Wang, Y., Chen, Z. G., & Shin, D. M. (2009). Advances of cancer therapy by nanotechnology. *Cancer Research and Treatment, 1.* https://doi.org/10.4143/crt.2009.41.1.1.

Wang, S., Zhao, Q., Han, N., Bai, L., Li, J., Liu, J., et al. (2015). Mesoporous silica nanoparticles in drug delivery and biomedical applications. *Nanomedicine: Nanotechnology, Biology, and Medicine, 11*(2), 313–327. https://doi.org/10.1016/j.nano.2014.09.014.

Warheit, D. B., Laurence, B. R., Reed, K. L., Roach, D. H., Reynolds, G. A. M., & Webb, T. R. (2004). Comparative pulmonary toxicity assessment of single-wall carbon nanotubes in rats. *Toxicological Sciences, 77*(1), 117–125. https://doi.org/10.1093/toxsci/kfg228.

Wüstneck, N., Wüstneck, R., Fainerman, V. B., Miller, R., & Pison, U. (2001). Interfacial behaviour and mechanical properties of spread lung surfactant protein/lipid layers. *Colloids and Surfaces B: Biointerfaces, 21*(1–3), 191–205. https://doi.org/10.1016/S0927-7765(01)00172-2.

Xia, T., Kovochich, M., Liong, M., Meng, H., Kabehie, S., George, S., et al. (2009). Polyethyleneimine coating enhances the cellular uptake of mesoporous silica nanoparticles and allows safe delivery of siRNA and DNA constructs. *ACS Nano, 3*(10), 3273–3286. https://doi.org/10.1021/nn900918w.

Yan, Y., Chan-Park, M. B., & Zhang, Q. (2007). Advances in carbon-nanotube assembly. *Small, 3*(1), 24–42. https://doi.org/10.1002/smll.200600354.

Yaron, P. N., Holt, B. D., Short, P. A., Lösche, M., Islam, M. F., & Dahl, K. N. (2011). Single wall carbon nanotubes enter cells by endocytosis and not membrane penetration. *Journal of Nanobiotechnology, 9.* https://doi.org/10.1186/1477-3155-9-45.

Yokoyama, M. (2005). Drug targeting with nano-sized carrier systems. *Journal of Artificial Organs, 8*(2), 77–84. https://doi.org/10.1007/s10047-005-0285-0.

Yu, S. H., & Possmayer, F. (2003). Lipid compositional analysis of pulmonary surfactant monolayers and monolayer-associated reservoirs. *Journal of Lipid Research, 44*(3), 621–629. https://doi.org/10.1194/jlr.M200380-JLR200.

Zeb, A., Qureshi, O. S., Kim, H. S., Kim, M. S., Kang, J. H., Park, J. S., et al. (2017). High payload itraconazole-incorporated lipid nanoparticles with modulated release property for oral and parenteral administration. *Journal of Pharmacy and Pharmacology, 69*(8), 955–966. https://doi.org/10.1111/jphp.12727.

Zeng, X. M., Martin, G. P., & Marriott, C. (1995). The controlled delivery of drugs to the lung. *International Journal of Pharmaceutics, 124*(2), 149–164. https://doi.org/10.1016/0378-5173(95)00104-Q.

Zhang, Q., Shen, Z., & Nagai, T. (2001). Prolonged hypoglycemic effect of insulin-loaded polybutylcyanoacrylate nanoparticles after pulmonary administration to normal rats. *International Journal of Pharmaceutics, 218*(1–2), 75–80. https://doi.org/10.1016/S0378-5173(01)00614-7.

Further reading

Kesharwani, P. (2019). *Nanotechnology-based targeted drug delivery systems for lung cancer* (pp. 161–192). Elsevier.

Sezer, A. D. (2014). *Application of nanotechnology in drug delivery* (pp. 1–50). InTechOpen.

CHAPTER 6

Transdermal drug delivery nanocarriers for improved treatment of skin diseases

Introduction

Skin diseases and skin disorders are emerging health issues affecting many people globally every day. Dermatological disorders are commonly caused by infectious pathogens, inflammatory situations, immune-mediated skin troubles, skin cancers, and wounds. Skin-related diseases associated with chronic or incurable skin lesions have a significant impact ranging from morbidity to mortality. Skin diseases due to infections and infestations have been a major health concern in developing countries, while skin cancers are more prevalent in the developed countries of the world (Basra & Shahrukh, 2009; Gupta, Agrawal, & Vyas, 2012). Dermatological illnesses are related to several infectious diseases that are still difficult to treat. These disorders vary depending on the pathogen involved, skin structure, and medical root cause of the patient. Most infectious diseases of the skin and hair follicles are due to bacterial, viral, or microbial infections. Moreover, genetic factors, weak immune systems, or contact with allergens also contribute to skin diseases (Sierra-Sánchez, Montero-Vilchez, Quiñones-Vico, Sanchez-Diaz, & Arias-Santiago, 2021). Chronic inflammatory skin problems such as psoriasis, eczema, atopic dermatitis vitiligo, or leg ulcers are not life-threatening although they can cause discomfort and disability and affect the patient's quality of life. These inflammatory conditions are characterized by infiltration of inflammatory T cells with increased production of cytokines in the lesion (Sigmundsdottir, 2010). Skin cancers such as malignant melanoma remain a major concern due to their higher mortality rate and proliferation of abnormal cells in the epidermis. Maintenance of skin is important for healthcare and aesthetic purposes, thus effective dermatological care is of paramount importance. Several innovations have been made for the treatment of skin problems. The clinical treatment of infectious skin diseases emphasizes diagnosing, identifying the pathogen, and selecting appropriate route of administration to release the payload in a controlled and timely manner (Dawson, Dellavalle, & Elston, 2012; Oliver, Pham, Li, Xu, & Boyer, 2021).

Nanocarriers for Organ-Specific and Localized Drug Delivery. https://doi.org/10.1016/B978-0-12-821093-2.00003-7

The defensive function of the skin is to protect the body against invasion of physical, microbial, and chemical exposure. It also provides immunity against microbes, impedes the transport of endogenous substances like water, thus maintaining the electrolytic balance, performing thermoregulation, as well as providing protection against ultraviolet radiation and free radicals. The skin has a large surface area and, due to easy accessibility, it has the potential for drug delivery applications. Most of the therapeutic agents, due to their impermeability, are incapable of penetrating through or into the skin, which results in non-invasive delivery of drugs (Gupta et al., 2012). Skin delivery offers a promising alternative to circumvent the limitations associated with oral and parenteral drug delivery systems. Skin delivery is categorized into topical (dermal) or transdermal drug delivery systems (TDDS). Transdermal drug delivery is an attractive approach to improve the therapeutic efficacy of drugs against skin diseases through controlled release of drugs via skin systematically. Drug delivery through the skin provides benefits of avoiding hepatic first-pass metabolism, better patient compliance, ease of use, effective drug release, tissue targeting, and reduces systemic side effects (Muzzalupo et al., 2011). Better understanding of the morphology of skin, its barrier properties, and transport route has to be explored properly for better treatment for skin diseases. The main objective of this chapter is to describe various approaches adopted for the transdermal drug delivery systems to treat skin-related disorders. Additionally, the current and future potentials to treat the skin lesions based on TDDS have been overviewed.

Skin anatomy

To better understand how the transdermal drug delivery system enhances therapeutic efficacy of drugs for the treatment of skin-related disorders, it is crucial to explore the structure and mechanism of the therapeutic agents' entry into skin. The skin covers the largest surface area of approximately 1.5–2 m^2 of the human body. Anatomically, it is a multilayered structure composed of many histological layers, i.e., epidermis, dermis, and subcutaneous connective tissues (Fig. 6.1).

Epidermis

Epidermis is the outermost avascular layer of the skin with 15–20 μm thickness and is divided into two regions, i.e., the non-viable stratum corneum and the viable epidermis (Walters, 2002). It is primarily made of three main cell groups, namely keratinocytes, melanocytes, and Langerhans cells. Epidermis mainly consists of keratinocytes (more than 90%) with no blood supply and therefore, any molecule permeating through the epidermis must diffuse across the dermal-epidermal junction to enter the systemic circulation of the body.

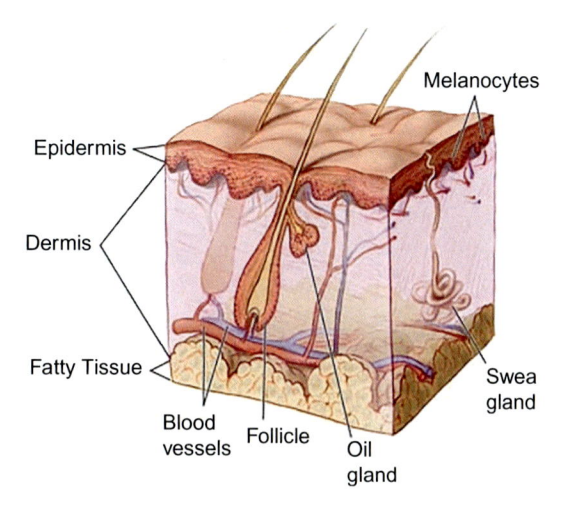

FIG. 6.1

Structure of the skin. *Data from Terese Winslow.* Skin anatomy. *National Cancer Institute https://visualsonline.cancer.gov/details.cfm? imageid=4604.*

Epidermis is composed of five distinctive layers: stratum corneum, stratum lucidum (SL), granular cell layer (stratum granulosum; SG), squamous cell layer (stratum spinosum; SS), and basal cell layer (stratum germinativum; SB). The viable epidermis consists of layers beneath the stratum corneum. Stratum corneum, the outermost epidermis, confers barrier properties against foreign material and is involved in homeostatic functions (Jain, Patel, Shah, Khatri, & Vora, 2017). This layer is transformed by keratinocytes, which consist of non-living corneocytes surrounded by a lamellar matrix formed by intercellular lipids assembled like brick and mortar. The stratum corneum is held together with stratum granulosum, to form a highly hydrophobic layer that reduces the passive permeation of molecules with over 500 kDa molecular weight (Bos & Meinardi, 2000). Thus, the architecture of stratum corneum is mainly responsible for limiting the penetration of transdermal drugs (Barua & Mitragotri, 2014). The other components like keratinocytes, melanocytes, Merkel cells, and Langerhans cells all perform key functions in the viable epidermis. Melanocytes are responsible for the melanin production and prevent skin from UV radiation. Merkel cells are responsible for sensory perception while Langerhans cells are responsive toward immune system of skin (Oliver et al., 2021).

Dermis

Dermis is the middle layer of the skin embedded between epidermis and subcutaneous tissues. This thick layer (1000–2000 μm) consists of fibrous and elastic tissues (collagen and elastin), glucoaminoglycans, salt, and water. Its key role is to provide support to the epidermis and thus strengthen the skin. The dermis is composed of papillary and reticular dermis layers (Losquadro, 2017). It also contains hair follicles, sweat glands, sebaceous glands, nerves, blood, and lymphatic vessels. The dermis has an extensive vascular network that connects to the systemic circulation through arterioles and venules. It is responsible for maintaining skin integrity, thermoregulation, and immunological responses. Moreover, the sweat glands and hair follicles connect to the outermost layer of skin crossing the stratum corneum and offer an appendageal route for skin penetration (Otberg et al., 2004; Walters, 2002).

Subcutaneous tissues

Subcutaneous tissues also known as hypodermis, located underneath the dermis, comprise adipocytes with connective tissues functioning as fat storage for the body. The subcutaneous tissues are the energy source and serve as a thermal barrier as well as mechanical cushion for the body (Jain, Gupta, Jain, & Bhola, 2007).

Human skin function

Because of its distinct structure and composition, skin is vulnerable to a variety of environmental stresses, including humidity, pH, and temperature (Schommer & Gallo, 2013). Thermal regulation, healing function, and barrier function are three fundamental parameters adopted by skin that result in considerable human body homeostasis and protection. Skin serves as first-line defense and has evolved to impede the xenobiotics and minimize the water loss. Thus, it poses low permeability toward the penetration of external molecules (Kassem & El-Alim, 2021). The lipid-rich matrix in upper strata is responsible for barrier function. The stratum corneum (SC) layer generally comprises cholesterol, ceramides, and fatty acids that are arranged in multi-lamellar bilayer. The lipid layer covalently binds to adjacent corneocyte for maintaining the barrier properties (Prausnitz & Langer, 2008; Prausnitz, Mitragotri, & Langer, 2004). Moreover, due to its flexibility, the barrier prevents skin from the external mechanical forces. The skin regulates the body temperature through two possible mechanisms: the blood flow and the sweat. Furthermore, physical properties (desquamation, pH) and metabolic enzymes residing in the interstitial gaps of the viable epidermis and hair follicles also influence the skin's protection, thus providing a hostile environment for foreign agents (Prausnitz et al., 2004; Sala, Diab, Elaissari, & Fessi, 2018). The defensive characteristic of the skin relies on the immunocompetent cells (Langerhans cells, Dendritic cells), melanocytes, hair follicles, sebaceous glands (fatty acids and lysozymes production), and natural feature of the skin (low pH, peeling, microflora) (Sala et al., 2018).

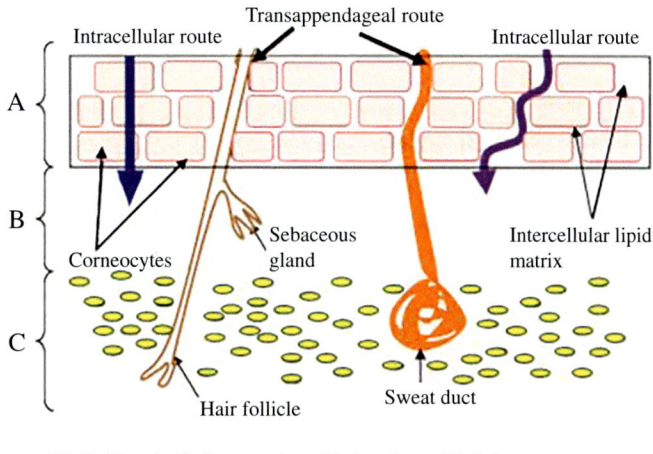

(A) Epidermis (S. Corneum)   (B) Dermis   (C) Subcutaneous layer

FIG. 6.2

Pathways of skin penetration. *Reproduced with permission from Nafisi, S., & Maibach, H. I. (2018). Skin penetration of nanoparticles. In Emerging nanotechnologies in immunology: The design, applications and toxicology of nanopharmaceuticals and nanovaccines (pp. 47–88). Elsevier. https://doi.org/10.1016/B978-0-323-40016-9.00003-8.*

Pathways of skin penetration

The phenomenon of percutaneous absorption of active compounds via skin occurs through two different pathways: transepidermal (intercellular or intracellular) and transappengageal (hair follicles, sebaceous glands, sweat glands) pathways (Fig. 6.2). The SC is primarily responsible for serving as a physical barrier against invading particles. The SC is considered as rate-limiting step for the delivery of therapeutic agents through skin. Indeed, the skin barrier takes 14 days for complete renewal in healthy people (Sala et al., 2018).

Transepidermal pathway

Transepidermal pathway encompasses intercellular and intracellular (transcellular) pathways. Drug molecule can diffuse either through intercellular or intracellular routes. In intercellular pathway, molecules transport between corneocytes cells and SC (Zhai & Zhai, 2014). Earlier, studies evidenced that intercellular lipids are the major obstacle in permeability of molecules through skin. Since, these cells are not assembled parallel to each other, that allows carrier to permeate through the lipid bilayer. The SC consists of brick (non-living corneocytes) and mortar (lipid mixture). Drug particles follow the constraint pathway around the corneocytes, the mechanism shows the transportation of hydrophilic molecules through the lipid head groups and lipophilic molecules penetrate through the lipid tails by forming nanometric spaces

(Prausnitz & Langer, 2008). The diffusion rate is dependent on the type of penetrating particle, volume, size, weight solubility, lipophilicity, and hydrogen-bonding ability. It is assumed that the particle exceeding 600 nm are not able to transport via skin. Whereas, the particle size less than 300 nm is desirable for the penetration through intercellular pathway (Kotla et al., 2017).

Transcellular pathway involves transverse substances intracellularly through corneocytes. Drugs diffusing by this pathway have to overcome various barriers in skin structure: hydrophilic (interior of cells) and lipophilic cell membrane, hydrophilic cellular keratin content, phospholidic cell barrier. Despite the penetrating particle by this pathway uses the imperfection in corneocytes that form gaps consisting of water. This route predominately favors the delivery of hydrophilic molecules (Ng & Lau, 2015).

Transappendageal pathway

The transappendageal pathway or "shunt" route permeates the drug molecules through the skin appendages via sweat glands and across the hair follicles and sebaceous glands, by passing through the SC (Bamba & Wepierre, 1993; Hueber, Wepierre, & Schaefer, 1992). In this pathway, hair follicle plays a main role due to its higher follicular distribution over the skin. However, appendages occupy small fraction (0.1%) of total skin area that is generally considered less significant to the overall permeation (Dąbrowska et al., 2018; Hueber et al., 1992). Several studies have demonstrated the importance of this pathway in molecule penetration through the skin. The follicular number, follicular volume, and opening diameter have significant impact on the extent of permeation in skin (Iqbal, Ali, & Baboota, 2018). Furthermore, due to the broad opening of hair follicles and the presence of sebum in the pilosebaceous gland duct, transit through hair follicles is a potential appendageal route that influences the successful usage of lipophilic drugs (Hwa, Bauer, & Cohen, 2011).

Transdermal drug delivery (TDD)

The skin is considered as an excellent site for the treatment of various dermatological disorders. People used to apply pharmaceutically active compounds externally in ancient times, and now there are many topical treatments available to treat local skin diseases. The dermal route was not previously found to be efficient way for delivering drugs to systemic circulation (Zhou et al., 2018). The success of TDS, on the other hand, has sparked interest in drug development to improve the treatment of skin clinical manifestations. TDD is an alternate means of delivering drugs to the systemic circulation through the skin, with benefits over traditional methods such as oral administration or needles (Arora, Prausnitz, & Mitragotri, 2008). In modern era, the TDD is initiated by the commercialization of scopolamine for the motion sickness.

TDD has been extensively researched, and a variety of drugs are available for clinical applications related to skin (Prausnitz & Langer, 2008).

Nowadays, approximately 70% of the drug candidates which are administered orally fail to show desired efficacy. TDD was founded with the aim of improving efficacy of such drugs (Marwah, Garg, Goyal, & Rath, 2016). As the skin has large surface area and has the potential for the topical applications. TDD presents unique potential for the delivery of drugs and vaccines that cannot be administered via the oral or parenteral routes. TDD enables to overcome first-pass hepatic metabolism, ensures sustain release of drug, and prolongs the residence time which is important parameter for drugs short with half-lives, reduces the side effects, and avoids fluctuation of the blood concentration. Despite these advantages, it is preferable route of administration of drugs due to the patient compliance, reduced side effects, the better clinical outcomes, and the non-invasive nature (Arora et al., 2008).

The ongoing demand for transdermal delivery derives from the advantages of transdermal route in comparison to other routes of drug administration. Despite its advantages, due to the nature of SC, which is highly impermeable and acts as a barrier for invaders from penetrating the skin, TDD's applications are limited to a broader range of drugs suited to administration by this route. To adopt this approach, the diffusional resistance produced by the membrane must be significant enough that the daily drug dose supplied from an acceptable size patch should be less than 10 mg or a few hundred Dalton, and the octanol-water partition coefficients favor lipids strongly (Benson, 2005; Prausnitz & Langer, 2008). Transdermal delivery of peptides and macromolecules, and hydrophilic drugs has posed significant hurdles, limiting the use of potent drugs by this administrative route (Naik, Kalia, & Guy, 2000). Therefore, there is a need to modify the method and device to deliver hydrophilic and high molecular weight drugs in controlled and reproducible manner.

Researchers are devoted to adopting important strategies in an attempt to improve transdermal delivery in order to overcome the impermeability of intact human skin (Lavon & Kost, 2004). These technologies include passive and active penetration enhancement and strategies to bypass the SC (Arora et al., 2008; Benson, 2005). Physical enhancement strategies could facilitate the transdermal delivery such as iontophoresis, electroporation, photochemical waves, and microneedle array.

Strategies to overcome skin barriers

One of the difficulties in developing the TDD system is overcoming the skin's barrier characteristics. Many studies have started to examine the most effective methods for maximizing the use of TDD systems for skin-impermeable drug candidates via passive or active methods. These strategies are based on two factors: (i) to enhance the skin's permeability and (ii) to provide driving force for drug transdermal applications (Alexander et al., 2012).

Passive penetration enhancement techniques

Passive penetration enhancement technique entails the manipulation of formulation and drug carrier to enhance their flux across the skin (Zhang, Tsai, Ramezanli, & Michniak-Kohn, 2013). Chemical enhancers, which raise the diffusion coefficient (increases diffusivity) and improve the solubility of pharmaceuticals in the skin, are one of the easiest ways used to optimize the TDS systems. Some more innovative TDD systems include super-loaded formulations (Naik et al., 2000; Zhao, Liu, Zhang, & Li, 2006), vesicular systems, and microemulsion. The chemical enhancers such as sulfoxides, fatty acids, and alcohols can be either used individually or in the form of synergistic mixture which causes temporary disorder in intercellular lipid matrix in the SC and reduces the skin resistance (Williams & Barry, 2012). Indeed, synergistic mixtures have proved to be potent for the skin permeabilization as compared to single chemicals. The advancement in this technology to improve TDD includes eutectic mixtures, vesicles, nanoparticles, and complexes (Alexander et al., 2012; Naik et al., 2000). As a result, it has been postulated that molecular simulation approach and its relevance include molecule penetration over the skin's lipophilic structure. Thus, the mechanism disrupts the lipid in the skin which facilitates the solubility and diffusion of hydrophilic therapeutic agents.

Microemulsion

Microemulsion is a dispersion consisting of oil, surfactant, co-surfactant, and water in appropriate ratios. It is thermodynamically stable, optically transparent, isotropic liquid solution. Moreover, it has potential characteristics due to its high solubilization of both lipophilic and hydrophilic drugs, ease of preparation, low viscosity, high drug loading capacity, and small droplet size ranging 10–100 nm, and thus has drawn the attention for the utilization of vehicle in drug delivery through different administrative routes. Recently, these versatile systems have gained the attention for transdermal administration (El Maghraby, 2008, 2010). It has proven to increase the systemic delivery of drugs via various methods: (i) structural composition favors to encapsulate greater amount of drug than conventional gels, creams, and lotions, (ii) the constituents and composition modify the diffusional barrier of the skin, and (iii) improve the thermodynamic activity of drug, thus resulting in the portioning into the skin (Malakar, Sen, Nayak, & Sen, 2011). Research work by Chudasama and co-worker reveals that the microemulsion is a promising vehicle for the TDD. The oil in water gel loaded with 1% itraconazole was formulated and used against fungal infections. The in-vitro results concluded that developed microemulsion have great potential for the delivery of itraconazole through transdermal route (Chudasama, Patel, Nivsarkar, Vasu, & Shishoo, 2011). Liu et al. evaluated terpene microemulsion consisting of polysorbate 80, limonene, ethanol, and aqueous phase entrapped curcumin. The study showed enhanced transdermal delivery of curcumin by reducing the diffusional barrier of the stratum corneum and the viscosity (Liu, Chang, & Hung, 2011). Likewise, microemulsion was applied to deliver several drugs such as fluconazole (Patel, Patel, Parikh, Solanki, & Patel, 2009), methicillin (Chhibber, Wadhwa, Chadha, Sharma, & Katare, 2015), luliconazole (Kansagra & Mallick, 2016), and quercetin (Caddeo et al., 2014) for the effective treatment of skin diseases.

Superloaded formulation

To achieve greater penetration rates across the skin, the degree of saturation is optimized for formulations and penetration enhancers. A reliable method to improve the drug permeation via skin is increasing the concentration of dissolved drug which retains the integrity of SC. Consequently, the relationship between drug permeability and drug content results in enhancement in drug flux which ultimately improves the thermodynamic activity. The excess amount of drug dissolved at its equilibrium attains the state of supersaturation which possesses high tendency to crystallize. As a result, crystallization is observed in patches with supersaturated drug concentration (Latsch, Selzer, Fink, & Kreuter, 2003). A wide range of supersaturated formulations is designed to improve drug delivery and skin permeation (Moser, Kriwet, Kalia, & Guy, 2001). There is a broad spectrum of polymers that can be utilized for enhancing the penetration across the skin and maintaining the supersaturation. Numerous polymers such as hydroxypropyl methylcellulose, hydroxypropyl cellulose, polyvinyl alcohol, polyvinyl pyrrolidone, polyethylene glycol or Eudragit have anti-nucleant properties (Valenta & Auner, 2004). Research conducted by Ghosh and co-worker used Propylene glycol (PG) as a co-solvent, and D-α-tocopheryl polyethylene glycol 1000 succinate (TPGS) as non-ionic solubilizers to enhance ibuprofen solubility. Polymeric stabilizers such as hydroxypropyl methylcellulose (HPMC) and polyvinylpyrrolidone (PVP) were used to inhibit the crystallization. In presence of PG, significant growth of ibuprofen crystals was observed. TPGS and poloxamer 407 when added into formulations, crystal growth appeared but had comparatively low size than PG. These findings concluded that TPGS and poloxamer 407 formed supersaturated solution and high permeation flux was observed by in-vitro permeation studies (Ghosh & Michniak-Kohn, 2012; Pham & Cho, 2017).

Vesicles

Vesicles are water-filled colloidal particles with an exterior membrane made up of amphiphilic molecules organized in a bilayer structure that can entrap both hydrophilic and lipophilic drug molecules (Honeywell-Nguyen & Bouwstra, 2005). The main objective of using vesicles in TDS systems is to function as a drug carrier for transporting entrapped drug molecules through the skin, as well as a penetration enhancer due to their nature. Furthermore, this vesicular carrier acts as a depot for the sustained release of therapeutically active compounds and provides a rate-limiting barrier for systemic absorption, hence providing controlled TDS system (Nair, 2019; Rajan, Jose, Mukund, & Vasudevan, 2011). Vesicle formulations are classified into two categories: (i) Rigid vesicles (Liposomes, niosomes) and (ii) elastic or ultra-deformable vesicles (transferosomes, ethosomes). According to literature, it has been concluded that rigid vesicles are not suitable for transdermal delivery as they cannot reach deeper into the skin and are likely to be trapped in the upper layer of stratum corneum (Cevc, 1997). Therefore, unique class of liposome "transferosomes" has paved the way to minimize the problems concomitant with conventional vesicles.

Conventional liposomes

Liposomes are microparticulate lipoidal vesicles and are phospholipid bilayer membranes that generate sphere containing hollow hydrophilic compartments. As a result, hydrophobic drugs can be carried in the envelope between the bilayers, while hydrophilic drugs can be entrapped in the core (Alavi, Karimi, & Safaei, 2017; Sharma & Sharma, 1997). Owing to their properties, they have achieved an immense interest in TDD. These systems can penetrate through the skin. The mechanism entails the permeation of vesicle carrier either through the fusion of lipids present in SC or by squeezing on the SC and facilitating the lipid molecule to penetrate. In addition, sebaceous gland could be a port of entry for liposomes which serves as reservoir for sustained drug release (Barua & Mitragotri, 2014). Liposomes are commonly prepared via thin-film hydration, reverse phase evaporation, and solvent injection techniques. Thin-film hydration method is the conventional technique used to design liposomes for TDD. In this method, lipids are dispersed in an organic solvent, followed by evaporation, thus leading to the dry thin film. Recently, liposomes are widely utilized in the treatment of skin diseases such as psoriasis, acne, atopic dermatitis, mycoses, etc. Through literature, it has been evidenced that the encapsulation of tretinoin in liposomal carrier has remarkable effect in the treatment of acne without any adverse symptoms (Patel, Misra, & Marfatia, 2000). Liposomes are considered effective for drug delivery since they are biodegradable, non-toxic, and capable of encapsulating both hydrophilic and lipophilic compounds (Hua, 2015). However, they are not suitable carriers for the drug delivery deeper into the skin layer.

Niosomes

Niosomes are non-ionic surfactants that self-assemble from a lamellar structure of amphiphilic molecules encircled by an aqueous compartment. The surfactants have both hydrophilic head and hydrophobic tail that aggregates or is self-assembled in various shapes (Auda, Fathalla, Fetih, El-Badry, & Shakeel, 2016). Numerous surfactants such as sorbitan esters and their derivatives, polyoxyethylene, polyglycerol, sugar or crown ether based, and additive excipients like cholesterol have been used. When niosomes are applied to the skin, they increase the permeation of drug across the skin that can be sufficiently explained via following mechanism. Because of the interaction between the vesicle and the skin, the non-ionic surfactant changes the barrier characteristics of SC and functions as a penetration enhancer. As the niosomes penetrate into the lipid matrix of skin SC, the vesicles increase the permeation and facilitate TDD. Moreover, reduced transepidermal water loss promotes SC hydration, which intercalates the niosomes in the gaps generated in its cellular structure, improving the drug's thermodynamic activity at the interface and, as a result, improving drug permeation. Niosomes are of prime interest for researchers for various skin-related diseases such as alopecia, skin cancer, acne, and psoriasis (Ghanbarzadeh, Khorrami, & Arami, 2015; Muzzalupo & Tavano, 2015).

Transferosomes

To further enhance the permeation capabilities of conventional vesicles, the liposomes are chemically and structurally modified to design special vesicular system. Transferosomes are a modified form of conventional vesicular carriers that contain an aqueous core surrounded by a lipid bilayer and an edge activator (EA) (Rai, Pandey, & Rai, 2017). The ultra-flexible, self-regulating, and self-optimizing vesicular carrier is made possible by the membrane flexibility, hydrophilicity, and ability to maintain the vesicle's integrity (Jadupati, Amites, & Kumar, 2012). Owing to their elastic nature, transferosomes can deform and squeeze themselves without any measurable loss through narrow pores or constriction of the skin that is comparatively smaller than the vesicle size. Contrary to conventional liposomes made from phosphatidylcholine or phospholipids, the modified transferosomes comprised of phospholipids and an edge activator (single-chain surfactant) (Opatha, Titapiwatanakun, & Chutoprapat, 2020). Typical edge activators used in the synthesis of transferosomes are sodium cholate, sodium deoxycholate, Span 60, 284 Span 65, Span 80, etc. (Jain et al., 2017). Edge activators have the ability to destabilize the membrane in an exceptional manner to increase the deformability in vesicle membrane, and when they are mixed with lipids in appropriate ratio, they result in the formation of deformable, ultra-flexible vesicular carrier with higher permeation capability. Hence, transferosomes are capable to overcome the obstacle faced by conventional liposomes and penetrate pores that are much smaller than their own diameters. Furthermore, they can effectively penetrate skin when applied under non-occlusive conditions, which necessitates the establishment of a transepidermal osmotic gradient via the skin (Cevc, 2003).

Ethosomes

Ethosomes are unique non-invasive vesicular system and deliver drugs to the deep skin layer or systemic circulation. These are soft, and malleable and exhibit superior delivery of active compounds. They mainly composed of phospholipids (phosphatidylcholine, phosphatidylserine) cholesterol, water, and high concentration of ethanol (Bodade, Shaikh, Kamble, & Chaudhari, 2013). It is thought that ethanol in ethosomes is the factor that disrupts the assembled bilayer of skin, and when incorporated in vesicle membrane, it improves the penetration ability of stratum corneum. Due to the presence of ethanol in high concentration, the lipid membrane is packed less tightly than conventional liposomes. It has been suggested that the ethanolic concentration is also responsible for deeper permeation and improves the drug's distribution ability in the SC (Satyam, Shivani, & Garima, 2010). The major advantage of ethosomal formulation is to deliver either hydrophilic or lipophilic drug under both occlusive and non-occlusive conditions (Carter, Narasimhan, & Wang, 2019).

Inclusion complexes

Inclusion complexes are structured molecular cages (host) that process an internal cavity where another active drug molecule (guest) can be entrapped. The most widely explored agents to tailor the inclusion complexes are cyclodextrins (Cal & Centkowska, 2008; Lopez, Collett, & Bentley, 2000). Cyclodextrines with a hydrophobic cavity and a hydrophilic exterior can generate inclusion complexes due to their structural features (Karande & Mitragotri, 2009). The cyclodextrin inclusion complexes are supposed to enhance the drug stability by preventing oxidation, hydrolysis, and degradation (Másson, Loftsson, Másson, & Stefánsson, 1999). Research shows that cyclodextrins are interesting TDD carriers for increasing the permeability via two mechanistic pathways. Cyclodextrins improve the drug's solubility, producing greater driving force to cross stratum corneum. The skin barrier properties are diminished by the interaction of cyclodextrins with lipid bilayer and significantly reduce the drug irritation (Lopez et al., 2000). Jain and co-worker reported that celecoxib entrapped in cyclodextrin complex enhances the skin permeation properties. Results showed that the system has remarkable in-vivo anti-inflammatory effects with sustained and prolonged release (Jain et al., 2007).

Eutectic method

Eutectic mixture is the combination of two or more components, which do not interact to form a new chemical moiety but at particular ratios inhibits the crystallization process for one another, resulting in lower melting point than individual component (Patel, Trivedi, Bhandari, & Shah, 2011). The process transforms the solid drugs into the oily state by depressing the melting point of the chemical entity below the ambient temperature (25°C). Literature reports that the melting point of drug is inversely proportional to its solubility and lipophilicity of lipids. Hence, decreasing the melting point of drug through the formation of eutectic mixture enhances its lipid solubility and results in increase of transdermal flux. It contributes to increase the transdermal flux in two ways: (i) to make the low melting point mixture that improves the active drug candidate's partition over the SC layer and (ii) the direct disturbance of the SC layer which improves the permeability of skin (Alexander et al., 2012). Yuan and co-worker reported the binary eutectic mixture of ibuprofen and ketoprofen that showed the depression in the melting point and disrupting crystallinity of the drug. The eutectic mixture increases the steady-state flux by lowering melting point and increases the solubility of the drug in lipid bilayer of skin. Therefore, it is suggested to have potential application for transdermal drug delivery system (Yuan & Capomacchia, 2010).

Coacervation effect

Coacervation (ion-pairing) is the separation of binary polyion mixture into two liquid phases; the denser phase rich in polyions, coacervate phase, and a dilute equilibrium phase. Coacervation phenomenon is based on the aggregation of oppositely charged ions through electrostatic interaction and effect of entropy tends to disperse them under specific conditions such as ionic strength, pH, molecular weight, and ion ratio. Ion-pair transport across the membrane has been studied as a driving force for permeation of ionizable drugs (Green & Hadgraft, 1987; Stott, Williams, & Barry, 1996).

Nanoparticles

Nanoparticles (NPs) have been extensively studied for the rapid development of nanomedicine to treat diseases. The pharmaceutical studies exploited NPs to improve therapeutic efficacy of numerous drugs. They are used as vehicles for (i) efficient delivery of poorly water-soluble drugs, (ii) targeted delivery at the site of action in a specified manner, (iii) transcytosis of drug molecules across the biological barriers, and (iv) co-delivery of drugs to improve the therapeutic efficacy and for combination therapy. Additionally, these systems have controlled morphology and chemical composition and allow the incorporation of several drugs (Farokhzad & Langer, 2009). Interestingly, NPs have gained attraction of scientists as transdermal delivery carriers because of their advantages over other TDD systems for the treatment of skin disorders. To design the NPs formulation for the skin delivery, it is viable to understand the interaction between the skin and NPS permeation mechanism into the skin. It is regarded that transdermal application of NPs formulations leads to the absorption of drug molecules through the transepidermal and transappendageal pathways. NPs can remain intact into the skin retaining their properties without

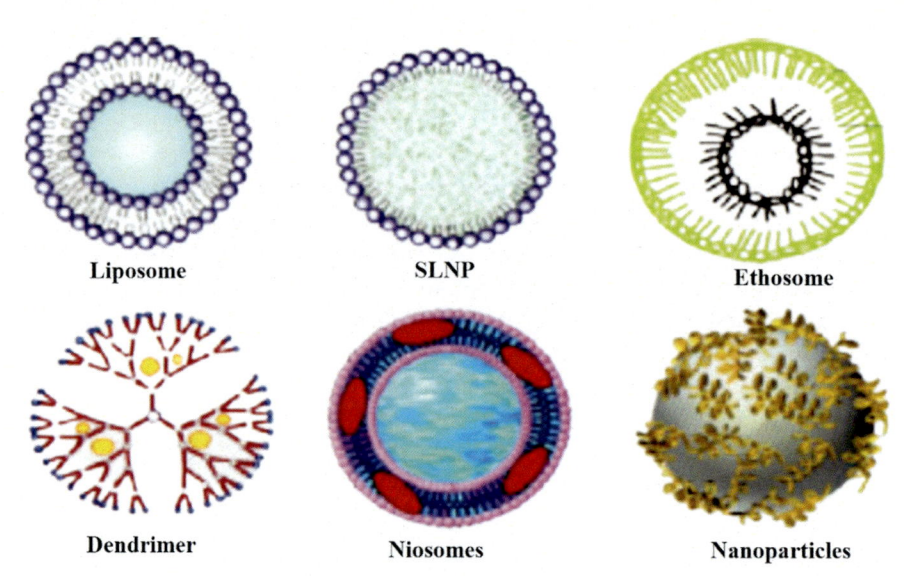

FIG. 6.3

Representative examples of nano-drug delivery systems for transdermal drug delivery. *From Angajala, G., & Subashini, R. (2014). Transdermal nanocarriers: New challenges and prospectives in the treatment of diabetes mellitus. Journal of Chemistry and Applied Biochemistry, 1(2), 1–10.*

degradation or they can degrade near the skin and the therapeutic active agent can permeate into the skin layers. The skin interaction with the NPs relies on few factors such as size, surface charge, properties of nanomaterial, encapsulating efficiency, lamellarity, and mode of application (Desai, Patlolla, & Singh, 2010). Although, the interaction of skin and other organisms is well established through blood vessels and lymph vessels, however, they do not promote the transport of NPs across the skin (Chen et al., 2020). Nonetheless, there is convincing evidence that irrespective of nanomaterial used, most particles do not cross the SC and transappendageal route appears to be the dominant pathway of NPs entry into skin.

The NPs have the potential to treat several skin problems including fungal infections, inflammatory disorders, rosacea, acne or atopic dermatitis skin, etc. (Papakostas, Rancan, Sterry, Blume-Peytavi, & Vogt, 2011). The NPs employed for TDD applications include metal nanoparticles, polymeric nanoparticles, dendrimers, nano emulsion to mention a few (Goyal, Macri, Kaplan, & Kohn, 2016), and are summarized in Fig. 6.3.

Polymeric nanoparticle

Polymeric nanocarriers are colloidal particles prepared from polymeric matrix like natural, synthetic, or semi-synthetic polymers with size range of 10–1000 nm. Polymer-based nanocarriers are widely used for drug delivery due to their biocompatibility and biodegradability and their efficient and targeted delivery of the pharmaceutically active agent to the diseased site. In the area of TDD, the polymeric nanoparticles attain the attraction due to the fact that they can overcome the obstacles concomitant with other lipid systems, as they prevent the drug denaturation and degradation, increase the concentration gradient for the better enhancement of percutaneous drugs (Zhou et al., 2018). The following are the mechanisms that aid drug absorption into skin through polymeric nanoparticles:

(i) Polymeric nanoparticle as a whole is quite large to spread passively deeper into the skin through SC directly. As a result, they are assembled on the skin's surface or in hair follicles, then transported to a drug reservoir for sustained drug release from the NPs. Because the drug diffuses at a concentration gradient in the active layer of the skin, this results in a higher concentration of drug (Watkinson, Bunge, Hadgraft, & Lane, 2013). (ii) The permeability of NPs predominantly depends on their physiochemical properties and skin's integrity, so any appropriate change observed in skin integrity could promote the percutaneous penetration of drug (Kimura, Kawano, Todo, Ikarashi, & Sugibayashi, 2012). (iii) Surface charge is an important factor affecting the permeability. The charge of NPs with highly cationic constituents tends to interact more strongly with the negatively charged skin, resulting in enhanced drug release from polymeric NPs (McConnell et al., 2016). (iv) The hair follicles facilitate the permeation of polymeric NPs to enter the skin following the follicular pathway (Desai et al., 2013). Nowadays, both natural and synthetic polymers are utilized in the preparation of polymeric NPS for TDD.

Natural polymeric nanoparticles

Natural polymeric nanoparticles are based on natural polymers such as chitosan, alginate, albumin, gelatin, etc. These polymers are extracted from various natural sources followed by purification. Chitosan, the N-deacetylated derivatives of chitin have been more frequently utilized in the TDD. It is a biodegradable, and cationic polymer comprised mainly of glucosamine units. Its antioxidant, anti-inflammatory, and antimicrobial properties make chitosan a suitable vehicle for delivering therapeutics to treat dermatological disorders (Wang, Du, Fan, Liu, & Wang, 2003). Moreover, at physiological pH, the primary amine groups of chitosan are protonated, and therefore chitosan is positively charged. The positive charge has tendency to form nanoparticles through cross-linking with anionic species in solution, and negative-charged drugs can be encapsulated by electrostatic interaction, therefore, facilitating the cellular internalization of drug-containing chitosan nanoparticles (Zhang et al., 2013). Tolentino et al. reported the polymeric nanoparticles of chitosan and hyaluronic acid for the delivery of clindamycin to the pilosebaceous structure to treat acne vulgaris. The results reveal that the targeting and penetrating potentials of clindamycin were improved, showing these nanocarriers as promising alternative to treat acne vulgaris (Tolentino, Pereira, Cunha-Filho, Gratieri, & Gelfuso, 2021).

Synthetic polymeric nanoparticles

Natural polymers are limited in their use in the preparation of NPs due to inconsistency and generally lack purity, compromising the NPs' repeatability and controlled release pattern. To circumvent the issue associated with natural polymers, synthetic polymers have been predominantly used in the synthesis of polymeric nanoparticles, which assures batch-to-batch reproducibility, better purity, and the drug release pattern can also be tuned. Most common methods employed for the synthesis of synthetic nanoparticles include solvent evaporation/extraction, nanoprecipitation, dispersion polymerization, and emulsification-diffusion (Escobar-Chávez et al., 2012). Synthetic NPs tend to accumulate in hair follicles after being used to deliver hydrophilic/lipophilic drugs. Drugs can either be adsorbed, encapsulated, or dispersed within the polymer matrix. Synthetic polymers used in drug delivery are either biodegradable or non-degradable polymers such as polylactide, poly(lactide co-glycolide), poly(ε-caprolactone) or poly acrylates, poly(methyl methacrylate), respectively. PLGA is the most common synthetic polymer capable of encapsulating lipophilic drugs, and it is widely used in both non-medicine devices and FDA-approved formulations due to its biocompatibility. Lactic acid and glycolic acid are the final degradation products of PLGA which eliminate from the body during the citric cycle (Panyam & Labhasetwar, 2003). A recent study conducted on psoriasis-like mouse model compared curcumin-loaded PLGA NP (50–150 nm) to curcumin hydrogel and observed that the former had apronounced performance. Tyrosine-derived polymeric nanospheres (TyroSpheres) is another degradable polymer that has been explored for TDDS. It is ABA-type triblock copolymer, which comprises hydrophilic poly(ethylene glycol) hydrophobic oligomers of suberic acid desaminotyrosyl-tyrosine esters. In an aqueous medium, TyroSpheres assemble in a hollow nanosphere exhibiting a hydrodynamic diameter of around 70 nm. TyroSpheres strongly associate with lipophilic drugs such as paclitaxel but do not form complexes with hydrophilic molecules (Sheihet, Piotrowska, Dubin, Kohn, & Devore, 2007). Due to this, TyroSpheres can be loaded with various lipophilic drugs for treating diseases like skin cancer and psoriasis (Sheihet, Dubin, Devore, & Kohn, 2005).

Dendrimers

Dendrimers are hyper-branched synthetic polymers and are highly applicable in TDDS due to their biocompatibility, monodispersity, and narrow hydrodynamic diameter. Dendrimers improve the transdermal flux by perturbing the skin layer to penetrate the molecule in deeper skin. Therefore, polymeric enhancers with amphiphilic properties have attracted increasing interest (Lee & Larson, 2008). PAMAM dendrimers can help hydrophobic drugs become more water-soluble and stable. These macromolecules, which have a hydrophilic envelope and a hydrophobic core, are likely to be excellent penetration enhancers because they meet the structural requirements of polymeric transdermal enhancers (Cheng, Xu, Ma, & Xu, 2008). Recently, several researchers have investigated the potential of dendrimers in the transdermal route of drug administration to treat tumors such as melanoma or squamous carcinoma (Dianzani et al., 2014).

Metal nanoparticles

The metal nanoparticles such as iron, gold, silver, titanium, and zinc oxides can be beneficial for skin targeting and are used to treat dermatological disorders. Drug molecules are loaded or modified on the surfaces of these particles (Baroli et al., 2007). Their diverse characteristics including stability, biocompatibility, customizable surface properties, and the capacity to be adjusted in a wide variety of sizes have been extensively studied. These rigid NPs encompass the passive penetration of NPs through skin SC lipid bilayer and hair follicles (Desai et al., 2010). According to Huang and colleagues, the interaction of Au-NPs with the skin barrier increases skin permeability and efficiently induces percutaneous absorption of the co-administered proteins (Huang et al., 2010). Gopukumar et al. studied the potential of nanosilver patch for the development of desirable wound dressings, concluding that AgNPs with high antioxidant, antibacterial, and cytotoxic activities are essential for the treatment of wounds via a radical scavenging mechanism (Gopukumar, Sana-Fathima, Alexander, Alex, & Praseetha, 2016). Rao et al. evaluated the potential of epirubicin (EPI) functionalized superparamagnetic iron-oxide nanoparticles (SPION). The covalent fabrication of the SPION results in EPI-SPION, a potential drug delivery carrier that follows magnetism for the targeted transdermal chemotherapy of skin tumors. In vitro transdermal studies investigated that the EPI-SPION can penetrate deep inside the skin driven by an external magnetic field. The magnetic-field-assisted SPION transdermal vector can circumvent the stratum corneum via follicular pathways (Rao et al., 2015).

Active permeation enhancement techniques

Active penetration enhancement techniques use physical enhancement methods to increase the transdermal flux (Fig. 6.4). These methods are discussed one by one in the coming section.

Iontophoresis

Iontophoresis is a method of introducing ions via the skin while using a tiny electric current to improve transdermal delivery (Priya, Rashmi, & Bozena,

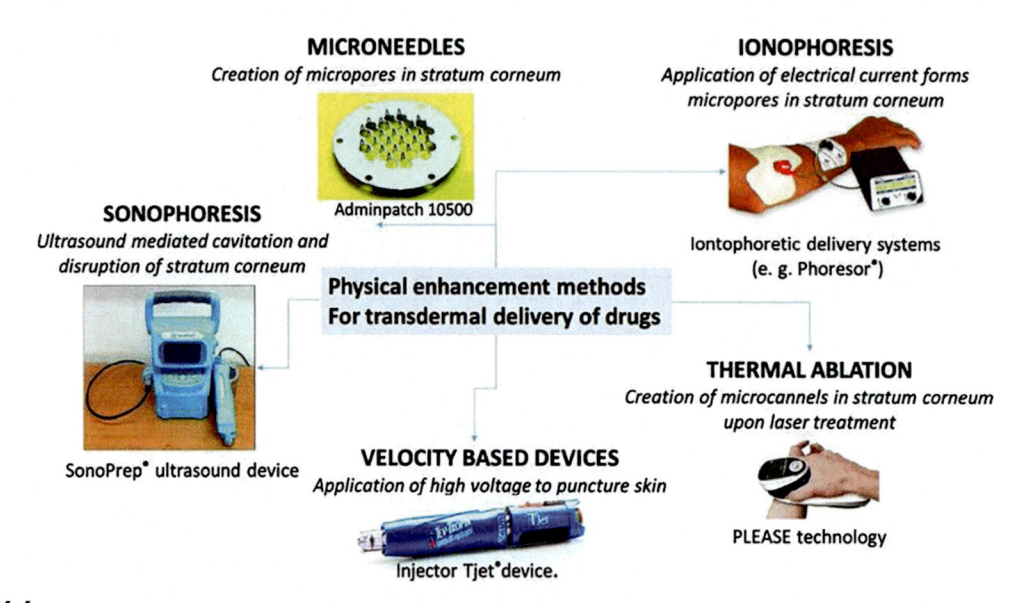

MICRONEEDLES
Creation of micropores in stratum corneum

Adminpatch 10500

IONOPHORESIS
Application of electrical current forms micropores in stratum corneum

Iontophoretic delivery systems
(e. g. Phoresor®)

SONOPHORESIS
Ultrasound mediated cavitation and disruption of stratum corneum

Physical enhancement methods For transdermal delivery of drugs

THERMAL ABLATION
Creation of microcannels in stratum corneum upon laser treatment

SonoPrep® ultrasound device

VELOCITY BASED DEVICES
Application of high voltage to puncture skin

Injector Tjet®device.

PLEASE technology

FIG. 6.4

Physical methods for enhancement of transdermal drug delivery. *Reproduced with permission from Szunerits, S., & Boukherroub, R. (2018). Heat: A highly efficient skin enhancer for transdermal drug delivery.* Frontiers in Bioengineering and Biotechnology, 6. *https://doi. org/10.3389/fbioe.2018.00015 under the terms of the Creative Commons Attribution License (CC BY).*

2006). Along with hydrophilic drugs, macromolecules including peptide and oligonucleotide may be effectively delivered to intact skin through iontophoresis. Transdermal iontophoresis achieves the delivery of several compounds in programmed manner (Costello & Jeske, 1995; Yan, Li, & Higuchi, 2005). Active therapeutic agents are delivered to the systemic circulation via blood capillaries from the drug solution/device entering the epidermis. Transdermal iontophoresis enhances the drugs skin permeation by the following mechanism: (i) the ion-electric field interaction provides an additional force that serves as a driving force to penetrate ions through the skin; (ii) flow of electric current; and (iii) electroosmosis produces motion of solvent that carries ionic or neutral charges with solvent stream (Pikal, 2001). An anodal iontophoretic device is made up of a current source, an electronic device to control the current, drug/ion in solution (positively charged), and an anodic and cathodic reservoir system in general (with the anode and cathode electrodes). The anode reservoir system contains the positively charged drugs/ions at the desired site of application, whereas the cathode is placed on a different site on the skin. When electric current is applied, all cations including the positively charged drug, move away from the anode and into the skin. Negatively charged ions in the body flow spontaneously from the body to the donor reservoir (Priya et al., 2006). The iontophoresis is applicable in TDD of anesthetic, retinoids, anti-inflammatory drugs, etc.

Sonophoresis

Sonophoresis is non-invasive ultrasound mediated technique to enhance the skin permeability of various pharmacologically active molecules including macromolecules and hydrophilic drugs under low-frequency sonophoresis. This technique operates at frequencies in the range of 20 kHz–16 MHz and intensities up to 14 W/cm^2 to enhance skin permeability. It is observed that transdermal flux is higher at low frequency (20–100 kHz) in contrast to high-frequency range (Lavon & Kost, 2004; Mitragotri, Blankschtein, & Langer, 1996). The proposed mechanism to increase the skin permeability via sonophoresis depends on two factors: (i) thermal effect and (ii) cavitation effect.

In thermal effect, when ultrasonic beam is passed through a medium, partial energy is absorbed. Tissues in the human body absorb ultrasound energy, which raises local temperatures. The rate of heat removal is dependent on the ultrasonic frequency, intensity, time of exposure, area of the ultrasound beam, and the rate of heat removal through blood flow or conduction. Consequently, the temperature of the skin rises that may enhance the permeability due to an increase in diffusivity of the skin (Merritt, Kremkau, & Hobbins, 1992; Nyborg, 2001). Cavitation process generates the cavities and creates the oscillation, expansion, and distortion of pre-existing gaseous bubbles in the solution. The violent formation and collapse of bubbles upon ultrasound propagation produces shock waves, which causes the structural alteration in the SC (Leighton, 2012; Park, Park, Seo, & Lee, 2014). These changes in the SC create channel for the facile diffusion of drug. Indeed, the safety of ultrasound application should be considerable while using this technique for TDD.

Electroporation

Electroporation is the creation of transient aqueous pathways by perturbation in lipid bilayer membranes using high voltage and short electric pulse. The electrical conductance and the permeability can be increased reversibly by many orders of magnitude depending on size. Moreover, electroporation can be achieved when the transmembrane voltage reaches a few hundred mV for electric field pulses typically of 10 µs to 100 ms duration (Prausnitz, 1999; Prausnitz, Bose, Langer, & Weaver, 1993). Electroporation is known to occur in metabolically inactive systems, such as synthetic lipid membranes and red blood cell ghosts as well as in living cells and tissues. Electroporation has proved beneficial in improving transdermal permeability of lipophilic molecules of various sizes, including proteins, peptides, oligonucleotides, and relatively tiny molecules, such as biopharmaceuticals with molecular weights more than 7 kDa (Zorec, Becker, Reberšek, Miklavčič, & Pavšelj, 2013).

Microneedles

Microneedles are needle-like structures used to improve skin permeation by providing a micron-scale channel across the lipid matrix of the skin, which acts as a point of entry for drugs, small molecules, proteins, and vaccines. Microneedles are in size of the micron range and lengths of up to 1 mm (Park, Choi, Seo, Choy, & Prausnitz, 2010; Van Der Maaden, Jiskoot, & Bouwstra, 2012). To permit TDD, these structures trigger piercing of the skin's stratum corneum. The substances used to modify the microneedles are polymers, silicon, and metals. Drug can be transported adopting various modes: (i) Drug patch applicability at the site of action by piercing the skin by solid microneedle, (ii) drug coated on microneedle and inserting them into skin for dissolution of drug within the skin, (iii) loading drug in polymeric microneedles, which on insertion into the skin releases drug in controlled manner, and (iv) injecting drugs through hollow microneedles (Saraf, Paliwal, & Saraf, 2011). Thus, this technique is an alternative approach to transport molecule through the SC which is the major obstacle for transdermal penetration.

Magnetophoresis

Magnetophoresis is a technique used to enhance the transdermal flux across the biological barrier under the influence of magnetic field. Previous work shows that magnetophoresis facilitates the transdermal delivery of benzoic acid,salbutamol sulfate, and terbutaline sulfate (Murthy, Sammeta, & Bowers, 2010). The phenomenon of magnetophoresis relies on magnetokinesis. It promotes TDD via magnetorepulsion and magnetohydrokinesis. (i) Magnetorepulsion occurs when diamagnetic drug molecules are driven away from the external magnetic field. (ii) Magnetohydrokinesis mediates drug transport through moving water/solvent across the SC under external magnetic field. The contribution of magnetohydrokinesis to transdermal drug transport may be more apparent when higher magnetic field strengths are applied or a more concentrated drug solution is used in donor compartment. In magnetophoresis, the octanol/water partition coefficient of drugs is found to increase under the influence of magnetic field(Wong, 2014).

Needleless jet injectors

A needleless jet injector is a device that delivers drugs, proteins, or peptides in the form of liquid or powder across skin at high velocity. In contrast to conventional injections, they provide needleless alternative which circumvents the undesired pain and offers patient compliance (Benson, 2005; Stachowiak et al., 2007).

Photomechanical waves

Photochemical waves are generated by high-power lasers, which have broadband (fast rise time), unipolar, compressive waves. Photochemical waves interact with cells and tissues differently from those of ultrasound, they have only positive pressure unlike ultrasound. These waves are the result of mechanical forces that excludes the cavitation generated by ultrasound (Lee, McAuliffe,

Flotte, Kollias, & Doukas, 1998). Erbium-doped yttrium aluminum garnet, Q-switched ruby, and carbon dioxide lasers have been used to increase skin permeability to drugs via ablation mechanism or lipid bilayer disruption in a process ranging from nanoseconds (100 ns) to a few microseconds (10 μs) with pressure amplitude in the hundreds of atmospheres (300–1000 bar). Skin permeability can be improved through instant evaporation of water by laser, thereby forming microchannels in skin for TDD. Four phases of reactions may occur: (i) heating (37–60°C), (ii) coagulation (60–65°C), (iii) drying (90–100°C), and (iv) vaporization (>100°C). The depth of the formed channels is a function of duration of exposure. The channel depth is limited to less than 200 μm to avoid pain and bleeding. The peak pressure, rise time, and duration of waves are governed by wavelength, pulse duration, and fluency of laser, as well as optical and mechanical properties of target material applied on the top surfaces of skin (Wong, 2014).

Combination systems

The advantages of the above-mentioned transdermal drug delivery systems have led to more innovation in this field. Researchers from several disciplines are collaborating to improve the delivery of drugs through the skin barriers. Various combinational therapies have been investigated to show synergistic effect including; chemical iontophoresis, chemical ultrasound, chemical electroporation, electroporation-iontophoresis, electroporation- ultrasound, and pressure waves chemicals. These techniques are promising alternatives to enhance the skin permeability to several drugs and therapeutically active molecules.

Application of TDDS for the treatment of skin diseases

Melanoma

The malignant mutation in the pigment-producing cell, the melanocyte, is what distinguishes a skin tumor. According to recent US statistics, melanoma accounts for roughly 1% of all skin malignant tumors, while basal cell carcinoma, which begins in the epidermis' basal layer, affects over 2.8 million people each year. However, squamous cell carcinoma that originates in the upper layer of skin is most prevalent. Moreover, cutaneous malignant melanoma represents the chronic form of skin cancer. This disease affects mostly the Caucasian population of both genders and once it becomes metastatic, it is difficult to treat. The metastatic stage increases the migration of tumor cells to the other organs. Therefore, early diagnosis of these cancers may have the chance to increase the survival rate (Domingues, Lopes, Soares, & Pópulo, 2018; Vishnubhakthula, Elupula, & Durán-Lara, 2017).

Transdermal drug delivery systems provide an alternative to treat melanoma with improved penetration as compared to the conventional treatments. Yu. et al. demonstrated that the ethosomal gel encapsulated with mitoxantrone is a promising transdermal delivery system for anti-melanoma therapy with improved permeation efficiency, cytotoxicity, anticancer effects (Yu, Du, Li, Fu, & Jin, 2015). Liao et al. explored the Epigallocatechin gallate (EGCG)-nanoethosomes, which are composed of phosphatidylcholine, EGCG, sugar esters, ethanol, and surfactant, for skin cancer treatment. The nanoethosomes loaded with docetaxel were compacted into niosomes and were used for the melanoma cell tumors treatment via transdermal route (Fig. 6.5) (Liao et al., 2016).

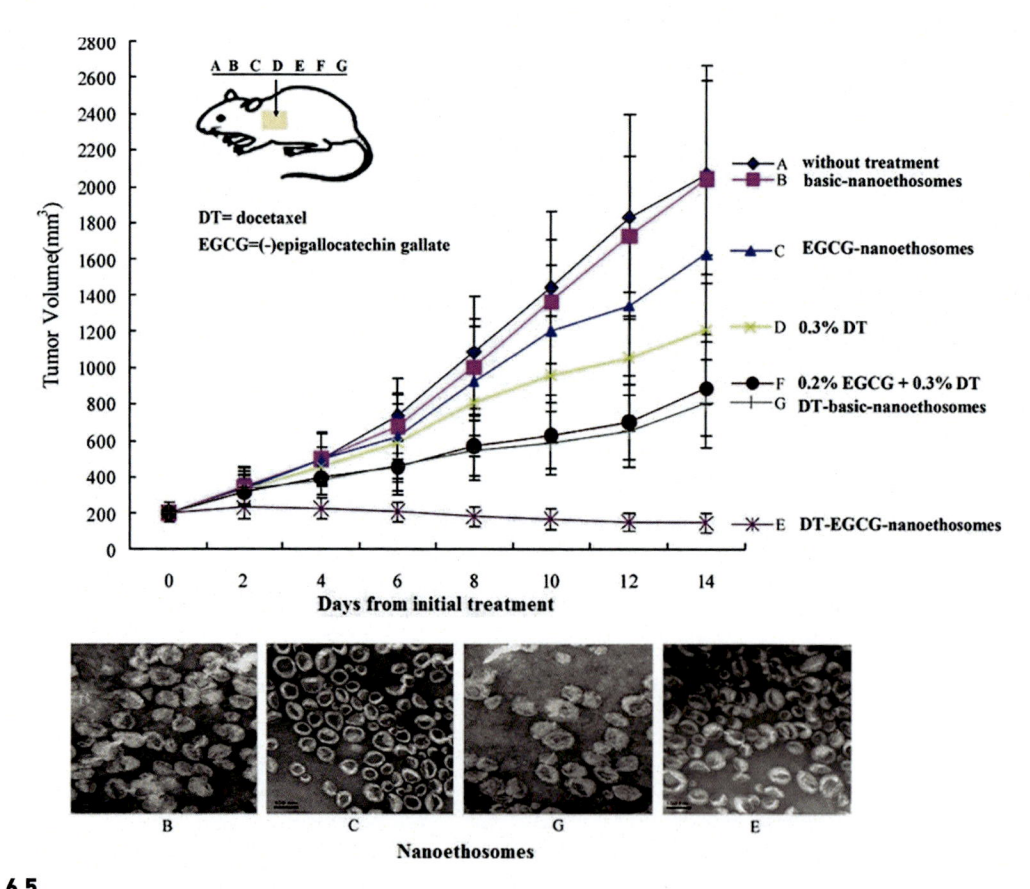

FIG. 6.5

Niosomes deliver docetaxel via transdermal route to improve the melanoma cell tumor. *From Liao, B., Ying, H., Yu, C., Fan, Z., Zhang, W., Shi, J., et al. (2016). (—)-Epigallocatechin gallate (EGCG)-nanoethosomes as a transdermal delivery system for docetaxel to treat implanted human melanoma cell tumors in mice.* International Journal of Pharmaceutics, 512(1), 22–31. https://doi.org/10.1016/j. ijpharm.2016.08.038.

Acne

Acne is the skin disorder, particularly in adolescents and young people. The root cause of acne is the pilosebaceous unit which has abnormalities in sebum production, bacterial proliferation, inflammation, and follicular epithelial desquamation. It is prevalent and leads to significant morbidity that is concomitant with scarring and has negative impact on the quality of life (Rivera, 2008; Tan, Schlosser, & Paller, 2018). Mostly used therapeutic agents for the treatment of acne are retinoid, antimicrobial agents, and hormonal drugs. Hend-Abd-Allah and co-worker proved that chitosan NPs loaded with nutraceutical can be opt for clinical treatment of acne lesion due to their anti-inflammatory and permeation capabilities (Abd-Allah, Abdel-Aziz, & Nasr, 2020).

Psoriasis

Psoriasis is skin disease that has elevated itchy plaques of red lesion and silvery scales. The main cause of psoriasis is not well established, though it is considered that inflammatory cells such as dendritic cells, T cells or keratinocytes over-proliferation could participate in this skin lesion. Therefore, the hyper proliferation state renders the differentiation keratinocyte causing formation of scales and skin barrier defects. The activation of cytokines further increases the chronic inflammation (Boakye et al., 2017). To overcome the barrier flaws, nanoparticles (NPs) in a TDD system have been used to treat psoriasis. Betamethasone dipropionate and calcipotriol loaded solid lipid nanoparticles were produced using a hot melt high shear homogenization process and then inserted into a Carbopol gel matrix to test their anti-psoriatic capabilities. The in vitro HaCaT cell line study reveals that solid lipid nanoparticles delayed the abrupt growth of keratinocytes, while in vivo mouse tail model showed that solid lipid nanoparticle gel significantly decreased the epidermal thickness and increased melanocyte count in comparison to commercial Daivobet ointment. Hence it could be used as an efficient carrier to treat psoriasis (Sonawane, Harde, Katariya, Agrawal, & Jain, 2014).

Wound healing

Wound healing is a primary response to injury through various stimuli that affect skin. In case of skin injury, several biochemical and organized complex cascades trigger that eventually result in the restoration of the affected area. Wound healing typically depends on four phases: Homeostasis, inflammation, proliferation, and remodeling. Different types of cells, cytokines, and proteins are involved in repairing tissues. Failure or delay in any one phase may result in prolonged healing or non-closure of wound (Rajendran, Kumar, Houreld, & Abrahamse, 2018; Witte & Barbul, 1997). Studies on wound healing have demonstrated the potential of Curcumin transdermal patches consisting of ethyl cellulose and polyvinylpyrrolidone formulation that result in enhanced permeation through the skin with effective systemic circulation. In-vivo wound contraction study has revealed drastic change in healing with reduced time, which can be attributed to the enhanced epithelialization. Moreover, histological data showed the well-organized angiogenesis and collagen fiber formation (Gadekar, Saurabh, Thakur, & Saurabh, 2012). A number of nanocarriers have been used for the application of wound healing including liposomes, polymeric nanoparticles, organic and inorganic nanoparticles, and so on (Wang, Lu, Yu, Huang, & Du, 2019). Cationic elastic liposomes loaded with growth factor complex significantly accelerated the wound closure rate in the diabetic mouse model, with the maximal shrinkage of wound size (58%) compared with the native growth factor complex (Choi et al., 2017). Daily application of baicalin transferosomes in mice model brought about complete skin restoration and inhibition of inflammatory markers such as oedema, TNF-α, and IL-1β (Manconi et al., 2018). Lipid-polymer hybrid nanoparticles formulation was reported to sustain release of the encapsulated drug (Norfloxacin) for 24 h with favorable skin permeation and reduced the frequency of application (Dave, Kushwaha, Yadav, & Agrawal, 2017). Despite the enormous potential of nano-DDSs, these systems also have exposed some limitations in researches such as lack of international standards and evaluation methods on their toxicology, biocompatibility, and targeting efficiency, as well as the undeniable restriction of industrial production for their complicated preparation procedures. However, it is an inevitable and unstoppable trend for researchers to further exploit the full potential of nano-DDSs, overcome the technical difficulties, and bring tangible benefits for the patients suffered from wounds, nano-DDSs are bound to constitute the most promising and cost-effective therapies to boost the wound healing and skin regeneration (Wang et al., 2019).

Conclusion

Transdermal delivery offers an effective substitute to other drug administration routes; specially for dermatological disorders which typically rely on site-specific delivery. It is non-invasive drug delivery system and is preferred due to: (i) avoidance of first-pass metabolism, (ii) greater patient compliance, (iii) ease of use, and (iv) increased bioavailability. Despite these benefits, its utilization is restricted due to the nature of SC that functions as a barrier for drug transport to the site of action. Nevertheless, continuous scientific efforts have resulted in the development of variety of transdermal drug delivery carriers/devices. These systems have potential to overcome the failure of drugs and biologically active molecules' penetration through skin via improving the permeation either by active or passive permeation enhancers. The literature described in chapter shows that various nanocarriers, microneedles, external stimuli responsive (ultrasound, laser, iontophoresis, etc.) have opened a new avenue for engineering of new enhanced therapies. In spite, of the successful development of transdermal delivery system to treat several skin diseases like acne, melanoma, dermatitis, inflammation, and so on, the stability, regulatory challenges and manufacturing processes are of prime concern to achieve the full potential clinical application of this route.

References

Abd-Allah, H., Abdel-Aziz, R. T. A., & Nasr, M. (2020). Chitosan nanoparticles making their way to clinical practice: A feasibility study on their topical use for acne treatment. *International Journal of Biological Macromolecules*, *156*, 262–270. https://doi.org/10.1016/j.ijbiomac.2020.04.040.

Alavi, M., Karimi, N., & Safaei, M. (2017). Application of various types of liposomes in drug delivery systems. *Advanced Pharmaceutical Bulletin*, *7*(1), 3–9. https://doi.org/10.15171/apb.2017.002.

Alexander, A., Dwivedi, S., Ajazuddin, Giri, T. K., Saraf, S., Saraf, S., et al. (2012). Approaches for breaking the barriers of drug permeation through transdermal drug delivery. *Journal of Controlled Release*, *164*(1), 26–40. https://doi.org/10.1016/j.jconrel.2012.09.017.

Arora, A., Prausnitz, M. R., & Mitragotri, S. (2008). Micro-scale devices for transdermal drug delivery. *International Journal of Pharmaceutics*, *364*(2), 227–236. https://doi.org/10.1016/j.ijpharm.2008.08.032.

Auda, S. H., Fathalla, D., Fetih, G., El-Badry, M., & Shakeel, F. (2016). Niosomes as transdermal drug delivery system for celecoxib: In vitro and in vivo studies. *Polymer Bulletin*, *73*(5), 1229–1245. https://doi.org/10.1007/s00289-015-1544-8.

Bamba, F. L., & Wepierre, J. (1993). Role of the appendageal pathway in the percutaneous absorption of pyridostigmine bromide in various vehicles. *European Journal of Drug Metabolism and Pharmacokinetics*, *18*(4), 339–348. https://doi.org/10.1007/BF03190183.

Baroli, B., Ennas, M. G., Loffredo, F., Isola, M., Pinna, R., & López-Quintela, M. A. (2007). Penetration of metallic nanoparticles in human full-thickness skin. *Journal of Investigative Dermatology*, *127*(7), 1701–1712. https://doi.org/10.1038/sj.jid.5700733.

Barua, S., & Mitragotri, S. (2014). Challenges associated with penetration of nanoparticles across cell and tissue barriers: A review of current status and future prospects. *Nano Today*, *9*(2), 223–243. https://doi.org/10.1016/j.nantod.2014.04.008.

Basra, M. K. A., & Shahrukh, M. (2009). Burden of skin diseases. *Expert Review of Pharmacoeconomics & Outcomes Research*, *9*(3), 271–283. https://doi.org/10.1586/erp.09.23.

Benson, H. A. E. (2005). Transdermal drug delivery: Penetration enhancement techniques. *Current Drug Delivery*, *2*(1), 23–33. https://doi.org/10.2174/1567201052772915.

Boakye, C. H. A., Patel, K., Doddapaneni, R., Bagde, A., Marepally, S., & Singh, M. (2017). Novel amphiphilic lipid augments the co-delivery of erlotinib and IL36 siRNA into the skin for psoriasis treatment. *Journal of Controlled Release*, *246*, 120–132. https://doi.org/10.1016/j.jconrel.2016.05.017.

Bodade, S. S., Shaikh, K. S., Kamble, M. S., & Chaudhari, P. D. (2013). A study on ethosomes as mode for transdermal delivery of an antidiabetic drug. *Drug Delivery*, *20*(1), 40–46. https://doi.org/10.3109/10717544.2012.752420.

Bos, J. D., & Meinardi, M. M. H. M. (2000). The 500 Dalton rule for the skin penetration of chemical compounds and drugs. *Experimental Dermatology*, *9*(3), 165–169. https://doi.org/10.1034/j.1600-0625.2000.009003165.x.

Caddeo, C., Díez-Sales, O., Pons, R., Fernàndez-Busquets, X., Fadda, A. M., & Manconi, M. (2014). Topical anti-inflammatory potential of quercetin in lipid-based nanosystems: In vivo and in vitro evaluation. *Pharmaceutical Research, 31*(4), 959–968. https://doi.org/10.1007/s11095-013-1215-0.

Cal, K., & Centkowska, K. (2008). Use of cyclodextrins in topical formulations: Practical aspects. *European Journal of Pharmaceutics and Biopharmaceutics, 68*(3), 467–478. https://doi.org/10.1016/j.ejpb.2007.08.002.

Carter, P., Narasimhan, B., & Wang, Q. (2019). Biocompatible nanoparticles and vesicular systems in transdermal drug delivery for various skin diseases. *International Journal of Pharmaceutics, 555,* 49–62. https://doi.org/10.1016/j.ijpharm.2018.11.032.

Cevc, G. (1997). Drug delivery across the skin. *Expert Opinion on Investigational Drugs, 6*(12), 1887–1937. https://doi.org/10.1517/13543784.6.12.1887.

Cevc, G. (2003). Transdermal drug delivery of insulin with ultradeformable carriers. *Clinical Pharmacokinetics, 42*(5), 461–474. https://doi.org/10.2165/00003088-200342050-00004.

Chen, M., Quan, G., Sun, Y., Yang, D., Pan, X., & Wu, C. (2020). Nanoparticles-encapsulated polymeric microneedles for transdermal drug delivery. *Journal of Controlled Release, 325,* 163–175. https://doi.org/10.1016/j.jconrel.2020.06.039.

Cheng, Y., Xu, Z., Ma, M., & Xu, T. (2008). Dendrimers as drug carriers: Applications in different routes of drug administration. *Journal of Pharmaceutical Sciences, 97*(1), 123–143. https://doi.org/10.1002/jps.21079.

Chhibber, T., Wadhwa, S., Chadha, P., Sharma, G., & Katare, O. P. (2015). Phospholipid structured microemulsion as effective carrier system with potential in methicillin sensitive Staphylococcus aureus (MSSA) involved burn wound infection. *Journal of Drug Targeting, 23*(10), 943–952. https://doi.org/10.3109/1061186X.2015.1048518.

Choi, J. U., Lee, S. W., Pangeni, R., Byun, Y., Yoon, I. S., & Park, J. W. (2017). Preparation and in vivo evaluation of cationic elastic liposomes comprising highly skin-permeable growth factors combined with hyaluronic acid for enhanced diabetic wound-healing therapy. *Acta Biomaterialia, 57,* 197–215. https://doi.org/10.1016/j.actbio.2017.04.034.

Chudasama, A., Patel, V., Nivsarkar, M., Vasu, K., & Shishoo, C. (2011). Investigation of microemulsion system for transdermal delivery of itraconazole. *Journal of Advanced Pharmaceutical Technology & Research, 2*(1).

Costello, C. T., & Jeske, A. H. (1995). Iontophoresis: Applications in transdermal medication delivery. *Physical Therapy, 75*(6), 554–563. https://doi.org/10.1093/ptj/75.6.554.

Dąbrowska, A. K., Spano, F., Derler, S., Adlhart, C., Spencer, N. D., & Rossi, R. M. (2018). The relationship between skin function, barrier properties, and body-dependent factors. *Skin Research and Technology, 24*(2), 165–174. https://doi.org/10.1111/srt.12424.

Dave, V., Kushwaha, K., Yadav, R. B., & Agrawal, U. (2017). Hybrid nanoparticles for the topical delivery of norfloxacin for the effective treatment of bacterial infection produced after burn. *Journal of Microencapsulation, 34*(4), 351–365. https://doi.org/10.1080/02652048.2017.1337249.

Dawson, A. L., Dellavalle, R. P., & Elston, D. M. (2012). Infectious skin diseases: A review and needs assessment. *Dermatologic Clinics, 30*(1), 141–151. https://doi.org/10.1016/j.det.2011.08.003.

Desai, P. R., Marepally, S., Patel, A. R., Voshavar, C., Chaudhuri, A., & Singh, M. (2013). Topical delivery of anti-TNFα siRNA and capsaicin via novel lipid-polymer hybrid nanoparticles efficiently inhibits skin inflammation in vivo. *Journal of Controlled Release, 170*(1), 51–63. https://doi.org/10.1016/j.jconrel.2013.04.021.

Desai, P., Patlolla, R. R., & Singh, M. (2010). Interaction of nanoparticles and cell-penetrating peptides with skin for transdermal drug delivery. *Molecular Membrane Biology, 27*(7), 247–259. https://doi.org/10.3109/09687688.2010.522203.

Dianzani, C., Zara, G. P., Maina, G., Pettazzoni, P., Pizzimenti, S., Rossi, F., et al. (2014). Drug delivery nanoparticles in skin cancers. *BioMed Research International, 2014,* 1–15. https://doi.org/10.1155/2014/895986. 895986.

Domingues, B., Lopes, J. M., Soares, P., & Pópulo, H. (2018). Melanoma treatment in review. *Immu-noTargets and Therapy, 7*.

El Maghraby, G. M. (2008). Transdermal delivery of hydrocortisone from eucalyptus oil microemulsion: Effects of cosurfactants. *International Journal of Pharmaceutics, 355*(1–2), 285–292. https://doi.org/10.1016/j.ijpharm.2007.12.022.

El Maghraby, G. M. (2010). Self-microemulsifying and microemulsion systems for transdermal delivery of indomethacin: Effect of phase transition. *Colloids and Surfaces B: Biointerfaces, 75*(2), 595–600. https://doi.org/10.1016/j.colsurfb.2009.10.003.

Escobar-Chávez, J. J., Rodríguez-Cruz, I. M., Domínguez-Delgado, C. L., Díaz-Torres, R., Revilla-Vázquez, A. L., & Aléncaster, N. C. (2012). Nanocarrier systems for transdermal drug delivery. In A. Demer Sezer (Ed.), *Recent advances in novel drug carrier systems* (pp. 201–240). IntechOpen.

Farokhzad, O. C., & Langer, R. (2009). Impact of nanotechnology on drug delivery. *ACS Nano, 3*(1), 16–20. https://doi.org/10.1021/nn900002m.

Gadekar, R., Saurabh, M. K., Thakur, G. S., & Saurabh, A. (2012). Study of formulation, characterisation and wound healing potential of transdermal patches of Curcumin. *Asian Journal of Pharmaceutical and Clinical Research, 5*(4), 225–230. http://www.ajpcr.com/Vol5Suppl4/1447.pdf.

Ghanbarzadeh, S., Khorrami, A., & Arami, S. (2015). Nonionic surfactant-based vesicular system for transdermal drug delivery. *Drug Delivery, 22*(8), 1071–1077. https://doi.org/10.3109/10717544.2013.873837.

Ghosh, I., & Michniak-Kohn, B. (2012). A comparative study of vitamin e TPGS/HPMC supersaturated system and other solubilizer/polymer combinations to enhance the permeability of a poorly soluble drug through the skin. *Drug Development and Industrial Pharmacy, 38*(11), 1408–1416. https://doi.org/10.3109/03639045.2011.653363.

Gopukumar, S., Sana-Fathima, T., Alexander, P., Alex, V., & Praseetha, P. (2016). Evaluation of antioxidant properties of silver nanoparticle embedded medicinal patch. *Nanomedicine & Nanotechnology Open Access, 1*.

Goyal, R., Macri, L. K., Kaplan, H. M., & Kohn, J. (2016). Nanoparticles and nanofibers for topical drug delivery. *Journal of Controlled Release, 240*, 77–92. https://doi.org/10.1016/j.jconrel.2015.10.049.

Green, P. G., & Hadgraft, J. (1987). Facilitated transfer of cationic drugs across a lipoidal membrane by oleic acid and lauric acid. *International Journal of Pharmaceutics, 37*(3), 251–255. https://doi.org/10.1016/0378-5173(87)90037-8.

Gupta, M., Agrawal, U., & Vyas, S. P. (2012). Nanocarrier-based topical drug delivery for the treatment of skin diseases. *Expert Opinion on Drug Delivery, 9*(7), 783–804. https://doi.org/10.1517/17425247.2012.686490.

Honeywell-Nguyen, P. L., & Bouwstra, J. A. (2005). Vesicles as a tool for transdermal and dermal delivery. *Drug Discovery Today: Technologies, 2*(1), 67–74. https://doi.org/10.1016/j.ddtec.2005.05.003.

Hua, S. (2015). Lipid-based nano-delivery systems for skin delivery of drugs and bioactives. *Frontiers in Pharmacology, 6*. https://doi.org/10.3389/fphar.2015.00219.

Huang, Y., Yu, F., Park, Y. S., Wang, J., Shin, M. C., Chung, H. S., et al. (2010). Co-administration of protein drugs with gold nanoparticles to enable percutaneous delivery. *Biomaterials, 31*(34), 9086–9091. https://doi.org/10.1016/j.biomaterials.2010.08.046.

Hueber, F., Wepierre, J., & Schaefer, H. (1992). Role of transepidermal and transfollicular routes in percutaneous absorption of hydrocortisone and testosterone: In vivo study in the hairless rat. *Skin Pharmacology, 5*(2), 99–107. https://doi.org/10.1159/000211026.

Hwa, C., Bauer, E. A., & Cohen, D. E. (2011). Skin biology. *Dermatologic Therapy, 24*(5), 464–470. https://doi.org/10.1111/j.1529-8019.2012.01460.x.

Iqbal, B., Ali, J., & Baboota, S. (2018). Recent advances and development in epidermal and dermal drug deposition enhancement technology. *International Journal of Dermatology, 57*(6), 646–660. https://doi.org/10.1111/ijd.13902.

Jadupati, M., Amites, G., & Kumar, N. A. (2012). Transferosomes: An opportunistic carrier for transdermal drug delivery system. *International Journal of Pharmacy IRJP, 3,* 35–38.

Jain, S. K., Gupta, Y., Jain, A., & Bhola, M. (2007). Multivesicular liposomes bearing celecoxib-β-cyclodextrin complex for transdermal delivery. *Drug Delivery, 14*(6), 327–335. https://doi.org/10.1080/10717540601098740.

Jain, S., Patel, N., Shah, M. K., Khatri, P., & Vora, N. (2017). Recent advances in lipid-based vesicles and particulate carriers for topical and transdermal application. *Journal of Pharmaceutical Sciences, 106*(2), 423–445. https://doi.org/10.1016/j.xphs.2016.10.001.

Kansagra, H., & Mallick, S. (2016). Microemulsion-based antifungal gel of luliconazole for dermatophyte infections: Formulation, characterization and efficacy studies. *Journal of Pharmaceutical Investigation, 46*(1), 21–28. https://doi.org/10.1007/s40005-015-0209-9.

Karande, P., & Mitragotri, S. (2009). Enhancement of transdermal drug delivery via synergistic action of chemicals. *Biochimica et Biophysica Acta - Biomembranes, 1788*(11), 2362–2373. https://doi.org/10.1016/j.bbamem.2009.08.015.

Kassem, A. A., & El-Alim, S. H. (2021). Vesicular nanocarriers: A potential platform for dermal and transdermal drug delivery. In V. Yata, S. Ranjan, N. Dasgupta, & E. Lichtfouse (Eds.), *Vol. 2. Nanopharmaceuticals: Principles and applications* (p. 155). Springer, Cham.

Kimura, E., Kawano, Y., Todo, H., Ikarashi, Y., & Sugibayashi, K. (2012). Measurement of skin permeation/penetration of nanoparticles for their safety evaluation. *Biological and Pharmaceutical Bulletin, 35*(9), 1476–1486. https://doi.org/10.1248/bpb.b12-00103.

Kotla, N. G., Chandrasekar, B., Rooney, P., Sivaraman, G., Larrañaga, A., Krishna, K. V., et al. (2017). Biomimetic lipid-based nanosystems for enhanced dermal delivery of drugs and bioactive agents. *ACS Biomaterials Science & Engineering, 3*(7), 1262–1272. https://doi.org/10.1021/acsbiomaterials.6b00681.

Latsch, S., Selzer, T., Fink, L., & Kreuter, J. (2003). Crystallisation of estradiol containing TDDS determined by isothermal microcalorimetry, X-ray diffraction, and optical microscopy. *European Journal of Pharmaceutics and Biopharmaceutics, 56*(1), 43–52. https://doi.org/10.1016/S0939-6411(03)00042-0.

Lavon, I., & Kost, J. (2004). Ultrasound and transdermal drug delivery. *Drug Discovery Today, 9*(15), 670–676. https://doi.org/10.1016/S1359-6446(04)03170-8.

Lee, H., & Larson, R. G. (2008). Lipid bilayer curvature and pore formation induced by charged linear polymers and dendrimers: The effect of molecular shape. *Journal of Physical Chemistry B, 112*(39), 12279–12285. https://doi.org/10.1021/jp805026m.

Lee, S., McAuliffe, D. J., Flotte, T. J., Kollias, N., & Doukas, A. G. (1998). Photomechanical transcutaneous delivery of macromolecules. *Journal of Investigative Dermatology, 111*(6), 925–929. https://doi.org/10.1046/j.1523-1747.1998.00415.x.

Leighton, T. (2012). *The acoustic bubble.* Academic press.

Liao, B., Ying, H., Yu, C., Fan, Z., Zhang, W., Shi, J., et al. (2016). (−)-Epigallocatechin gallate (EGCG)-nanoethosomes as a transdermal delivery system for docetaxel to treat implanted human melanoma cell tumors in mice. *International Journal of Pharmaceutics, 512*(1), 22–31. https://doi.org/10.1016/j.ijpharm.2016.08.038.

Liu, C. H., Chang, F. Y., & Hung, D. K. (2011). Terpene microemulsions for transdermal curcumin delivery: Effects of terpenes and cosurfactants. *Colloids and Surfaces B: Biointerfaces, 82*(1), 63–70. https://doi.org/10.1016/j.colsurfb.2010.08.018.

Lopez, R. F. V., Collett, J. H., & Bentley, M. V. L. B. (2000). Influence of cyclodextrin complexation on the in vitro permeation and skin metabolism of dexamethasone. *International Journal of Pharmaceutics, 200*(1), 127–132. https://doi.org/10.1016/S0378-5173(00)00365-3.

Losquadro, W. D. (2017). Anatomy of the skin and the pathogenesis of nonmelanoma skin cancer. *Facial Plastic Surgery Clinics of North America, 25*(3), 283–289. https://doi.org/10.1016/j.fsc.2017.03.001.

Malakar, J., Sen, S. O., Nayak, A. K., & Sen, K. K. (2011). Development and evaluation of microemulsions for transdermal delivery of insulin. *ISRN Pharmaceutics, 2011*. https://doi.org/10.5402/2011/780150, 780150 (Hindawi).

Manconi, M., Manca, M. L., Caddeo, C., Valenti, D., Cencetti, C., Diez-Sales, O., et al. (2018). Nanodesign of new self-assembling core-shell gellan-transfersomes loading baicalin and in vivo evaluation of repair response in skin. *Nanomedicine: Nanotechnology, Biology, and Medicine, 14*(2), 569–579. https://doi.org/10.1016/j.nano.2017.12.001.

Marwah, H., Garg, T., Goyal, A. K., & Rath, G. (2016). Permeation enhancer strategies in transdermal drug delivery. *Drug Delivery, 23*(2), 564–578. https://doi.org/10.3109/10717544.2014.935532.

Másson, M., Loftsson, T., Másson, G., & Stefánsson, E. (1999). Cyclodextrins as permeation enhancers: Some theoretical evaluations and in vitro testing. *Journal of Controlled Release, 59*(1), 107–118. https://doi.org/10.1016/S0168-3659(98)00182-5.

McConnell, K. I., Shamsudeen, S., Meraz, I. M., Mahadevan, T. S., Ziemys, A., Rees, P., et al. (2016). Reduced cationic nanoparticle cytotoxicity based on serum masking of surface potential. *Journal of Biomedical Nanotechnology, 12*(1), 154–164. https://doi.org/10.1166/jbn.2016.2134.

Merritt, C. R. B., Kremkau, F. W., & Hobbins, J. C. (1992). Diagnostic ultrasound: Bioeffects and safety. *Ultrasound in Obstetrics and Gynecology, 2*(5), 366–374. https://doi.org/10.1046/j.1469-0705.1992.02050366.x.

Mitragotri, S., Blankschtein, D., & Langer, R. (1996). Transdermal drug delivery using low-frequency sonophoresis. *Pharmaceutical Research, 13*(3), 411–420. https://doi.org/10.1023/A:1016096626810.

Moser, K., Kriwet, K., Kalia, Y. N., & Guy, R. H. (2001). Enhanced skin permeation of a lipophilic drug using supersaturated formulations. *Journal of Controlled Release, 73*(2–3), 245–253. https://doi.org/10.1016/S0168-3659(01)00290-5.

Murthy, S. N., Sammeta, S. M., & Bowers, C. (2010). Magnetophoresis for enhancing transdermal drug delivery: Mechanistic studies and patch design. *Journal of Controlled Release, 148*(2), 197–203. https://doi.org/10.1016/j.jconrel.2010.08.015.

Muzzalupo, R., & Tavano, L. (2015). Niosomal drug delivery for transdermal targeting: Recent advances. *Research and Reports in Transdermal Drug Delivery, 23*. https://doi.org/10.2147/RRTD.S64773.

Muzzalupo, R., Tavano, L., Cassano, R., Trombino, S., Ferrarelli, T., & Picci, N. (2011). A new approach for the evaluation of niosomes as effective transdermal drug delivery systems. *European Journal of Pharmaceutics and Biopharmaceutics, 79*(1), 28–35. https://doi.org/10.1016/j.ejpb.2011.01.020.

Naik, A., Kalia, Y. N., & Guy, R. H. (2000). Transdermal drug delivery: Overcoming the skin's barrier function. *Pharmaceutical Science & Technology Today, 3*(9), 318–326. https://doi.org/10.1016/S1461-5347(00)00295-9.

Nair, S. S. (2019). Strategies to improve the potential of transdermal devices by enhancing the skin permeation of therapeutic entities. *Journal of Drug Delivery and Therapeutics, 9*(3-s), 972–976.

Ng, K. W., & Lau, W. M. (2015). Skin deep: The basics of human skin structure and drug penetration. In *Percutaneous penetration enhancers chemical methods in penetration enhancement: Drug manipulation strategies and vehicle effects* (pp. 3–11). Springer Berlin Heidelberg. https://doi.org/10.1007/978-3-662-45013-0_1.

Nyborg, W. L. (2001). Biological effects of ultrasound: Development of safety guidelines. Part II: General review. *Ultrasound in Medicine and Biology, 27*(3), 301–333. https://doi.org/10.1016/S0301-5629(00)00333-1.

Oliver, S., Pham, T. T. P., Li, Y., Xu, F. J., & Boyer, C. (2021). More than skin deep: Using polymers to facilitate topical delivery of nitric oxide. *Biomaterials Science, 9*(2), 391–405. https://doi.org/10.1039/d0bm01197e.

Opatha, S. A. T., Titapiwatanakun, V., & Chutoprapat, R. (2020). Transfersomes: A promising nanoencapsulation technique for transdermal drug delivery. *Pharmaceutics, 12*(9).

Otberg, N., Richter, H., Schaefer, H., Blume-Peytavi, U., Sterry, W., & Lademann, J. (2004). Variations of hair follicle size and distribution in different body sites. *Journal of Investigative Dermatology, 122*(1), 14–19. https://doi.org/10.1046/j.0022-202X.2003.22110.x.

Panyam, J., & Labhasetwar, V. (2003). Biodegradable nanoparticles for drug and gene delivery to cells and tissue. *Advanced Drug Delivery Reviews, 55*(3), 329–347. https://doi.org/10.1016/S0169-409X(02)00228-4.

Papakostas, D., Rancan, F., Sterry, W., Blume-Peytavi, U., & Vogt, A. (2011). Nanoparticles in dermatology. *Archives of Dermatological Research, 303*(8).

Park, J. H., Choi, S. O., Seo, S., Choy, Y. B., & Prausnitz, M. R. (2010). A microneedle roller for transdermal drug delivery. *European Journal of Pharmaceutics and Biopharmaceutics, 76*(2), 282–289. https://doi.org/10.1016/j.ejpb.2010.07.001.

Park, D., Park, H., Seo, J., & Lee, S. (2014). Sonophoresis in transdermal drug deliverys. *Ultrasonics, 54*(1), 56–65. https://doi.org/10.1016/j.ultras.2013.07.007.

Patel, V. B., Misra, A., & Marfatia, Y. S. (2000). Topical liposomal gel of tretinoin for the treatment of acne: Research and clinical implications. *Pharmaceutical Development and Technology, 5*(4), 455–464. https://doi.org/10.1081/PDT-100102029.

Patel, M. R., Patel, R. B., Parikh, J. R., Solanki, A. B., & Patel, B. G. (2009). Effect of formulation components on the in vitro permeation of microemulsion drug delivery system of fluconazole. *AAPS PharmSciTech, 10*(3), 917–923. https://doi.org/10.1208/s12249-009-9286-2.

Patel, H. J., Trivedi, D. G., Bhandari, A. K., & Shah, D. A. (2011). Penetration enhancers for transdermal drug delivery system: A review. *Journal of Pharmaceutics and Cosmetology, 1*(2), 67–80.

Pham, C. V., & Cho, C. W. (2017). Application of d-α-tocopheryl polyethylene glycol 1000 succinate (TPGS) in transdermal and topical drug delivery systems (TDDS). *Journal of Pharmaceutical Investigation, 47*(2), 111–121. https://doi.org/10.1007/s40005-016-0300-x.

Pikal, M. J. (2001). The role of electroosmotic flow in transdermal iontophoresis. *Advanced Drug Delivery Reviews, 46*(1–3), 281–305. https://doi.org/10.1016/S0169-409X(00)00138-1.

Prausnitz, M. R. (1999). A practical assessment of transdermal drug delivery by skin electroporation. *Advanced Drug Delivery Reviews, 35*(1), 61–76. https://doi.org/10.1016/S0169-409X(98)00063-5.

Prausnitz, M. R., Bose, V. G., Langer, R., & Weaver, J. C. (1993). Electroporation of mammalian skin: A mechanism to enhance transdermal drug delivery. *Proceedings of the National Academy of Sciences, 90*(22), 10504–10508. https://doi.org/10.1073/pnas.90.22.10504.

Prausnitz, M. R., & Langer, R. (2008). Transdermal drug delivery. *Nature Biotechnology, 26*(11), 1261–1268. https://doi.org/10.1038/nbt.1504.

Prausnitz, M. R., Mitragotri, S., & Langer, R. (2004). Current status and future potential of transdermal drug delivery. *Nature Reviews Drug Discovery, 3*(2), 115–124. https://doi.org/10.1038/nrd1304.

Priya, B., Rashmi, T., & Bozena, M. (2006). Transdermal iontophoresis. *Expert Opinion on Drug Delivery, 3*(1), 127–138. https://doi.org/10.1517/17425247.3.1.127.

Rai, S., Pandey, V., & Rai, G. (2017). Transfersomes as versatile and flexible nano-vesicular carriers in skin cancer therapy: The state of the art. *Nano Reviews & Experiments*, 1325708. https://doi.org/10.1080/20022727.2017.1325708.

Rajan, R., Jose, S., Mukund, V. B., & Vasudevan, D. T. (2011). Transferosomes—A vesicular transdermal delivery system for enhanced drug permeation. *Journal of Advanced Pharmaceutical Technology & Research, 2*(3).

Rajendran, N. K., Kumar, S. S. D., Houreld, N. N., & Abrahamse, H. (2018). A review on nanoparticle based treatment for wound healing. *Journal of Drug Delivery Science and Technology, 44,* 421–430. https://doi.org/10.1016/j.jddst.2018.01.009.

Rao, Y. F., Chen, W., Liang, X. G., Huang, Y. Z., Miao, J., Liu, L., et al. (2015). Epirubicin-loaded superparamagnetic iron-oxide nanoparticles for transdermal delivery: Cancer therapy by circumventing the skin barrier. *Small, 11*(2), 239–247. https://doi.org/10.1002/smll.201400775.

Rivera, A. E. (2008). Acne scarring: A review and current treatment modalities. *Journal of the American Academy of Dermatology, 59*(4), 659–676. https://doi.org/10.1016/j.jaad.2008.05.029.

Sala, M., Diab, R., Elaissari, A., & Fessi, H. (2018). Lipid nanocarriers as skin drug delivery systems: Properties, mechanisms of skin interactions and medical applications. *International Journal of Pharmaceutics, 535*(1–2), 1–17. https://doi.org/10.1016/j.ijpharm.2017.10.046.

Saraf, S., Paliwal, S., & Saraf, S. (2011). Drug delivery microneedles: From micromachining to transdermal. *International Journal of Current Biomedical and Pharmaceutical Research, 1,* 80–87.

Satyam, G., Shivani, S., & Garima, G. (2010). Ethosomes: A novel tool for drug delivery through the skin. *Journal of Pharmacy Research, 3*(4), 688–691.

Schommer, N. N., & Gallo, R. L. (2013). Structure and function of the human skin microbiome. *Trends in Microbiology, 21*(12), 660–668. https://doi.org/10.1016/j.tim.2013.10.001.

Sharma, A., & Sharma, U. S. (1997). Liposomes in drug delivery: Progress and limitations. *International Journal of Pharmaceutics, 154*(2), 123–140. https://doi.org/10.1016/S0378-5173(97)00135-X.

Sheihet, L., Dubin, R. A., Devore, D., & Kohn, J. (2005). Hydrophobic drug delivery by self-assembling triblock copolymer-derived nanospheres. *Biomacromolecules, 6*(5), 2726–2731. https://doi.org/10.1021/bm050212u.

Sheihet, L., Piotrowska, K., Dubin, R. A., Kohn, J., & Devore, D. (2007). Effect of tyrosine-derived triblock copolymer compositions on nanosphere self-assembly and drug delivery. *Biomacromolecules, 8*(3), 998–1003. https://doi.org/10.1021/bm060860t.

Sierra-Sánchez, Á., Montero-Vilchez, T., Quiñones-Vico, M. I., Sanchez-Diaz, M., & Arias-Santiago, S. (2021). Current advanced therapies based on human mesenchymal stem cells for skin diseases. *Frontiers in Cell and Developmental Biology, 9.* https://doi.org/10.3389/fcell.2021.643125.

Sigmundsdottir, H. (2010). Improving topical treatments for skin diseases. *Trends in Pharmacological Sciences, 31*(6), 239–245. https://doi.org/10.1016/j.tips.2010.03.004.

Sonawane, R., Harde, H., Katariya, M., Agrawal, S., & Jain, S. (2014). Solid lipid nanoparticles-loaded topical gel containing combination drugs: An approach to offset psoriasis. *Expert Opinion on Drug Delivery, 11*(12), 1833–1847. https://doi.org/10.1517/17425247.2014.938634.

Stachowiak, J. C., von Muhlen, M. G., Li, T. H., Jalilian, L., Parekh, S. H., & Fletcher, D. A. (2007). Piezoelectric control of needle-free transdermal drug delivery. *Journal of Controlled Release, 124*(1–2), 88–97. https://doi.org/10.1016/j.jconrel.2007.08.017.

Stott, P. W., Williams, A. C., & Barry, B. W. (1996). Characterization of complex coacervates of some tricyclic antidepressants and evaluation of their potential for enhancing transdermal flux. *Journal of Controlled Release, 41*(3), 215–227. https://doi.org/10.1016/0168-3659(96)01328-4.

Tan, A. U., Schlosser, B. J., & Paller, A. S. (2018). A review of diagnosis and treatment of acne in adult female patients. *International Journal of Women's Dermatology, 4*(2), 56–71. https://doi.org/10.1016/j.ijwd.2017.10.006.

Tolentino, S., Pereira, M. N., Cunha-Filho, M., Gratieri, T., & Gelfuso, G. M. (2021). Targeted clindamycin delivery to pilosebaceous units by chitosan or hyaluronic acid nanoparticles for improved topical treatment of acne vulgaris. *Carbohydrate Polymers, 253.*

Valenta, C., & Auner, B. G. (2004). The use of polymers for dermal and transdermal delivery. *European Journal of Pharmaceutics and Biopharmaceutics, 58*(2), 279–289. https://doi.org/10.1016/j.ejpb.2004.02.017.

Van Der Maaden, K., Jiskoot, W., & Bouwstra, J. (2012). Microneedle technologies for (trans)dermal drug and vaccine delivery. *Journal of Controlled Release, 161*(2), 645–655. https://doi.org/10.1016/j.jconrel.2012.01.042.

Vishnubhakthula, S., Elupula, R., & Durán-Lara, E. F. (2017). Recent advances in hydrogel-based drug delivery for melanoma cancer therapy: A mini review. *Journal of Drug Delivery*, 1–9. https://doi.org/10.1155/2017/7275985.

Walters, K. A. (2002). *Dermatological and transdermal formulations. Vol. 119.* Informa Health Care.

Wang, Q., Du, Y., Fan, L., Liu, H., & Wang, X. (2003). Structures and properties of chitosan-starch-sodium benzoate blend films. *Wuhan University Journal (Natural Science Edition), 6*, 1–13.

Wang, W., Lu, K. J., Yu, C. H., Huang, Q. L., & Du, Y. Z. (2019). Nano-drug delivery systems in wound treatment and skin regeneration. *Journal of Nanobiotechnology, 17*(1), 1–15. https://doi.org/10.1186/s12951-019-0514-y.

Watkinson, A. C., Bunge, A. L., Hadgraft, J., & Lane, M. E. (2013). Nanoparticles do not penetrate human skin—A theoretical perspective. *Pharmaceutical Research, 30*(8), 1943–1946. https://doi.org/10.1007/s11095-013-1073-9.

Williams, A. C., & Barry, B. W. (2012). Penetration enhancers. *Advanced Drug Delivery Reviews, 64*, 128–137. https://doi.org/10.1016/j.addr.2012.09.032.

Witte, M. B., & Barbul, A. (1997). General principles of wound healing. *Surgical Clinics of North America, 77*(3), 509–528. https://doi.org/10.1016/S0039-6109(05)70566-1.

Wong, T. W. (2014). Electrical, magnetic, photomechanical and cavitational waves to overcome skin barrier for transdermal drug delivery. *Journal of Controlled Release, 193*, 257–269. https://doi.org/10.1016/j.jconrel.2014.04.045.

Yan, G., Li, S. K., & Higuchi, W. I. (2005). Evaluation of constant current alternating current iontophoresis for transdermal drug delivery. *Journal of Controlled Release, 110*(1), 141–150. https://doi.org/10.1016/j.jconrel.2005.09.006.

Yu, X., Du, L., Li, Y., Fu, G., & Jin, Y. (2015). Improved anti-melanoma effect of a transdermal mitoxantrone ethosome gel. *Biomedicine and Pharmacotherapy, 73*, 6–11. https://doi.org/10.1016/j.biopha.2015.05.002.

Yuan, X., & Capomacchia, A. C. (2010). Physicochemical studies of binary eutectic of ibuprofen and ketoprofen for enhanced transdermal drug delivery. *Drug Development and Industrial Pharmacy, 36*(10), 1168–1176. https://doi.org/10.3109/03639041003695071.

Zhai, Y., & Zhai, G. (2014). Advances in lipid-based colloid systems as drug carrier for topic delivery. *Journal of Controlled Release, 193*, 90–99. https://doi.org/10.1016/j.jconrel.2014.05.054.

Zhang, Z., Tsai, P. C., Ramezanli, T., & Michniak-Kohn, B. B. (2013). Polymeric nanoparticles-based topical delivery systems for the treatment of dermatological diseases. *Wiley Interdisciplinary Reviews: Nanomedicine and Nanobiotechnology, 5*(3), 205–218. https://doi.org/10.1002/wnan.1211.

Zhao, X., Liu, J. P., Zhang, X., & Li, Y. (2006). Enhancement of transdermal delivery of theophylline using microemulsion vehicle. *International Journal of Pharmaceutics, 327*(1–2), 58–64. https://doi.org/10.1016/j.ijpharm.2006.07.027.

Zhou, X., Hao, Y., Yuan, L., Pradhan, S., Shrestha, K., Pradhan, O., et al. (2018). Nano-formulations for transdermal drug delivery: A review. *Chinese Chemical Letters, 29*(12), 1713–1724. https://doi.org/10.1016/j.cclet.2018.10.037.

Zorec, B., Becker, S., Reberšek, M., Miklavčič, D., & Pavšelj, N. (2013). Skin electroporation for transdermal drug delivery: The influence of the order of different square wave electric pulses. *International Journal of Pharmaceutics, 457*(1), 214–223. https://doi.org/10.1016/j.ijpharm.2013.09.020.

CHAPTER 7

Targeted nano drug delivery systems for renal disorders

Introduction

The kidneys are important organs that filter blood and eliminate waste fluids and products from the body. Renal disorders can disrupt fluid and electrolyte balance, resulting in serious complications. However, because renal disorders can go unnoticed until they have progressed to late stages, therefore they are sometimes referred to as silent killers (Huang, Ma, Li, Han, & Lin, 2021). Chronic renal disorders are estimated to affect roughly 11% of the population (Webster, Nagler, Morton, & Masson, 2017). According to the US Centers for Disease Control and Prevention, approximately 37 million Americans have chronic renal disorders, with roughly 0.7 million needing life-sustaining dialysis or a kidney transplant. Glomerulonephritis, tubulointerstitial fibrosis, diabetic nephropathy, and nephrotic syndrome are all common renal disorders. The natural progression of these disorders may lead to renal failure, which imposes a significant financial burden on sufferers (Centers for Disease Control and Prevention, 2021; Chen, Peng, & Zhang, 2020).

Currently used drugs like hormones and nonsteroidal anti-inflammatory drugs must be administered in high quantities into the kidney as part of current renal disorders treatment. Because the drugs are distributed to areas other than the kidneys without discrimination, these treatments are associated with systemic side effects (Zhou, Sun, & Zhang, 2014). These drugs may not work on specific target cells even within the kidney, especially in pathological conditions such as glomerular sclerosis, tubulointerstitial fibrosis, and other functional complications (Stridh, Palm, & Hansell, 2012). As a result, there is an unmet need for kidney-targeted drug delivery systems that can reduce systemic adverse effects, improve the therapeutic index and the efficacy of drugs used to treat renal disorders. The fast developments and applications of nanotechnology in drugs delivery are projected to have a dramatic impact on human health (Wojtynek & Mohs, 2020). One of the primary applications of nanotechnology is the delivery of anticancer drugs, some of which have been approved for cancer treatment and others are now undergoing clinical studies

167

(Klochkov et al., 2021; Zhang, Li, Gao, Chen, & Liu, 2019). Nanocarriers have the ability to efficiently load therapeutic molecules such as small drug molecules, ribonucleic acids, and peptides in order to achieve prolonged or regulated release throughout the body (Williams, Jaimes, & Heller, 2016). Moreover, the physicochemical features of nanocarriers can significantly alter their function, enabling them to exhibit enhanced biodistribution, pharmacokinetics, and long-circulation, resulting in the effect of targeted drug delivery to specific cells, tissues, or organs (Anselmo & Mitragotri, 2017; Jo, Kim, Lee, & Kim, 2015). Although research into developing kidney-targeted drug delivery systems began in the 1990s, it was Haas et al. who addressed the significance and plausibility of kidney-targeted drug delivery for the first time in 2002 (Haas, Moolenaar, Meijer, & De Zeeuw, 2002). Subsequent advancements in this field of research have been aided significantly by the rapid growth of nanotechnology and macromolecular carrier technologies concurrently. Ongoing research on kidney-targeted drug delivery systems has enabled the planning of cell-targeted administration of both new and established drugs (Liu et al., 2019). This chapter discusses recent advancements in tailored nano drug delivery systems for the treatment of a variety of renal disorders. The chapter will improve our understanding of the various targets that can be effectively used for drug delivery in the treatment and management of various renal disorders.

Targeted nanocarriers for renal tubulointerstitial diseases

Renal tubules include the collecting ducts, the loop of Henle, proximal tubule, and the distal tubule. Tubulointerstitial injuries to the kidneys, such as tubular atrophy and interstitial fibrosis or inflammation, are typical consequences of kidney disease and play a significant role in the evolution of chronic kidney disease and end-stage kidney disease. They can develop as a result of vascular and glomerular disorders, or as a result of primary tubulointerstitial diseases, such as acute kidney damage (Huang et al., 2021; Ramos et al., 2015). As a result, nanocarriers that deliver drugs to prevent reduce renal inflammation and fibrosis, cell death, and/or enhance tubule regeneration could be extremely beneficial.

Recently, Poly(lactic-*co*-glycolic acid) (PLGA) conjugated to polyethylene glycol (PEG) nanocarriers modified with kidney-targeting proteins (KTP) were used for targeted delivery of anti-inflammatory drug Asiatic acid for the treatment of chronic kidney diseases. The KTP modified PLGA-PEG targeted nanocarrier system achieved a 30-fold enhanced targeted delivery of Asiatic acid to renal tubules, thus leading to effective treatment of chronic kidney diseases (He et al., 2020). In another study, for specific renal targeting capabilities, hydrocaffeic acid-containing catechol-derived low molecular weight chitosan and metal ions-based nanocarriers were designed. The nanocarriers, when loaded with the antifibrosis drug emodine, slowed the progression of kidney fibrosis in ureter-obstruction mice (Qiao et al., 2014). In another work, hydrophobically modified glycol chitosan (HGC) polymeric nanomicelles were loaded with tacrolimus and exhibited efficient selective transport to the kidney, reducing systemic adverse effects considerably (Kim et al., 2020).

Several RNAi-based approaches to treat acute kidney injury have been proposed since proximal tubule cells collect siRNA. Nanocarriers-mediated siRNA transport can boost siRNA molecule accumulation within proximal tubule cells, making combination therapies or controlled-release modes of action easier (Morishita et al., 2015; Zheng et al., 2006). According to a study, nanocarriers with a size of 400 nm were able to localize preferentially in the kidney compared to other organs, and selectively accumulate in the renal proximal tubules, implying that such nanocarriers could be used for targeted delivery of antifibrotic compounds directly to proximal tubular cells, increasing renal specificity and minimizing adverse and teratogenic effects (Williams et al., 2015). In a cisplatin-induced model of acute kidney injury, carbon nanotubes containing siRNAs against Trp53, Mep1b, and Ctr1 were found to reduce renal fibrosis, injury, and immunological infiltration (Alidori et al., 2016). Another study looked at the effects of delivering an exogenous miRNA (miR-146a) via polyethylenimine nanocarriers in a mouse model of unilateral ureteral obstruction-induced kidney fibrosis. TGFβ1 and Smad mRNA expression, which are responsible for the transcription of profibrotic genes, were suppressed by these positively charged nanocarriers, which significantly reduced renal fibrosis (Morishita et al., 2015).

Targeted nanocarriers for glomerular diseases

The glomerulus is responsible for selective filtration of blood components across the capillary wall and for the construction of an ultra-filtrate that is modified secondary by the renal tubular system. Glomerular damage is defined in patients with chronic renal disease by an increase in albuminuria and a decrease in glomerular filtration rate (GFR) (Jha & Garcia-Garcia, 2013). Around 40% of patients with end-stage renal disease have IgA nephropathy, a prevalent kind of glomerulonephritis (Dillon, 2001). Nephrotic syndrome, rapidly progressing renal failure, nephritic syndrome, AKI, diabetic/hypertensive nephropathy, and isolated hematuria or proteinuria are further symptoms of glomerular disorders. Glomerular disease can occur exclusively in the kidney or as a result of a systemic condition (Hartman, Lai, & Patterson, 2007; Scindia, Deshmukh, & Bagavant, 2010).

The first line of treatment for glomerular disorders is systemic immunosuppressive drugs including corticosteroids, mycophenolate mofetil, tacrolimus, celastrol, and cyclophosphamide. Since the 1950s, this type of therapy has been used to halt the progression of disease by modifying systemic cofactors. They are not, however, universally successful, and treatment is frequently hampered by devastating side effects and extensive extra renal toxicity. It is hoped that by delivering these therapeutic drugs directly to the glomerulus, administration will be facilitated, systemic side effects will be minimized, and patients' prognoses will improve (Scindia et al., 2010; Tanna, Tam, & Pusey, 2013). Currently, only a few strategies are available for targeting glomeruli. Similar to the method used to target renal tubules, the shape, size, and, most importantly, the molecular recognition moiety are critical in glomeruli-specific delivery. Notably, damaged glomeruli expose normally sequestered epitopes, which facilitates antibody-mediated targeted delivery. Certain surface molecules are upregulated in diseased states and thus serve as receptors for carriers. Wang et al. conducted an exhaustive evaluation of renal targeting peptide and antibody ligands (Wang, Masehi-Lano, & Chung, 2017). In this section, we briefly discuss targeted drug delivery systems that are used to specifically deliver drugs for the treatment of glomerular disorders.

Wang et al. developed kidney-targeted rhein (RH)-loaded liponanoparticles (KLPPR) using polyethyleneimine-based cores and lipid layers modified with kidney-targeting protein (KTP). They accomplished this by targeting the glomerular endothelial cell barrier and basement membrane with CLPVASC, an elastin-like polypeptide (Wang et al., 2017). In diabetic nephropathy, the particle demonstrated good kidney-targeted distribution (Wang et al., 2017). Bruni et al. attempted to target podocytes using ultrasmall colloidal polymeric nanocarriers. A polymer with or without a hydrophobic polycaprolactone (PCL) core and a hydrophilic PEG shell was used as the carrier. PEG Dexamethasone was encapsulated, and podocyte healing was demonstrated in vitro (Bruni et al., 2017). Pollinger et al. and Colombo et al. developed poly (N-2-hydroxypropyl) methacrylamide (PHMAM), PCL, and liposome-based nanocarriers containing the cycloRGD peptide (Colombo et al., 2017; Pollinger et al., 2012). In vitro, the nanocarriers were capable of binding to the v3αvβ3 integrin receptor on podocytes. Increased accumulation of RGD liposomes in the fibrotic kidney was seen in an in vivo investigation using unilateral ureteral obstruction mice. Celastrol loaded in these nanocarriers ameliorated inflammatory responses in glomeruli while causing no clear injury to other organs (Zhou et al., 2014). Celastrol-loaded albumin nanoparticles have been proven to be beneficial in the treatment of mesangioproliferative glomerulonephritis. Celastrol encapsulated in these nanocarriers has shown increased ability to repair renal damage and decreased toxicity to the heart, brain, and liver in animal models. Similarly, Wang et al. recently developed a PEG-modified cationic liposome coated with octa-arginine (R8) and measuring around 110 nm in diameter. In animal models, the liposomal nanocarriers were loaded with both p38α MAPK and p65 siRNA and accumulated mostly in the kidney, with the maximum uptake by glomerular mesangial cells (Wang et al., 2020).

Targeted nanocarriers for renal cell carcinoma

Renal cell carcinoma is a type of malignancy characterized by a very diverse epithelium that originates in the renal tubules. Renal cell carcinoma is the third most prevalent type of cancer of the urinary system, occurring at an incidence rate of approximately 5–10 occurrences per 100,000 persons and accounting for approximately 2%–3% of all malignant tumors (Yuan et al., 2016). While chemotherapy is a primary modality of cancer treatment, its efficacy is hampered by drug resistance. For instance, sunitinib is presently the recommended first-line treatment for advanced renal cell carcinoma, and patients will inevitably develop resistance to sunitinib. Furthermore, clinical results for patients with renal cell carcinoma continue to be dismal. As a result, rapid advancements in nanotechnology may enable the development of more effective therapeutic techniques for the treatment of renal cell carcinoma (He et al., 2020). The following section describes the nanocarrier systems developed and reported for the targeted drug delivery of renal cells carcinoma.

Various liposomal drug delivery approaches have been examined in preclinical models of renal cell carcinoma, but only a few clinical trials have been conducted. The majority of these studies used conventional renal disease therapy in combination with antibody-targeting methods. In a phase II clinical research, EGylated-liposomal doxorubicin was used to treat patients with resistant renal cell carcinoma but demonstrated no efficacy and was less toxic (Skubitz, 2002). Additionally, polymeric nanocarriers have been studied in preclinical models for the delivery of the anti-angiogenic drug tetraiodothyroacetic acid to tumor cells in the chorioallantoic membrane of chicks and in mice xenografts (Yalcin et al., 2009). Liu et al. investigated the in vitro applicability of several drugs as delivery systems for the treatment of renal cell carcinoma, including sorafenib-loaded PLGA nanoparticles, liposomes based 1,2-dipalmitoyl-sn-glycero-3-phosphocholine (DPPC), and chitosan-coated DPPC liposomes. At lower doses, sorafenib-loaded chitosan-coated DPPC liposomes and PLGA nanoparticles appeared to kill more renal cell carcinoma cells than sorafenib alone (Liu et al., 2019). Additionally, XL184 liposomes inhibited tumor activity in renal carcinoma cells by reducing the phosphorylation of AKT, Met, and mitogen-activated protein kinase pathways (Kulkarni, Vijaykumar, Natarajan, Sengupta, & Sabbisetti, 2016).

Sunitinib is currently the recommended first-line treatment for renal cells carcinoma; nevertheless, patients acquire resistance to this drug inexorably, resulting in therapy failure and a dismal prognosis (Stone, 2016). Nanotechnology's rapid advancement has resulted in the development of novel strategies for the treatment of drug-resistant renal carcinoma cells. Yang et al. have shown that cuprous oxide nanoparticles can significantly reduce renal cell carcinoma tumor development in vitro and in animals with minimal renal damage, reversing sunitinib resistance. The mechanism may involve the downregulation of copper chaperone proteins antioxidant 1 copper chaperone and copper chaperone for superoxide dismutase in renal cell carcinoma, affecting copper trafficking to the mitochondria and endoplasmic reticulum and thus initiating mitochondrial and endoplasmic reticulum stress-induced apoptosis via activation of caspase-12, caspase-9, and caspase-3 (Yang et al., 2017). The utilization of PEG-conjugated antibodies coupled to nanoshells ensures biocompatibility when delivering drugs to tumor cells. Pannerec-Varna et al. synthesized PEGylated gold nanoshells having wider distribution in various cellular components of human kidney carcinoma. The distribution kinetics advanced from intravascular flow to intratumoral cells 24 h later, and no toxicity was identified in 6-month-old mice treated with PEGylated gold nanoshells. This groundbreaking study shed new light on the application of large PEGylated gold nanoshells by enhancing the duration of targeted hyperthermia or topical medication delivery to cancer cells (Pannerec-Varna et al., 2013).

Conclusion

Numerous hybrids, nanoparticluate-based kidney-targeted drug delivery systems are being studied at the moment. Though the majority of them are in the early stages of development, several of them have demonstrated significant therapeutic benefit and significantly reduced adverse effects when loaded with drugs in diseased animal models. However, some hurdles appear to be identifying a safe and effective carrier capable of increasing the effectiveness of targeting and intracellular delivery of drug payload into target renal cells and tissues. With a better understanding of the kidney's structure and function, and continued optimization of carriers and targeting ligands, drug delivery systems targeted specifically to the renal cells and tissues will develop into powerful tools for the treatment of kidney diseases.

References

Alidori, S., Akhavein, N., Thorek, D. L. J., Behling, K., Romin, Y., Queen, D., et al. (2016). Targeted fibrillar nanocarbon RNAi treatment of acute kidney injury. *Science Translational Medicine, 8*(331). https://doi.org/10.1126/scitranslmed.aac9647.

Anselmo, A. C., & Mitragotri, S. (2017). Impact of particle elasticity on particle-based drug delivery systems. *Advanced Drug Delivery Reviews, 108*, 51–67. https://doi.org/10.1016/j.addr.2016.01.007.

Bruni, R., Possenti, P., Bordignon, C., Li, M., Ordanini, S., Messa, P., et al. (2017). Ultrasmall polymeric nanocarriers for drug delivery to podocytes in kidney glomerulus. *Journal of Controlled Release, 255*, 94–107. https://doi.org/10.1016/j.jconrel.2017.04.005.

Centers for Disease Control and Prevention. (2021). *Chronic kidney disease in the United States, 2021.* US Department of Health and Human Services.

Chen, Z., Peng, H., & Zhang, C. (2020). Advances in kidney-targeted drug delivery systems. *International Journal of Pharmaceutics, 587*, 119679. https://doi.org/10.1016/j.ijpharm.2020.119679.

Colombo, C., Li, M., Watanabe, S., Messa, P., Edefonti, A., Montini, G., et al. (2017). Polymer nanoparticle engineering for podocyte repair: From in vitro models to new nanotherapeutics in kidney diseases. *ACS Omega, 2*(2), 599–610. https://doi.org/10.1021/acsomega.6b00423.

Dillon, J. J. (2001). Treating IgA nephropathy. *Journal of the American Society of Nephrology, 12*(4), 846–847.

Haas, M., Moolenaar, F., Meijer, D. K. F., & De Zeeuw, D. (2002). Specific drug delivery to the kidney. *Cardiovascular Drugs and Therapy, 16*(6), 489–496. https://doi.org/10.1023/A:1022913709849.

Hartman, H. A., Lai, H. L., & Patterson, L. T. (2007). Cessation of renal morphogenesis in mice. *Developmental Biology, 310*(2), 379–387. https://doi.org/10.1016/j.ydbio.2007.08.021.

He, J., Chen, H., Zhou, W., Chen, M., Yao, Y., Zhang, Z., et al. (2020). Kidney targeted delivery of asiatic acid using a FITC labeled renal tubular-targeting peptide modified PLGA-PEG system. *International Journal of Pharmaceutics, 584*, 119455. https://doi.org/10.1016/j.ijpharm.2020.119455.

Huang, X., Ma, Y., Li, Y., Han, F., & Lin, W. (2021). Targeted drug delivery systems for kidney diseases. *Frontiers in Bioengineering and Biotechnology, 9*. https://doi.org/10.3389/fbioe.2021.683247.

Jha, V., & Garcia-Garcia, G. (2013). Global kidney disease—Authors' reply. *The Lancet, 382*.

Jo, D. H., Kim, J. H., Lee, T. G., & Kim, J. H. (2015). Size, surface charge, and shape determine therapeutic effects of nanoparticles on brain and retinal diseases. *Nanomedicine: Nanotechnology, Biology, and Medicine, 11*(7), 1603–1611. https://doi.org/10.1016/j.nano.2015.04.015.

Kim, C. S., Mathew, A. P., Uthaman, S., Moon, M. J., Bae, E. H., Kim, S. W., et al. (2020). Glycol chitosan-based renal docking biopolymeric nanomicelles for site-specific delivery of the immunosuppressant. *Carbohydrate Polymers, 241*, 116255–116279. https://doi.org/10.1016/j.carbpol.2020.116255.

Klochkov, S. G., Neganova, M. E., Nikolenko, V. N., Chen, K., Somasundaram, S. G., Kirkland, C. E., et al. (2021). Implications of nanotechnology for the treatment of cancer: Recent advances. *Seminars in Cancer Biology, 69*, 190–199. https://doi.org/10.1016/j.semcancer.2019.08.028.

Kulkarni, A. A., Vijaykumar, V. E., Natarajan, S. K., Sengupta, S., & Sabbisetti, V. S. (2016). Sustained inhibition of cMET-VEGFR2 signaling using liposome-mediated delivery increases efficacy and reduces toxicity in kidney cancer. *Nanomedicine: Nanotechnology, Biology, and Medicine, 12*(7), 1853–1861. https://doi.org/10.1016/j.nano.2016.04.002.

Liu, C. P., Hu, Y., Lin, J. C., Fu, H. L., Lim, L. Y., & Yuan, Z. X. (2019). Targeting strategies for drug delivery to the kidney: From renal glomeruli to tubules. *Medicinal Research Reviews, 39*(2), 561–578. https://doi.org/10.1002/med.21532.

Morishita, Y., Imai, T., Yoshizawa, H., Watanabe, M., Ishibashi, K., Muto, S., et al. (2015). Delivery of micro RNA-146a with polyethylenimine nanoparticles inhibits renal fibrosis in vivo. *International Journal of Nanomedicine, 10*.

Pannerec-Varna, M., Ratajczak, P., Bousquet, G., Ferreira, I., Leboeuf, C., Boisgard, R., et al. (2013). In vivo uptake and cellular distribution of gold nanoshells in a preclinical model of xenografted human renal cancer. *Gold Bulletin, 46*(4), 257–265. https://doi.org/10.1007/s13404-013-0115-8.

Pollinger, K., Hennig, R., Breunig, M., Tessmar, J., Ohlmann, A., Tamm, E. R., et al. (2012). Kidney podocytes as specific targets for cyclo(RGDfC)-modified nanoparticles. *Small, 8*(21), 3368–3375. https://doi.org/10.1002/smll.201200733.

Qiao, H., Sun, M., Su, Z., Xie, Y., Chen, M., Zong, L., et al. (2014). Kidney-specific drug delivery system for renal fibrosis based on coordination-driven assembly of catechol-derived chitosan. *Biomaterials, 35*(25), 7157–7171. https://doi.org/10.1016/j.biomaterials.2014.04.106.

Ramos, A. M., González-Guerrero, C., Sanz, A., Sanchez-Niño, M. D., Rodríguez-Osorio, L., Martín-Cleary, C., et al. (2015). Designing drugs that combat kidney damage. *Expert Opinion on Drug Discovery, 10*(5), 541–556. https://doi.org/10.1517/17460441.2015.1033394.

Scindia, Y. M., Deshmukh, U. S., & Bagavant, H. (2010). Mesangial pathology in glomerular disease: Targets for therapeutic intervention. *Advanced Drug Delivery Reviews, 62*(14), 1337–1343. https://doi.org/10.1016/j.addr.2010.08.011.

Skubitz, K. M. (2002). Phase II trial of pegylated-liposomal doxorubicin (Doxil™)in renal cell cancer. *Investigational New Drugs, 20*(1), 101–104. https://doi.org/10.1023/A:1014428720551.

Stone, L. (2016). Exosome transmission of sunitinib resistance. *Nature Reviews Urology*, 297. https://doi.org/10.1038/nrurol.2016.88.

Stridh, S., Palm, F., & Hansell, P. (2012). Renal interstitial hyaluronan: Functional aspects during normal and pathological conditions. *American Journal of Physiology - Regulatory, Integrative and Comparative Physiology, 302*(11), R1235–R1249. https://doi.org/10.1152/ajpregu.00332.2011.

Tanna, A., Tam, F. W. K., & Pusey, C. D. (2013). B-cell-targeted therapy in adult glomerulonephritis. *Expert Opinion on Biological Therapy, 13*(12), 1691–1706. https://doi.org/10.1517/14712598.2013.851191.

Wang, J., Masehi-Lano, J. J., & Chung, E. J. (2017). Peptide and antibody ligands for renal targeting: Nanomedicine strategies for kidney disease. *Biomaterials Science, 5*(8), 1450–1459. https://doi.org/10.1039/c7bm00271h.

Wang, Y., Wu, Q., Wang, J., Li, L., Sun, X., Zhang, Z., et al. (2020). Co-delivery of p38α MAPK and p65 siRNA by novel liposomal glomerulus-targeting nano carriers for effective immunoglobulin a nephropathy treatment. *Journal of Controlled Release, 320*, 457–468. https://doi.org/10.1016/j.jconrel.2020.01.024.

Webster, A. C., Nagler, E. V., Morton, R. L., & Masson, P. (2017). Chronic kidney disease. *The Lancet, 389*(10075), 1238–1252. https://doi.org/10.1016/S0140-6736(16)32064-5.

Williams, R. M., Jaimes, E. A., & Heller, D. A. (2016). Nanomedicines for kidney diseases. *Kidney International, 90*(4), 740–745. https://doi.org/10.1016/j.kint.2016.03.041.

Williams, R. M., Shah, J., Ng, B. D., Minton, D. R., Gudas, L. J., Park, C. Y., et al. (2015). Mesoscale nanoparticles selectively target the renal proximal tubule epithelium. *Nano Letters, 15*(4), 2358–2364. https://doi.org/10.1021/nl504610d.

Wojtynek, N. E., & Mohs, A. M. (2020). Image-guided tumor surgery: The emerging role of nanotechnology. *Wiley Interdisciplinary Reviews: Nanomedicine and Nanobiotechnology, 12*(4). https://doi.org/10.1002/wnan.1624.

Yalcin, M., Bharali, D. J., Lansing, L., Dyskin, E., Mousa, S. S., Hercbergs, A., et al. (2009). Tetraiodothyroacetic acid (Tetrac) and tetrac nanoparticles inhibit growth of human renal cell carcinoma xenografts. *Anticancer Research, 29*(10), 3825–3831.

Yang, Q., Wang, Y., Yang, Q., Gao, Y., Duan, X., Fu, Q., et al. (2017). Cuprous oxide nanoparticles trigger ER stress-induced apoptosis by regulating copper trafficking and overcoming resistance to sunitinib therapy in renal cancer. *Biomaterials*, 72–85. https://doi.org/10.1016/j.biomaterials.2017.09.008.

Yuan, Z. X., Mo, J., Zhao, G., Shu, G., Fu, H. L., & Zhao, W. (2016). Targeting strategies for renal cell carcinoma: From renal cancer cells to renal cancer stem cells. *Frontiers in Pharmacology*, 7. https://doi.org/10.3389/fphar.2016.00423.

Zhang, Y., Li, M., Gao, X., Chen, Y., & Liu, T. (2019). Nanotechnology in cancer diagnosis: Progress, challenges and opportunities. *Journal of Hematology & Oncology*, *12*(1), 1–13. https://doi.org/10.1186/s13045-019-0833-3.

Zheng, X., Zhang, X., Sun, H., Feng, B., Li, M., Chen, G., et al. (2006). Protection of renal ischemia injury using combination gene silencing of complement 3 and caspase 3 genes. *Transplantation*, *82*(12), 1781–1786. https://doi.org/10.1097/01.tp.0000250769.86623.a3.

Zhou, P., Sun, X., & Zhang, Z. (2014). Kidney–targeted drug delivery systems. *Acta Pharmaceutica Sinica B*, 37–42. https://doi.org/10.1016/j.apsb.2013.12.005.

Further reading

He, M.-H., Chen, L., Zheng, T., Tu, Y., He, Q., Fu, H.-L., et al. (2018). Potential applications of nanotechnology in urological cancer. *Frontiers in Pharmacology*, 9.

Liu, J., Boonkaew, B., Arora, J., Mandava, S. H., Maddox, M. M., Chava, S., et al. (2015). Comparison of sorafenib-loaded poly (lactic/glycolic) acid and DPPC liposome nanoparticles in the in vitro treatment of renal cell carcinoma. *Journal of Pharmaceutical Sciences*, *104*(3), 1187–1196. https://doi.org/10.1002/jps.24318.

Wang, G., Li, Q., Chen, D., Wu, B., Wu, Y., Tong, W., et al. (2019). Kidney-targeted rhein-loaded liponanoparticles for diabetic nephropathy therapy via size control and enhancement of renal cellular uptake. *Theranostics*, *9*(21).

Zhou, J., Li, R., Zhang, J., Liu, Q., Wu, T., Tang, Q., et al. (2021). Targeting interstitial myofibroblast-expressed integrin αvβ3 alleviates renal fibrosis. *Molecular Pharmaceutics*, *18*(3), 1373–1385. https://doi.org/10.1021/acs.molpharmaceut.0c01182.

CHAPTER 8

Designing of nanocarriers for enhancing drugs absorption through gastrointestinal tract

Introduction

Oral drug delivery remains the favored route of drug ingestion because of its simplicity and convenience. This route of drug administration is associated with the highest level of patient compliance as it offers the greatest degree of flexibility in dosage plans and allows convenient self-administration. Oral drug administration is also of great interest due to the physiological reasons and conditions that the drug faces in the gastrointestinal (GI) tract. The GI tract is composed of absorptive enterocytes (epithelial cells) that offer an extensive surface area (up to 400 m^2) for drug absorption (Pridgen, Alexis, & Farokhzad, 2015). The GI tract also comprises some other types of absorptive cells such as Paneth cells, endocrine cells, mucin-secreting goblet cells, and specialized M cells which take part in drug absorption. The specialized M cells in Peyer's patches are also responsible for the transport of antigen via dendritic cells (Pawar et al., 2014). However, several hydrophilic and hydrophobic drugs (e.g., polyene antibiotics, aminoglycosides, taxanes, etc.) have low bioavailability when ingested orally because of their mediocre physicochemical (stability, solubility) and biopharmaceutical (metabolic stability, permeability) properties (Fig. 8.1) (Gao et al., 2013; Leo et al., 2010). Moreover, most of the new chemical substances identified by drug discovery methods exhibit low aqueous solubility and inadequate permeability. Almost 70% of the newly identified substances are rejected during the pre-clinical process owing to their low oral bioavailability (Gao et al., 2013).

The intestinal enterocytes (i.e., epithelial cells) have unique physiology and structure in the GI tract. They act as the primary barrier against the exogenous entities that are administrated orally. Therefore, a comprehensive understanding of the physiology, structure, function as a barrier, and characteristics of the intestinal epithelial cells is very important and essential for developing an effective nanocarrier for oral drug delivery. One of the foremost characteristics of these intestinal cells is their reliability due to the incidence of tight junction (TJs) proteins, these proteins set up connections among nearby cells via the intercellular space (Vllasaliu, Fowler, Garnett, Eaton, & Stolnik, 2011).

Nanocarriers for Organ-Specific and Localized Drug Delivery. https://doi.org/10.1016/B978-0-12-821093-2.00010-4

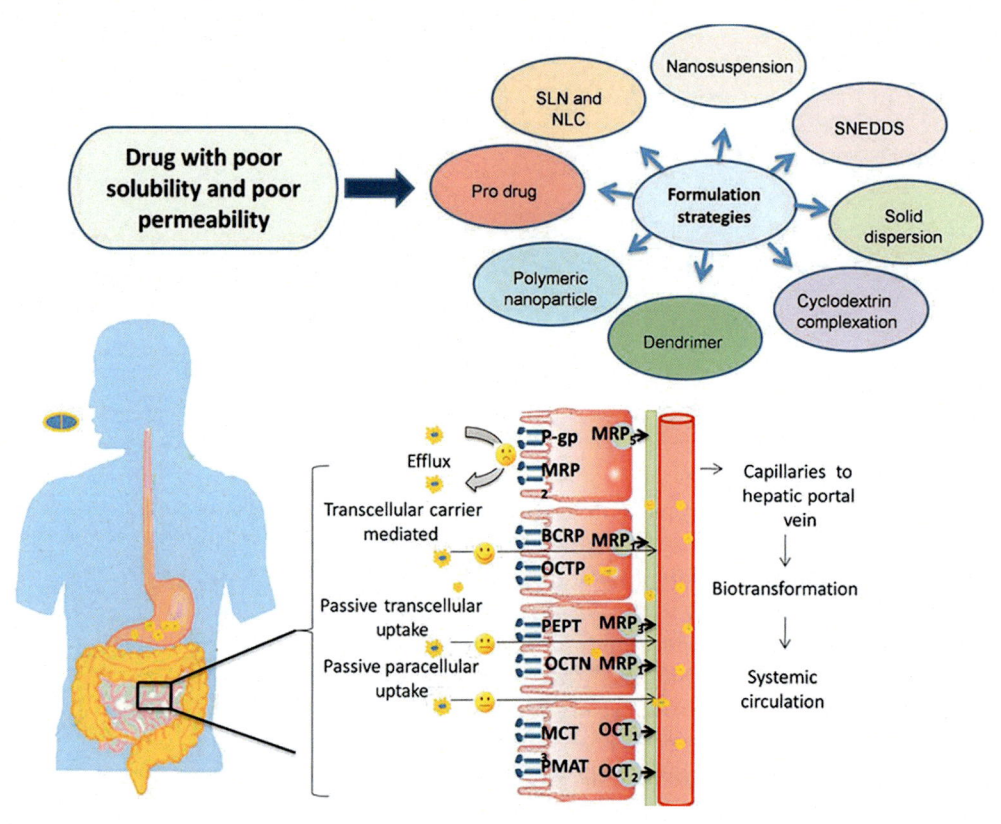

FIG. 8.1

Strategies for enhancing drug absorption through the gastrointestinal tract. *From Shekhawat, P. B., & B. Pokharkar, V. (2017).*
Understanding peroral absorption: Regulatory aspects and contemporary approaches to tackling solubility and permeability hurdles. Acta
Pharmaceutica Sinica B, 7(3), 260–280. https://doi.org/10.1016/j.apsb.2016.09.005.

The absorptive cells (enterocytes) and the secretory cells (i.e., Paneth cells and
goblet cells) are the most abundant intestinal epithelial cells. The absorptive
cells have some typical characteristics because of the microvilli at their apical
surfaces which make a dense and well-organized brush-border facilitating
the transport of nutrients. Whereas, the primary function of goblet secretory
cells is to discharge mucin (glycosylated proteins) which are actively participat-
ing in the mucosal host defense. Similarly, the Paneth cells also secrete many
antimicrobial substances to provide protection against microorganisms in the
small intestine (Lueschow & McElroy, 2020). However, the specialized M cells
in the Peyer's patches have some exceptional features, for example, large numbers of mitochondria, lack of mucus layer at surfaces, and small numbers of
cytoplasmic lysosomes (Azzali, 2003).

The progressive research in the field of nanotechnology has offered innovative
applications of predesigned nanomaterials in several other fields of science

owing to their exceptional and exclusive properties (Kuzum, Yu, & Philip Wong, 2013). The versatility of nanomaterials can be expanded by manipulating their chemical composition, surface, shape, and size. In the last two decades, controlling and directing the chemical properties of inorganic and organic materials at the nanoscale has been effectively used for developing nanocarriers for different routes of administration (Lombardo, Kiselev, & Caccamo, 2019). Nanomedicines or drug delivery through nanocarriers have opened new perspectives to address the flaws of the conventional dosage regimen (Hamidi, Shahbazi, & Rostamizadeh, 2012). In this connection, numerous drug delivery systems (DDSs) have been developed that demonstrated exceptional potential in therapeutic agents' delivery for treating different diseases while exhibiting marginal side effects and minimizing toxicity to other tissues of the body. Formulation into nanocarriers reduces the susceptibility of drugs in the GI tract environment, improves drug solubility and absorption, provides protection against enzymatic degradation, and offers controlled release in the GI tract. For example, some polymers, i.e., chitosan when coated on nanocarriers significantly increase the bioadhesivity of nanocarriers, reversibly open TJs, decline cellular integrity, and reduce transepithelial electrical resistance (TEER) in monolayers of Caco-2 cell lines, resulting in high permeability of the nanocarriers (Faralli, Shekarforoush, Ajalloueian, Mendes, & Chronakis, 2019). Even though extensive progress in oral drug delivery has been made, there are still remaining some shortcomings that need to be considered to improve the effectiveness of oral drug delivery. For instance, several nano formulations are highly sensitive to the pH variations in the GI tract, resulting in pH-induced oxidation or even hydrolysis and release of the drug molecules (Broesder et al., 2020). Besides, the mucus layer secreted from the surface of the epithelial cells five to six time a day act as an absorption barrier for nanocarriers (Zhang et al., 2021). Moreover, the nanocarriers may also be degraded by some specific enzymatic barriers (Cao et al., 2019). Keeping in mind the needs of commercial availability and clinical uses, it is obligatory to design smart nanocarriers or nanomedicines that can deliver the therapeutic agents to the desired site and can keep maintaining the functional features of therapeutic agents in the body. This scenario can be achieved only by addressing the harsh biological milieu that is formed by degrading chemicals and enzymes and protecting the therapeutic agents from modification in these harsh conditions (Petri et al., 2007).

Biological barriers to oral drug delivery

Orally taken all digested ingredients pass through three major biological environments in the GI tract irrespective of its target or absorption mechanism. These three main environments are the lumen, mucus, and tissue of GI (Fig. 8.2). The stomach, a harsh acidic and enzymatic environment, is the first

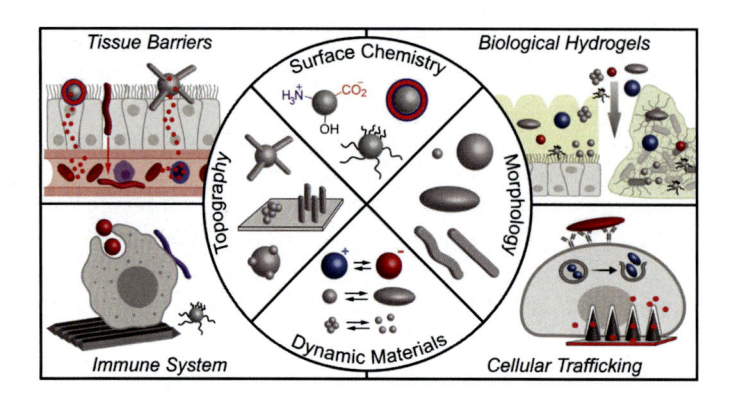

FIG. 8.2

Strategies to design drug nanocarrier to overcome biological barriers to oral drug delivery. Drug nanocarrier surface chemistry, topography, morphology, and trigger responsive behaviors can be modified to overcome biological barriers. *From Finbloom, J. A., Sousa, F., Stevens, M. M., & Desai, T. A. (2020). Engineering the drug carrier biointerface to overcome biological barriers to drug delivery.* Advanced Drug Delivery Reviews, 167, *89–108. https://doi.org/10.1016/j.addr.2020.06.007.*

biological barrier faced by the drug administrated orally. The strong acidic environment (pH 1–2.5) inside the stomach denatures most of the acid-sensitive drug molecules resulting in the low effectiveness of the drug (Bar-Zeev, Assaraf, & Livney, 2016; Moroz, Matoori, & Leroux, 2016). Besides the harsh acidic environment, some gastric enzymes, i.e., gelatinase and pepsin can also degrade drug molecules inside the stomach. In this connection, pH-responsive hydrogels have been prepared for protecting drug molecules from the harsh acidic environment of the stomach. These hydrogels protect drug molecules not only from the strong acidic condition but also provide shielding against gastric enzymes. For example, Yamagata et al. reported that sensitive drugs, i.e., insulin can be preserved against stomach acidic and enzymatic fluids when encapsulated in pH-responsive hydrogel microparticles (Yamagata et al., 2006). Similarly, Cerchiara et al. also reported a microencapsulation system containing a pH-responsive feature that effectively protects drug molecules from both acidic and enzymatic environments of the stomach (Cerchiara et al., 2016). Moreover, pancreatic enzymes are also important for digestion which are produced in the pancreas and secreted into the GI tract. Pancreatic enzymes include peptidases, amylase, trypsin, and lipase which digest peptides, starch, proteins, and fats, respectively. They are particularly abundant at the duodenum (entrance of the small intestine) and can also easily decompose biomolecules like nucleic acids resulting in gastric residential instability of biomolecules (O'Neill, Bourre, Melgar, & O'Driscoll, 2011). Despite the harsh enzymatic degradation, drug molecules also face some strong mechanical stresses inside the lumen. Mechanical stresses inside the lumen include peristaltic movement of the GI muscles, osmatic pressure along the GI tract, and the shear stresses generated inside the lumen by the flowing of gastric juice (Choi, Kim, Kang, & Montemagno, 2014; Ensign, Cone, & Hanes, 2012). The flow rate of gastric fluids may also reduce the contact time of drug molecules with the epithelial layer and leads to decreasing the absorption rate of drug molecules (O'Neill et al., 2011).

The second main biological environment of the GI tract is the mucus that directly interacts with all the digested ingredients. Mucus is a viscous, sticky, and elastic layer coated on the inner side of the entire GI tract responsible for the abduction of exotic moieties, particularly hydrophobic molecules, reducing their contact time with the epithelial layer. The abducted moieties are then discarded and thus the mucus layer is acting as one of the major parts of the immune system (mucosal immunity) (Liu et al., 2018; Shan et al., 2015). Mucus is mainly composed of proteoglycans coated mucin protein molecules and water, developing a negative charge on the mucus. Other chemical and biological moieties such as salts, antibodies, carbohydrates, cellular remnants, and bacteria are also found in mucus (Leal et al., 2018; Zhang et al., 2018).

The physicochemical properties of the drug molecules or the nanocarriers will decide their absorption sites and their mechanism of how they cross through the enterocytes (intestinal epithelial cells) in the GI tract. The two main pathways or routes for the absorption of drug molecules or nanocarriers are the transcellular route (transcytosis) and the paracellular route (diffusion via the spaces among epithelial cells). In the transcellular route, the drug molecules or the nanocarriers pass through the cell membrane of the epithelial cells and enter the enterocytes. Whereas the paracellular route offers a passage for small hydrophobic molecules only. The role of this route is limited in drug absorption owing to its thin physiology and covering a small fraction of the entire epithelium (Boegh, García-Díaz, Müllertz, & Nielsen, 2015). The major barrier in the paracellular route for oral drug delivery is the tight junctions (TJs) (Bein, Eventov-Friedman, Arbell, & Schwartz, 2018; Zhaeentan et al., 2018). One of the foremost characteristics of these intestinal cells is their reliability due to the incidence of TJs proteins, these proteins set up connections among nearby cells via the intercellular space.

Absorption mechanisms of nanocarriers in intestine

The transportation of nanocarriers across the intestinal membrane is a complex and dynamic process as the passage of nanocarriers and release of the drug via different functional routes occur in parallel. The absorption of nanocarriers in the intestine may occur through one or more pathways at the same time. The main pathways or routes for the absorption of drug molecules or nanocarriers are the transcellular route (transcytosis), the paracellular route (diffusion via the spaces among epithelial cells), and nanocarriers mediated transportation such as peptide transporters (influx mechanisms) and P-glycoprotein (efflux mechanisms) as shown in Fig. 8.3. One of the interesting features of oral drug

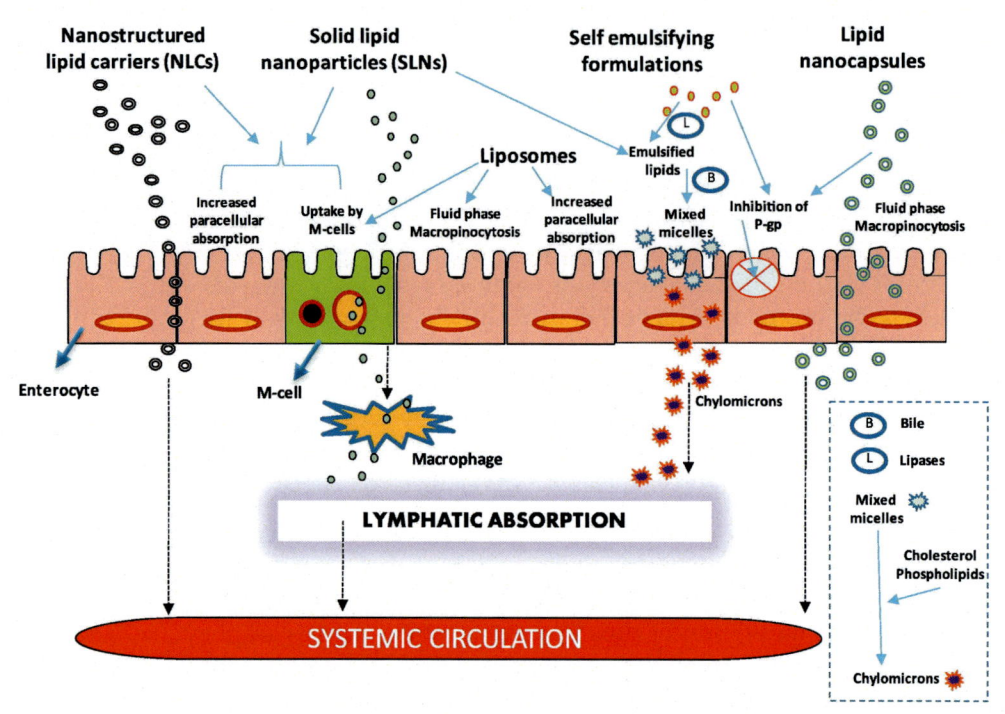

FIG. 8.3

The transportation mechanism of nanocarriers across the intestinal membrane. *From Ghadi, R., & Dand, N. (2017). BCS class IV drugs: Highly notorious candidates for formulation development. Journal of Controlled Release, 248, 71–95. https://doi.org/10.1016/j.jconrel.2017.01.014.*

delivery systems is that a nano-formulation ingested orally can pass through the GI tract, stick and absorb into the mucus layer, pass along the intestinal epithelium, enter the portal vein, and finally become part of peripheral circulation. The physical and chemical features of the nano-formulations are directly responsible to adopt the absorption sites and their routes to cross the enterocytes (intestinal epithelial cells) in the GI tract. The two main routes for the absorption of nano-formulations are the transcellular route and the paracellular route.

Transcellular transport

A passive transcellular transport (via passive diffusion) pathway is the most important route for lipophilic drug absorption through the oral route (Fig. 8.4). This route is more effective for drugs that have a low molecular weight (<700 Da), are highly permeable, and have a steep concentration gradient at the cell membrane. It also functions as a barrier for supramolecular or

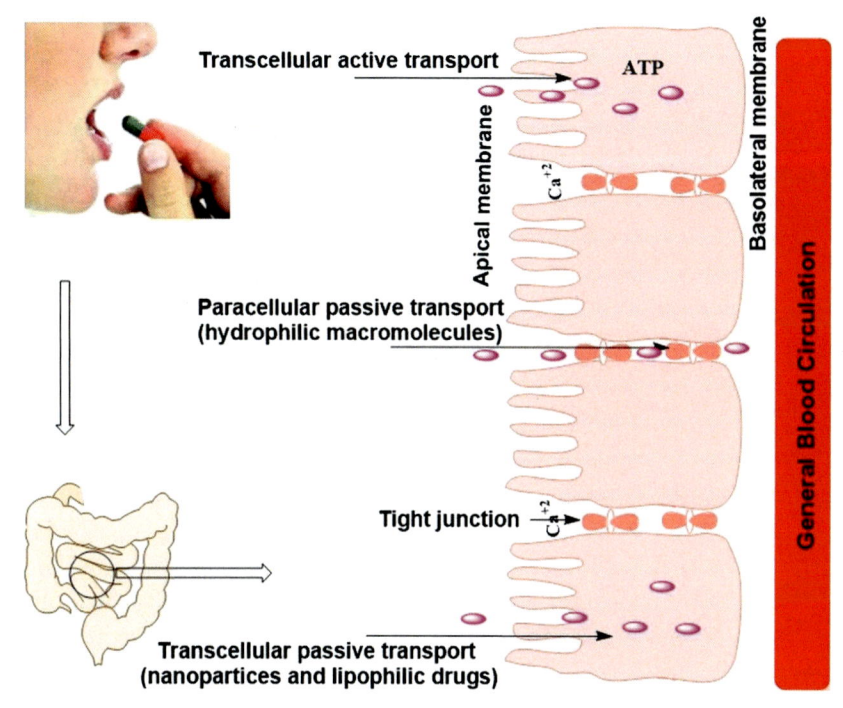

FIG. 8.4

Transport mechanism for absorption of drug molecules through intestine. *From Ibrahim, Y. H.-E. Y., Regdon, G., Hamedelniel, E. I., & Sovány, T. (2020). Review of recently used techniques and materials to improve the efficiency of orally administered proteins/peptides. DARU Journal of Pharmaceutical Sciences, 28(1), 403–416. https://doi.org/10.1007/s40199-019-00316-w.*

large-sized molecules which are taken up by M-cells of Peyer's patches or enterocytes (Homayun, Lin, & Choi, 2019). The transportation of nanocarriers, nano-formulations, supramolecular structures, or macromolecules largely happens through active transcellular transport. This type of transport occurs through different energy-dependent mechanisms such as caveolin-mediated endocytosis, clathrin-mediated endocytosis, micropinocytosis, and phagocytosis (Su, Jin, Li, Huang, & Li, 2020). The active transcellular transport of nanocarriers or macromolecules starts with the endocytic process, followed by conveying via the cells, and finally release at the basolateral sides (Salah, Abouelfetouh, Pan, Chen, & Xie, 2020). Several studies have demonstrated that easy active transcellular transport of nanocarriers through the endocytosis process is largely affected by factors such as size, surface charges, materials composing the nanocarriers, and even the ligand moieties anchored to the nanocarriers (Agrahari, Burnouf, Burnouf, & Agrahari, 2019). Hence, optimizing the physical and chemical features of the nanocarriers is very important for enhancing endocytosis. When muco-adhesivity of the nanocarriers is increased by coating with polyacrylic acid, polyvinyl alcohol, chitosan, or vitamin E succinylated PEG, the possibilities of uptake and absorption of these nanocarriers by epithelial cells in the small intestine are also enhanced (Babadi, Dadashzadeh, Osouli, Daryabari, & Haeri, 2020). Moreover, the use of lectin targeting ligands has expanded the specific-receptor-mediated transcytosis (Russell-Jones, 1996).

Paracellular transport

Hydrophilic molecules have a low affinity toward the lipophilic constituents of epithelial cells and cannot freely diffuse via the transcellular route. Therefore, the paracellular route is the only pathway for the absorption of hydrophobic molecules in the absence of an appropriate membrane transporter (Fig. 8.4). However, TJs work as a barrier for the absorption of hydrophilic molecules through the paracellular route and allow only less than 200 Da molecular weight drug molecules through this pathway (Xu, Shrestha, Préat, & Beloqui, 2021). Hence, the paracellular route can hardly be contributed to the absorption of drug-loaded nanocarriers into the bloodstream. To encounter this problem, along with the bio-adhesive polymers (i.e., methylcellulose and PEG) some absorption enhancers such as calcium chelators, anionic polymers, and cationic polymers (i.e., chitosan) can be used to open the TJs reversibly by non-covalent interactions with the mucus (Dan, Samanta, & Almoazen, 2020). Yeh et al. reported that the treatment of Caco-2 cells with chitosan resulted in the decrease of TJ strength because of the redistribution of claudin 4 (a transmembrane protein in TJs) and increase the paracellular permeability (Yeh et al., 2011). However, the intercellular spaces amid the epithelial cells also restrict the absorption of nanocarriers into the bloodstream through the paracellular route, and absorption of nanocarriers cannot entirely occur through this pathway (Vahdatpour, Mamaghani, Goloujeh, Maheri-Sis, & Mahmoodpour, 2016). As a matter of fact, paracellular permeability is associated with the release of drug molecules from the disintegrated nanocarriers.

Design considerations for nanocarriers to a specific section of the GI tract

An advantage of nanocarrier formulation is its potential for offering localized and/or targeted drug delivery. Though the term "targeted" sparks the vision of nanocarriers that actively search for a specific target and selectively; however, "targeting" in the GI tract is referred to different strategies employed for extending residence time, decreasing the ratio of drug degradation/release that occurs, and/or enhancing the interaction of nanocarrier formulations with specific tissues and cells in different sections of the GI tract.

Stomach targeting

For stomach targeting, a nanocarrier delivery system must address many physiological barriers such as gastric pH, gastric motility, and gastric mucus. The retention of drug molecules in the stomach is the most important factor for stomach targeting but is the most difficult to accomplish. Usually, microscopic delivery systems, for example, hydrogels, microparticles, and tablets are used for extending gastric retention (Adebisi & Conway, 2011, 2015). The general belief is to use a property like density or geometry to slow down gastric emptying and allow the drug release in the gastric environment for longer periods (Garg, Kumar, Rath, & Goyal, 2014; Prajapati, Jani, Khutliwala, & Zala, 2013). These microscopic dosages increase the retention time of the drug molecules in the stomach but in most cases, they may not be optimum for those drugs that have low absorption in gastric mucus.

Stomach targeting can be advantageous for drugs that are pH-dependent soluble (e.g., verapamil), pH-sensitive and degradable in the intestinal environment (e.g., captopril), and mainly absorbed in the stomach (e.g., metronidazole) (More, Gavali, Doke, & Kasgawade, 2018). Most nanocarriers designed to target the stomach are coated with muco-adhesive materials (e.g., chitosan) to extend gastric retention of nanocarriers (Abd El Hady, Mohamed, Soliman, & El-Sabbagh, 2019; Frank et al., 2020). Sometimes polyanions are also complexed with chitosan to prevent its rapid dissolution in the acidic pH of the stomach. Chang et al. reported amoxicillin-loaded nanocarriers for the treatment of *Helicobacter pylori* infections which were composed of a complex of poly-glutamic acid polyelectrolyte and chitosan (Chang et al., 2010). They also synthesized a cross-linked hydrogel from calcium ions, gelatin, and alginate to absorb the amoxicillin-loaded nanocarriers. They observed that after the incorporation of nanocarriers into hydrogel, the release of amoxicillin from nanoformulation was significantly lower at pH 1.2. For the evaluation of the therapeutic potential of these nanocarriers, fluorescently labeled nanocarriers were treated with *H. pylori*-infected gastric adenocarcinoma cells' monolayers. They found that nanocarriers successfully interact with *H. pylori*-infected sites of the monolayers.

Small intestine targeting

The small intestine's role in the absorption of nutrients, vitamins, electrolytes, drugs, and several other small molecules is very important as the end absorption of all these ingredients takes place here (Bornhorst, Gouseti, Wickham, & Bakalis, 2016; Pawar et al., 2014). Therefore, targeting the small intestine means to get increased nanocarriers uptake to ultimately make systemic drug delivery more effective and convenient. The small intestine is composed of specialized cells, receptors, specialized cell membrane transporters, and adopt dedicated transport mechanisms to absorb chemical moieties with diverse chemical structures (Qi et al., 2020). The aforesaid physiological aspects of the small intestine must be considered while designing nanocarriers for small intestine targeting (Madni et al., 2021). Furthermore, nanocarriers must endure the harsh gastric condition to reach the small intestine except they are encapsulated in a capsule that detours the stomach. To maintain nanocarriers' integrity in the gastric conditions and allow them to reach the intestine, enteric polymers have been developed that are labile to neutral pH but insensitive to acidic pH or gastric conditions (Ofridam et al., 2021). Therefore, enteric polymers-coated nanocarriers are employed for targeting drug to the small intestine passively. Enterocytes are the leading cell type in the intestine and targeting enterocytes leads to the maximum delivery of nanocarriers to the intestine and enhances absorption into the circulation system (Xu, Liu, et al., 2020). In this connection, PEG-coated nanocarriers have been found more interactive with small intestine villi and offer more consistent coverage of the surface of the epithelial cells as compared to unmodified nanocarriers (Maisel, Ensign, Reddy, Cone, & Hanes, 2015). Moreover, nanocarriers coated with enzymes such as papain or thiols have also been used for improving mucus penetration (Müller, Perera, König, & Bernkop-Schnürch, 2014; Wilcox, Van Rooij, Chater, Pereira De Sousa, & Pearson, 2015). Enzymes (e.g., papain) have the ability to actively degrade mucus and increase the interaction of nanocarriers with the epithelial cells, while the incorporation of thiols into nanocarriers reduces the cross-linking of mucus to improve nanocarriers penetration (Köllner et al., 2015; Pereira De Sousa et al., 2015). Besides, the size, morphology, surface charge, enteric polymer density, and materials of nanocarriers also direct their ability to absorb through mucus, hence, these physical aspects should also be kept in mind while designing nanocarriers (Ensign et al., 2013; Yildiz, McKelvey, Marsac, & Carrier, 2015).

Intestinal lymphatic targeting

The lymphatic system is a part of the circulatory and immune system composed of tonsils, Peyer's patches, thymus, spleen, lymph nodes, and lymphatic vessels. The intestinal lymphatic targeting means to increase the relative amount of therapeutic agents that makes part of the lymphatic system. Targeting the lymphatic system is very decisive in managing and predicting diseases such as AIDS, tuberculosis, metastatic cancers, filariasis, and several other chronic inflammatory diseases (Ahammed et al., 2017; Managuli, Raut, Reddy, & Mutalik, 2018). It can also help to bypass the first-pass metabolism as the mesenteric lymph bypass the liver organ and joins directly with the systemic circulation. So, targeting the lymphatic system is very effective for increasing the oral bioavailability of drugs like steroids that are very susceptible to first-pass metabolism (Nabi,

Rehman, Baboota, & Ali, 2019; Pandya, Giram, Bhole, Chang, & Raut, 2021; Plaza-Oliver, Santander-Ortega, & Lozano, 2021).

Comprehensive knowledge about the fundamental function of the lymphatic system is very essential to better understand the aspects of intestinal lymph-targeted nanocarriers. Currently, many studies have been reported on the subject of designing, developing, and evaluating lymph-targeted ligands and prodrugs. These prodrugs and ligands are also very useful for designing and developing nanocarriers targeted to the lymphatic system. As the capillaries of the lymphatic system are much more amenable and permeable to large-sized particle transport as compared to the nearby blood capillaries (Han et al., 2014; Trevaskis, Kaminskas, & Porter, 2015). So, exocytosed lipoproteins are particularly transferred to the mesenteric lymph and, ultimately, enter the systemic circulation. Several studies have been conducted to find out the physico-chemical requirements of different lipids for increasing intestinal lymphatic transport. It has been observed that fatty acids of 14 or greater carbon chains are primarily shuttled through lymphatic transport (Porter & Charman, 2001). Besides, studies designed about the prodrugs have revealed that triglyceride-based lipid prodrugs show better lymphatic transport as compared to the monoacylglycerol-based lipid prodrug (Trevaskis, Charman, & Porter, 2008; Trevaskis et al., 2015). Similarly, phospholipids such as lysophosphati-dylcholine and phosphatidylcholine can also improve the lymphatic transport of vitamins and drugs. All these studies have apprised the principles of design and development for lipid nanocarriers containing phospholipids or fatty acid glycerides and targeting the intestinal lymphatic system.

Colon targeting

The colon "targeting" means to adopt such approaches that result in minimum drug degradation and absorption in the GI tract before reaching the colon, which can enhance drug delivery to colorectal tissues. Colon targeting is of great interest in addressing the bowel ailments such as amebiasis (parasitic diseases), colonic dysmotility, diverticulitis, irritable bowel syndrome, colon cancer, and inflammatory bowel diseases (Alhakamy et al., 2020; Goyal, Singh, Kumar, & Gupta, 2021; Sarangi, Rao, Parcha, & Upadhyay, 2020). It has also been used as a good approach for increasing the macromolecules' systemic absorption due to diminished P-glycoprotein expression, decreased CYP3A4 activity, lower proteolytic activity, and prolonged transit time in the colon in contrast to the small intestine (Hua, Marks, Schneider, & Keely, 2015; Vinarov et al., 2021). Nanocarriers designed for colon targeting should be capable to mini-mize drug release in the GI tract, despite the harsh acidic conditions, facing a wide pH range, and the presence of different digestive enzymes. In addition, nanocarriers should be responsive to release drug upon its entry to the colon.

pH-responsive polymers have usually been used for designing nanocarriers to target the colon. The colon pH remains in the range of 6.2–7.2 at different sections of the colon. In this connection, different pH-responsive methacrylic acid copolymers, such as Eudragit FS 30D (Luo et al., 2020; Moghimipour et al., 2018), Eudragit S100 (El-Maghawry, Tadros, Elkheshen, & Abd-Elbary, 2020; Sood et al., 2019), and Eudragit L100 (Dong et al., 2019) that dissolve in the range of pH 6–7 are commonly utilized for designing drug delivery systems for targeting the colon. However, it should be kept in mind that some colonic ailments, i.e., inflammatory bowel diseases are directly associated with variations in the colonic pH (Zeeshan, Ali, Khan, Khan, & Weigmann, 2019). The colonic pH of ulcerative colitis patients usually ranges from 2.7 to 5.5 while the average colonic pH of Crohn's disease patients is 5.3. Fortunately, few Eudragit polymers such as Eudragit E100 and Eudragit L100-55 are also available that dissolve in the range of pH from 2.5 to 5.5 (Ofridam, Lebaz, Gagnière, Mangin, & Elaissari, 2020). Thus, it may be needed to coat the nanocarriers with additional coatings of such suitable polymers to avoid premature drug release in the GI tract before reaching the colon. Some studies have also evaluated the potential of lectins as ligands for nanocarriers to achieve colon targeting (Moulari, Béduneau, Pellequer, & Lamprecht, 2014). Other targeting ligands which are commonly used for targeting goblets cells, M cells, and enterocytes can also be employed to achieve colon targeting of nanocarriers.

Nanocarriers for enhancing intestinal drug absorption

Enhancing the transportation of drug molecules across the gastrointestinal tract has been one of the key factors for oral drug delivery. With the advancement of nanomedicines, different types of nanosized tunable systems have been developed as an interesting approach potentially able to address the unmet challenge of enhancing transportation of therapeutic agents across the gastrointestinal tract (Agrahari et al., 2019; Homayun et al., 2019). Nanocarriers' technology among these systems is rapidly advancing. Nanocarriers are nano-sized (1–100 nm) objects that are considered as complete units in terms of properties and transport. The nanocarriers based on polymeric nanoparticles (NPs), liposomes, hydrogels, micelles, polymer drug conjugates, dendrimers, liposomes, solid lipid nanoparticles, and spontaneously emulsifying systems are more popular in oral drug delivery (Mishra, Patel, & Tiwari, 2010). The high expectation from these nanocarriers is linked to their easy multi-functionalization along with their capability to carry payloads and cross the gastrointestinal tract. The drugs or therapeutic agents delivery to systemic circulation across the gastrointestinal tract via nanocarriers are wholly dependent on the biomimetic and physiochemical characteristics of the nanocarriers and do not depend on the physical or chemical properties of the drug which is encapsulated inside the nanocarriers.

Polymeric nanoparticles

Polymeric NPs have some exceptional advantages for use as drug carriers, including protecting drug molecules from harsh environments (chemical modification and enzymatic degradation), enhanced drug solubility, controlled and sustained drug release, and increasing the therapeutic index by improving drug bioavailability (Cano et al., 2019, 2020). Polymeric NPs are synthesized in one of the two morphological forms, i.e., nanospheres and nanocapsules (Schaffazick, Pohlmann, Dalla-Costa, & Guterres, 2003). Nanospheres are made of a continuous polymeric framework in which the drug molecule can be adsorbed onto their surface or retained within the framework. Whereas nanocapsules possess an oily core which provides a medium to dissolve drugs and the oily core is surrounded by a polymeric shell controlling the drug release from the core (Crucho & Barros, 2017; Szczęch & Szczepanowicz, 2020).

Polymeric nanoparticles have generally brought significant outcomes to oral drug delivery by improving tolerability, specificity, efficacy, stability, and absorption of the therapeutic agent, or even more precisely, delivering poorly water-soluble molecules, targeting specific regions of the GI tract, and transporting macromolecules across GI barriers, resulting in improved pharmacokinetic properties and therapeutic index of their payload (Crucho & Barros, 2017; El-Say & El-Sawy, 2017). To accomplish this, the polymeric nanocarriers must exhibit several distinguishing characteristics, including sufficient stability, nontoxicity, biocompatibility (non-immunogenicity), biodegradability, minimal degradation of the loaded drug, adaptability to a variety of drugs, and resistance to reticuloendothelial system (RES) uptake (Taghipour-Sabzevar, Sharifi, & Moghaddam, 2019).

Hydrogels

In the field of targeted drug delivery, polymeric hydrogels are a class of nanocarriers that are particularly well suited for increasing oral bioavailability due to their tunable physical and chemical properties such as smartness, softness, low toxicity, high affinity for water, excellent biodegradability, and biocompatibility (Lavrador, Esteves, Gaspar, & Mano, 2021; Liarou et al., 2018). Hydrogels are well-characterized hydrophilic gels composed of cross-linked materials that have the ability to absorb large contents of water without dissolving (Croitoru et al., 2020). In this connection, increased efforts have been made to develop new and emerging types of hydrogels possessing the capability of controlled drug release kinetics. As a result, hydrogels composed of acrylic moieties containing polymers have drawn considerable interest for oral drug delivery due to their safety, biocompatibility, ease of manufacture, and most importantly, they are mucoadhesive (Bhullar, Rani, Kumari, & Sud, 2021). Because of their mucoadhesivity, they can effectively interact with the mucosa lining of the intestine, increasing the drugs' oral bioavailability. For instance, Carballido and colleagues investigated an oral drug delivery application for poly(magnesium acrylate) (PAMgA) hydrogel (Cheddadi, López-Cabarcos, Slowing, Barcia, & Fernández-Carballido, 2011). They found the PAMgA hydrogels as excellent carriers for oral administration due to their low cytotoxicity and adequate biocompatibility.

Newly designed and manipulated hydrogel-based nanostructures are tuned to deliver drugs to targeted regions of the GI tract. Patel and Amiji reported a cationic pH-sensitive hydrogel for targeting the stomach and observed promising results in the treatment of *H. pylori*-induced peptic ulcers (Patel & Amiji, 1996). These nano-hydrogels act in the stomach's acidic environment through their swelling properties and antibiotic release. Additionally, the potential of hydrogels for colon-specific protein delivery is being extensively investigated, owing to the colon's lower proteolytic activity. In practice, these carriers degrade or exhibit swelling properties in response to colonic enzymes or microflora, resulting in colon-specific drug delivery. There are numerous types of unique hydrogels available today, made from natural (dextran and carrageenan) or synthetic (polyanhydrides and polyesters) polymers (Amani et al., 2021; Lerma, Garcés, & Palencia, 2020; Patil et al., 2019; Pettinelli et al., 2020). For instance, the most frequently used polymers for hydrogel preparation include alginate, chitosan, polyvinyl alcohol (PVA), polyvinyl pyrrolidone (PVP), polyethylene oxide (PEO), and poly-*N* isopropylacrylamide (Hamidi, Azadi, & Rafiei, 2008).

Micelles

The amphiphilic copolymers produce polymeric micelles upon aggregation in the aqueous medium. Polymeric micelles are spherical nanostructures composed of a hydrophobic core and a hydrophilic shell and are highly stable in suspension form (Mishra et al., 2010). The stability of the polymeric micelles' system can be further improved by crosslinking the core or the shell chains. The core of the polymeric micelles is used as a reservoir for drug molecules holding or encapsulation. Furthermore, the polymeric micelles can be decorated with special tunable features to make them responsive to stimuli, i.e., temperature, light, pH, ultrasound, hypoxia, overexpression of a specific enzyme, etc., and triggering a sustain and control release of encapsulated drug molecules (Kabanov et al., 1992). Pluronic-type block copolymer composed of propylene oxide and ethylene oxide are among the most used polymers for preparing the polymeric micelles.

Micelles are generally preferred because of their excellent attributes including enhanced drug solubility, small particle size, low toxicity, and an extended circulation time (Bose, Roy Burman, Sikdar, & Patra, 2021). The micelles must be able to resist dissociation and degradation in the harsh environment of the GI tract to ensure a successful drug delivery and absorption. While they can be administered orally, micelles are more susceptible to changes in pH and the presence of digestive enzymes and bile salts in the gut. So, adequate drug retention in micelles and micelle stability is necessary for proper drug delivery because the drug must be delivered in the solubilized state to the absorption site, which can be accomplished by first transferring the drug to the absorption site and then releasing it in a sustained manner. It is imperative to gain a deep understanding of how drug-loaded micelles interact with the intestinal membrane (Trivedi & Kompella, 2010). Although the corona of micelle appears hydrophilic, it is possible that they can limit the interaction with the intestinal mucosa. They can, however, be internalized through an energy-dependent process such as fluid-phase pinocytosis, which may be the most probable and major route. Despite that, there are disagreements about exactly how the mechanism works. In Caco-2 experiments, it hasn't been found that micelles come into contact with cellular membranes, possibly due to the micelles' hydrated shells blocking the way. However, it has been proved that in the presence of nano micelles the transepithelial electrical resistance does not change (Collnot et al., 2006). Zhang and colleagues prepared a typical micellar system of PEG3000-PLA3000 containing a fluorescent probe Coumarin-6 and assessed its permeability mechanism in MDCK epithelial cell line to evaluate its permeability potential across the epithelial barriers (Zhao et al., 2012). After introducing different inhibitors for macropinocytosis (amiloride), endosome maturation (monensin), cholesterol depletion (nystatin and MβCD), and clathrin-mediated endocytosis (hypertonic sucrose, potassium depletion, and chlorpromazine), they concluded that MβCD is the most imperative transport inhibitor. The hypertonic sucrose was categorized as a second and then followed by nystatin and monensin. After treatment with MβCD (inhibition of caveolae-mediated endocytosis and clathrin-independent endocytosis), the micelles integrated poorly into MDCK cells, but the cell cytoskeleton remained intact.

Dendrimers

Dendrimers are well-ordered, tree-shaped multiple branched homogenous synthetic molecules with a central core (Ahlawat, Henriquez, & Narayan, 2018). Their unique features such as rigidity, low polydispersity, predictable molecular weight, monodispersed phase, nanosized range, and easy to be modified render them excellent candidates to be used for developing nanocarriers (Zhang et al., 2017; Zhao et al., 2015). They are mostly used as carriers for hydrophobic drugs to a specific area because of their potential for accommodating hydrophobic molecules within their nanostructures. Despite the excellent properties of the dendrimers, there are also some limitations associated with them, such as complications in accomplishing targeted delivery, batch-to-batch variation, and high cost of production. Some safety issues associated with cationic dendrimers, i.e., positively charged nanocarriers strongly interact with the anionic cell surface which is negatively charged, resulting in high toxicity. These interactions create nanopores in the cell membrane and provoke the probable leakage of cellular content, resulting in cell death (Janaszewska, Lazniewska, Trzepiński, Marcinkowska, & Klajnert-Maculewicz, 2019).

Numerous studies have demonstrated the potential of dendrimers to improve the oral bioavailability of poorly water-soluble drugs. For instance, surface-modified G3 PAMAM increases propranolol's solubility while reducing the impact of P-gp efflux transporters, thereby increasing drug bioavailability (Sorroza-Martínez, Ruiu, González-Méndez, & Rivera, 2021). Another interesting study was reported by Ghandehari and colleagues who discovered that 7-ethyl-10-hydroxycampothecin (SN38), a poorly water-soluble drug, conjugated to PAMAM dendrimers showed promising results against hepatic colorectal cancer metastases when administered orally (Goldberg, Vijayalakshmi, Swaan, & Ghandehari, 2011). Conjugates of G3.5-SN38 were shown to increase transepithelial transport of SN38 in the apical to basolateral direction, demonstrating the capability of this vehicle for oral delivery. Also, the authors expressed the critical importance of using a strong linker with the capacity to accept intact drug dendrimer complexes, which then enabled them to be broken down in the bloodstream. Similarly, diethylene glycol was demonstrated to be an appropriate linker for the oral delivery of naproxen, a poorly water-soluble drug, because the PAMAM dendrimer could maintain its chemical stability while releasing the drug, which greatly enhanced the drug's oral bioavailability (Najlah, Freeman, Attwood, & D'Emanuele, 2007).

Lipid or oil based drug carriers

Liposomes

Liposomes are small artificial lipid bilayer spherical-shaped vehicles that are composed of cholesterol and amphiphilic lipids or natural phospholipids. They can hold a variety of molecules, including therapeutic agents such as proteins, nucleic acids, drugs, and vaccines. They are mostly biocompatible, amphiphilic in nature, can be easily modified, and enhance circulation time. These characteristics make them exceptional and efficient carriers for the delivery of both types of drugs, whether lipophilic or hydrophilic (Ahlawat et al., 2018). Liposomes are mostly used for targeted drug delivery because of their ability to easy modification, easily bypassing the parenchyma cells, and delivering drugs to the desired sites (Barchet & Amiji, 2009). In spite of the unique properties of the liposomes, there are some limitations also associated with them, such as unstable to keep stored for a long period, leakage of the held drugs during storing for a long time, short half-life, batch-to-batch variation, low solubility, and high cost of production in some cases. Some safety issues such as hepatotoxicity and immunogenic responses are raised upon liposome introduction (Wolfram et al., 2015). They also induce the overexpression of pro-inflammatory cytokines leading to inflammation.

There are numerous reports indicating that orally administered drug-loaded liposomes have better bioavailability and elimination half-life as well as reduced side effects compared to the free drug (Fricker et al., 2010; Riehemann et al., 2009). This advantage is thought to be due to the drug being physically entrapped within the liposome or conjugated to its constituent components, which results in effective solubilization, offers protection against enzymatic degradation, and remains inactive in the GI tract and bloodstream (Hofheinz, Gnad-Vogt, Beyer, & Hochhaus, 2005; Lilia Romero & Morilla, 2011). Apart from water-soluble drugs, the prospective use of liposomes for oral administration of drugs with low stability, poorly water-solubility, and low absorption have also been extensively studied (Chen et al., 2009; van Hoogevest, Liu, & Fahr, 2011). Despite some disagreements about liposomes' ability to protect sensitive drugs from degradation in the harsh environment of the GI tract, there is widespread agreement regarding the resistance of multi-lamellar vesicles (MLVs) (composed of cholesterol and phospholipids with a phase transition temperature higher than 37°C) in the GI tract (Chakraborty, Shukla, Mishra, & Singh, 2009). Another issue regarding the liposome absorption study relates to differences in results between similar studies utilizing the same liposomes, which are assessed in different cell lines. Indeed, this demonstrates the significance and effect of the liposome's intracellular fate on the cell line characteristics and cellular uptake pathway. In non-polarized NIH-3T3 cells, octaarginine (R8)-modified liposomes are primarily taken up via macropinocytosis, with an innate ability to escape efficiently from subsequent lysosomal degradation and endosomes (Khalil, Kogure, Futaki, & Harashima, 2006).

Solid lipid nanoparticles (SLNs)

Solid lipid nanoparticles (SLNs) contain a solid core made of hydrophobic lipid and its primary function is to hold drugs in the dispersed or dissolved form (Kaur, Bhandari, Bhandari, & Kakkar, 2008). They are composed of biocompatible lipids, i.e., waxes, fatty acids, and triglycerides. The small size of SLNs (40–200 nm) provides them with the advantage of escaping from the reticuloendothelial system (Pardeshi et al., 2012). In the manufacturing process, the drug is first mixed in melted lipid and then dispersed in the aqueous surfactants by micro-emulsification or high-pressure homogenization. The SLNs have advantages over other NPs systems as they are biocompatible, exhibit high drug entrapment efficiency, capable of providing an uninterrupted release of the drug for weeks, and can be stored for long periods (Mishra et al., 2010). Moreover, their composition can be modified and controlled to incorporate special properties to their surface to limit their uptake via the reticuloendothelial system and target the brain cells (Blasi, Giovagnoli, Schoubben, Ricci, & Rossi, 2007). Surfactants are key components of these nanocarriers because they provide stability to the SLNs. SLNs are better than liposomes in terms of drug protection, physical stability, and the ability to control the release of loaded drugs. These benefits come with a few drawbacks, including insufficient drug loading capacity and the possibility of drug expulsion. Khuller and colleagues demonstrated the potential of SLNs to be used for oral delivery of anti-tuberculosis (TB) drugs, intending to shorten the long duration of therapy and reduce the high dose of TB drugs (rifampicin, pyrazinamide, and isoniazid) (Pandey, Sharma, & Khuller, 2005). In vivo experiments demonstrated that therapeutic drug concentrations can be maintained for up to 10 days in the organs and 8 days in the plasma when administrated through SLNs, compared to 1–2 days for free drugs.

Spontaneous emulsifying systems (SES)

SES are oil-in-water isotropic emulsions that have revolutionized oral drug delivery with significant inter- or intra-individual variability in high susceptibility to hydrolysis, slow dissolution rate, and pharmacokinetic parameters, mainly due to their unique structure composed of drug, surfactant(s), and a mixture of oil (Avasarala, Dinakaran, Kakaraparthy, & Jayanti, 2019). Other substances such as solubilizers cosurfactants, and/or co-emulsifiers may be added to enhance SES formation and improve stability as well as drug encapsulation. In addition to the benefits of conventional nano-emulsions, SES comprised of appropriate components can help in reducing P-gp efflux in the gastrointestinal tract (Betageri, Kovvasu, Kunamaneni, & Joshi, 2019). Furthermore, the SES' anhydrous nature offers a variety of different ways to encapsulate it, including soft or hard gelatin capsules, as well as other forms, such as tablets and pellets, resulting in easy oral administration. One disadvantage of the SES is the sudden increase in drug plasma concentration (because of the rapid gastric emptying of emulsions, which are like solutions), as well as an increased risk of adverse side effects if the drug has a low therapeutic index. To overcome this, controlled release tablets/pellets of the SES are intended to control the administered drug's high peak plasma concentrations.

Along with the more commonly used O/W SES for drugs with low solubility, double emulsions water-in-oil-in-water (W/O/W) are well suited for drugs with high water-solubility and low permeability; however, there are some worries about their stability. A novel self-dual emulsifying pidotimod has recently been developed by Zhu and colleagues, adding some specific hydrophilic surfactants to W/O emulsions (Qi, Wang, Zhu, Hu, & Zhang, 2011). Pidotimod is a synthetic dipeptide with immunological and biological activity on both the innate and adaptive immune responses. The resulting emulsions are stable up to 6 months at 25°C, exhibit improved absorption compared to the parent pidotimod solution (~2.5-fold), and have minimal local damage. The polarity and particle size of the loaded drugs determine the release pattern of the loaded drugs in these systems.

Strategies for intestinal transport improvement of drug nanocarriers

Despite significant efforts to develop oral nano-drug delivery systems, a complicated real concern has drawn the researchers' attention toward the entrapment of nano vehicles in mucus and their rapid elimination. The continuous mucus secretion and shedding (a prophylactic process that prevents penetration in the exposed epithelial surfaces) causes rapid elimination of nano vehicles, which overcomes the controlled release of loaded drugs and the effectiveness of nanoparticles. Mucus is a complex hydrogel made up of a variety of components including antibodies, bacteria, cell debris, carbohydrates, lipids, salts, and most critically mucins, a group of proteins that give mucus its dynamic viscoelastic feature, which is critical for effective shielding. Although nanocarriers can compensate for poor stability and solubility, mucus acts as a barrier, preventing drug-loaded nanocarriers from permeabilizing and being absorbed (Liarou et al., 2018). To optimize nanoparticle transport through the colon, a variety of direct and indirect solutions are currently available for better understanding the main properties of nanoparticles that are most important for mucosal absorption (Cao et al., 2019).

Mucoadhesive nanoparticles

The majority of nanoparticles directly transit in intact form through the GI tract after oral administration, resulting in insufficient payload release and preventing the realization of a high drug concentration in the GI tract, leading to limited oral bioavailability and poor efficacy (Yan et al., 2020). In recent years, several materials' mucoadhesive qualities have been used as a non-specific approach to extend nanocarriers' residence time in the GI tract by enhancing mucus association and reducing direct transit and fecal removal. Hydrogen-bonding, van der Waals contacts, hydrophobic interactions, electrostatic interactions (positively charged materials), polymer entanglement, and polymer chain interpenetration with mucins are all involved in this strategy (Xu, Shrestha, Préat, & Beloqui, 2020). In addition to the benefits listed earlier, the likelihood of unintentional surface adhesion is a drawback when using mucoadhesive compounds in oral medication delivery systems. Furthermore, trapping in the weakly adhering mucus layer may occur, leaving the mucociliary system vulnerable to fast mucociliary clearance before gradual penetration into the firmly adhering mucus layer (the physiological mucus layer turnover time, which ranges between 50 and 270 min, plays a limiting function). As a result, there is a need for operational endeavors to overcome these flaws by utilizing the benefits of nanoparticle surface modification or utilizing particular lipid-based or pH-responsive formulations. Des Rieux and coworkers have demonstrated that the surface chemistry of nanoparticles has a significant impact on intestinal uptake by showing that nanosized aminated particles have better cellular uptake than carboxylated particles, which they attribute to the interaction of tertiary amines with anionic components of mucus (Rieux et al., 2005). The most important polymers with mucoadhesive characteristics include chitosan (CS), Poly-lactic acid (PLA), Polyglycolic acid (PLGA), poly(sebacic acid), and poly-acrylic acid (PAA).

Permeation enhancer nanoparticles

The drawbacks of mucoadhesive carriers, as well as unsatisfactory absorption results have encouraged researchers to design engineered muco-inert mucus-penetrating nanocarriers which make them capable of penetrating the mucus barrier and releasing the drug near epithelial cells. These nanocarriers take advantage of penetration enhancer materials in a variety of ways, including mucus viscosity reduction, mucus rheology, and improved membrane fluidity through interactions with proteins or lipids in epithelial cells (Sabir et al., 2021; Sladek et al., 2020). The disruption of intracellular barriers, and intercellular lipids, hindering the effects of different enzymes inside the mucosal surface, interfering with TJ products (notably desmosomes), and modifying the drug's partition coefficient, and interacting with calcium ions to increase the thermo-dynamic activity of the drug have also been studied for enhanced permeation of nanocarriers. The most common permeation enhancers employed in nanos-tructures are surfactants, chelators, and fatty acids (Franz-Montan, de Araújo, de Morais Ribeiro, de Melo, & de Paula, 2017). The hydrophobic nature of the mucus is due to the non-glycosylated part of the glycoproteins and the involvement of lipid molecules, a molecule with prominent features like small size, highly hydrophilic, and lack of mucoadhesive hydrophobic surface area are preferred for ideal mucus penetration.

Coated nanoparticles

Surface coating of nanoparticles with PEG and CS uncharged hydrophilic mate-rials are other strategies for surface modification, resulting in decreased hydro-phobicity, higher zeta potential, and enhance the stability of the mucus, and transportation across the mucus via particular interactions between the intesti-nal epithelium and the nanoparticles' surface (Yang, Gao, Dagnæs-Hansen, Jakobsen, & Kjems, 2017). PEGylation can lead to bigger particles and impaired epithelial transport, but they can still move within the mucus pores through low-viscosity channels, as well as other unexplored processes, without chang-ing the structure of mucus. Larger particles, on the other hand, may aid in greater drug encapsulation and release, resulting in improved therapeutic effi-cacy (Almalik, Alradwan, Kalam, & Alshamsan, 2017). For determining the optimal size and encapsulation efficiency of mucosal delivery, the friction forces should be considered using the Stokes-Einstein equation, as they are important considerations. Nanoparticles with diverse features must be produced because the thicknesses of mucus in different locations are variable in the gastrointestinal tract. For example, the most effective technique for improving Peyer's patch uptake is to combine M-cell adhesive and mucus penetrating substances in nano-formulation.

Mucolytic nanoparticles

The next method for oral nanoparticle delivery involves altering the mucus lining's barrier characteristics. Mucinex (N-acetyl-L-cysteine; NAC), a regularly used mucolytic agent, can create a mucus-free surface, which increases the number of nanoparticles exposed to underlying epithelial cells and improves the penetration rate. By breaking disulfide bonds, NAC reduces the cross-linking of mucin fibers, resulting in a reduction in rheology and mucus thickness (Storms & Miller, 2018). As a result, the administration of such drugs in combination with targeting moieties to increase nanoparticles absorption in the intestine could be considered. Despite the benefits, few studies have found that removing the mucus barrier significantly increases bacterial translocation, indicating that mucus plays a crucial defensive role in protecting deep-seated epithelial cells from contamination. Mucus undermining may also cause epithelial cell destruction as a result of exposure to acids, enzymes, or other GI harmful secreted chemicals (Ensign et al., 2012). Various mucolytic agents are currently being developed in order to aid nanoparticles' mucosal transport. Pilocarpine, nacystelyn, thymosin-4, and gelsolin are a few examples.

Targeted nanoparticles

Given the preceding sections and the assumption that the majority of nanoparticles are retained on the mucus surface and do not reach the enterocytes. Amoebiasis, Crohn's disease, colorectal cancer, ulcerative colitis, and Inflammatory bowel disease are among the disorders that can be treated with nanoparticles (Cao et al., 2019). It is possible to deliver drugs to the area of interest and to localize it at the target site by grafting ligands like glycoproteins, antibodies, peptides, microorganism-derived adhesive molecules (invasins and flagellin) or carbohydrates to the surface of nano-vehicles to reduce their toxicity, enhance bioavailability, and alter the pharmacokinetic behavior of the delivered drug (Gou et al., 2019; Shen et al., 2019; Shi et al., 2018; Xiao et al., 2018). For example, the limitation of colorectal cancer chemotherapy is due to non-selectivity for cancer and cause toxicity. This can be overcome by leveraging specific interactions of cancerous cells' surface receptors and the bound ligands to nanocarriers. The development of novel efficient orally delivered colon-targeted drug delivery systems could help cancer patients get better treatment. In this regard, Jain and Jain designed hyaluronic acid conjugated (HA-conjugated) CS nanoparticles containing 5-fluorouracil for the successful treatment of colon tumors (Jain & Jain, 2008). They used nanoparticles to target overexpressed HA receptors like RHAMM (CD168) and CD44. By reducing the administered dose and time of medication, this approach was able to increase effective and safe treatments by reducing systemic side effects. The scientists demonstrated that HA-connected CS nanoparticles engaged with the HT-29 cancer cells receptors (CD44), resulting in greater absorption and cellular uptake than uncoupled nanoparticles. To increase oral drug absorption, this method could be used for the treatment of other disorders. The permeability of dextran-g-PEO-C16 loaded cyclosporine (CyA) micelles was tested by Francis in Caco-2 cells by conjugating vitamin B12 intrinsic factor and then attaching it to intrinsic factor receptors expressed on the surface of enterocytes (Francis, Cristea, & Winnik, 2005). When compared to naked micelles, receptor-mediated endocytosis increased the apical to basolateral transport of CyA in vitamin B12-tagged micelles by twofold.

Conclusions and future perspectives

Nowadays, numerous promising nanocarriers systems have been developed to improve the pharmacological and biological performance of a variety of orally administered drugs. Though their application and beneficial effects are now undeniable, their long-term safety must be considered over time. Reducing the dose that is required to elicit a therapeutic effect, as well as preserving the biological and physicochemical stability of orally administered drugs, are future enhancements to be implemented. Indeed, with prudent applications of nanotechnology, nanocarriers-based systems have the potential to significantly improve the quality, efficacy, and safety of drugs, thereby paving the way for a bright future in the treatment of various diseases. By getting a better understanding of the factors that influence oral absorption, there is a way to develop effective and safe nanosystems that enhance therapeutic delivery, relieving many of the issues associated with oral therapy and, in the end, benefiting patients. Acquiring a comprehensive understanding of the biological fate and penetration mechanism of ingested nanocarriers is exceedingly difficult due to intra- and interindividual variations in gastric pH, emptying time, GI tract composition, GI flora, and mucus layer thickness. For this reason, the development of in vitro physiologically relevant models is absolutely essential in order to obtain further understanding of the nanocarriers' gastrointestinal absorption and to aid in the development of better nano-formulations before performing experiments in vivo. Despite the extensive use and acceptance of available in vitro testing models, developing cost-effective, high-throughput, and more predictive cell culture-based models is one of the most significant tasks currently confronting scientists working on oral nanocarriers drug delivery. Additionally, one of the primary goals of developing in vitro models is to correlate the results obtained using various drug formulations to the in vivo drug profile. Developing a model that is capable of correlating in vitro and in vivo data efficiently helps in reducing the formulation development time, expenditure, and improving product quality. Along with the extensive array of established in vitro models, in vivo experiments are critical due to the drawbacks of in vitro assays, such as their variable sensitivity and the brief duration of cytotoxicity experiments (12–48 h). However, a significant number of biological reactions which usually persist for more than 48 h in vivo, implying that in vitro testing is inadequate and does not fully mimic the in vivo environment and thus in vivo testing should be performed instead. Numerous factors, including the patient-friendly oral formulations, the ease with which nanocarriers can be fabricated and scaled up, the cost of materials, the necessity for innovative instrumentation, the in vitro and in vivo efficacy of large-scale batches, and the reproducibility, all play a part in whether the proposed nanotechnology-based oral formulation could be commercialized.

References

Abd El Hady, W. E., Mohamed, E. A., Soliman, O. A. E. A., & El-Sabbagh, H. M. (2019). In vitro–in vivo evaluation of chitosan-PLGA nanoparticles for potentiated gastric retention and anti-ulcer activity of diosmin. *International Journal of Nanomedicine, 14*.

Adebisi, A., & Conway, B. R. (2011). Gastroretentive microparticles for drug delivery applications. *Journal of Microencapsulation, 28*(8), 689–708. https://doi.org/10.3109/02652048.2011.590613.

Adebisi, A. O., & Conway, B. R. (2015). Modification of drug delivery to improve antibiotic targeting to the stomach. *Therapeutic Delivery, 6*(6), 741–762. https://doi.org/10.4155/tde.15.35.

Agrahari, V., Burnouf, P. A., Burnouf, T., & Agrahari, V. (2019). Nanoformulation properties, characterization, and behavior in complex biological matrices: Challenges and opportunities for brain-targeted drug delivery applications and enhanced translational potential. *Advanced Drug Delivery Reviews, 148*, 146–180. https://doi.org/10.1016/j.addr.2019.02.008.

Ahammed, V., Narayan, R., Paul, J., Nayak, Y., Roy, B., Shavi, G. V., et al. (2017). Development and in vivo evaluation of functionalized ritonavir proliposomes for lymphatic targeting. *Life Sciences, 183*, 11–20. https://doi.org/10.1016/j.lfs.2017.06.022.

Ahlawat, J., Henriquez, G., & Narayan, M. (2018). Enhancing the delivery of chemotherapeutics: Role of biodegradable polymeric nanoparticles. *Molecules, 23*(9), 2157. https://doi.org/10.3390/molecules23092157.

Alhakamy, N. A., Fahmy, U. A., Ahmed, O. A. A., Caruso, G., Caraci, F., Asfour, H. Z., et al. (2020). Chitosan coated microparticles enhance simvastatin colon targeting and pro-apoptotic activity. *Marine Drugs, 18*(4). https://doi.org/10.3390/md18040226.

Almalik, A., Alradwan, I., Kalam, M. A., & Alshamsan, A. (2017). Effect of cryoprotection on particle size stability and preservation of chitosan nanoparticles with and without hyaluronate or alginate coating. *Saudi Pharmaceutical Journal, 25*(6), 861–867. https://doi.org/10.1016/j.jsps.2016.12.008.

Amani, N., Javar, H. A., Dorkoosh, F. A., Rouini, M. R., Amini, M., Sharifzadeh, M., et al. (2021). Preparation and pulsatile release evaluation of teriparatide-loaded multilayer implant composed of polyanhydride-hydrogel layers using spin coating for the treatment of osteoporosis. *Journal of Pharmaceutical Innovation, 16*(2), 337–358. https://doi.org/10.1007/s12247-020-09453-1.

Avasarala, H., Dinakaran, S. K., Kakaraparthy, R., & Jayanti, V. R. (2019). Self-emulsifying drug delivery system for enhanced solubility of asenapine maleate: Design, characterization, in vitro, ex vivo and in vivo appraisal. *Drug Development and Industrial Pharmacy, 45*(4), 548–559. https://doi.org/10.1080/03639045.2019.1567758.

Azzali, G. (2003). Structure, lymphatic vascularization and lymphocyte migration in mucosa-associated lymphoid tissue. *Immunological Reviews, 195*, 178–189. https://doi.org/10.1034/j.1600-065X.2003.00072.x.

Babadi, D., Dadashzadeh, S., Osouli, M., Daryabari, M. S., & Haeri, A. (2020). Nanoformulation strategies for improving intestinal permeability of drugs: A more precise look at permeability assessment methods and pharmacokinetic properties changes. *Journal of Controlled Release, 321*, 669–709. https://doi.org/10.1016/j.jconrel.2020.02.041.

Barchet, T. M., & Amiji, M. M. (2009). Challenges and opportunities in CNS delivery of therapeutics for neurodegenerative diseases. *Expert Opinion on Drug Delivery, 6*(3), 211–225. https://doi.org/10.1517/17425240902758188.

Bar-Zeev, M., Assaraf, Y. G., & Livney, Y. D. (2016). β-casein nanovehicles for oral delivery of chemotherapeutic drug combinations overcoming P-glycoprotein-mediated multidrug resistance in human gastric cancer cells. *Oncotarget, 7*(17), 23322–23334. https://doi.org/10.18632/oncotarget.8019.

Bein, A., Eventov-Friedman, S., Arbell, D., & Schwartz, B. (2018). Intestinal tight junctions are severely altered in NEC preterm neonates. *Pediatrics and Neonatology, 59*(5), 464–473. https://doi.org/10.1016/j.pedneo.2017.11.018.

Betageri, G. V., Kovvasu, S. P., Kunamaneni, P., & Joshi, R. (2019). Self-emulsifying drug delivery systems and their marketed products: A review. *Asian Journal of Pharmaceutics, 13*(02), 73–84. https://doi.org/10.22377/ajp.v13i02.3102.

Bhullar, N., Rani, S., Kumari, K., & Sud, D. (2021). Amphiphilic chitosan/acrylic acid/thiourea based semi-interpenetrating hydrogel: Solvothermal synthesis and evaluation for controlled release of organophosphate pesticide, triazophos. *Journal of Applied Polymer Science, 138*(25), 50595. https://doi.org/10.1002/app.50595.

Blasi, P., Giovagnoli, S., Schoubben, A., Ricci, M., & Rossi, C. (2007). Solid lipid nanoparticles for targeted brain drug delivery. *Advanced Drug Delivery Reviews, 59*(6), 454–477. https://doi.org/10.1016/j.addr.2007.04.011.

Boegh, M., García-Díaz, M., Müllertz, A., & Nielsen, H. M. (2015). Steric and interactive barrier properties of intestinal mucus elucidated by particle diffusion and peptide permeation. *European Journal of Pharmaceutics and Biopharmaceutics, 95*, 136–143. https://doi.org/10.1016/j.ejpb.2015.01.014.

Bornhorst, G. M., Gouseti, O., Wickham, M. S., & Bakalis, S. (2016). Engineering digestion: Multiscale processes of food digestion. *Journal of Food Science, 81*(3).

Bose, A., Roy Burman, D., Sikdar, B., & Patra, P. (2021). Nanomicelles: Types, properties and applications in drug delivery. *IET Nanobiotechnology, 15*(1), 19–27. https://doi.org/10.1049/nbt2.12018.

Broesder, A., Kosta, A. M. M. A. C., Woerdenbag, H. J., Nguyen, D. N., Frijlink, H. W., & Hinrichs, W. L. J. (2020). pH-dependent ileocolonic drug delivery, part II: Preclinical evaluation of novel drugs and novel excipients. *Drug Discovery Today, 25*(8), 1374–1388. https://doi.org/10.1016/j.drudis.2020.06.012.

Cano, A., Ettcheto, M., Chang, J. H., Barroso, E., Espina, M., Kühne, B. A., et al. (2019). Dual-drug loaded nanoparticles of Epigallocatechin-3-gallate (EGCG)/Ascorbic acid enhance therapeutic efficacy of EGCG in a APPswe/PS1dE9 Alzheimer's disease mice model. *Journal of Controlled Release, 301*, 62–75. https://doi.org/10.1016/j.jconrel.2019.03.010.

Cano, A., Sánchez-López, E., Ettcheto, M., López-Machado, A., Espina, M., Souto, E. B., et al. (2020). Current advances in the development of novel polymeric nanoparticles for the treatment of neurodegenerative diseases. *Nanomedicine, 15*(12), 1239–1261. https://doi.org/10.2217/nnm-2019-0443.

Cao, S. J., Xu, S., Wang, H. M., Ling, Y., Dong, J., Xia, R. D., et al. (2019). Nanoparticles: Oral delivery for protein and peptide drugs. *AAPS PharmSciTech, 20*(5), 1–11. https://doi.org/10.1208/s12249-019-1325-z.

Cerchiara, T., Abruzzo, A., Parolin, C., Vitali, B., Bigucci, F., Gallucci, M. C., et al. (2016). Microparticles based on chitosan/carboxymethylcellulose polyelectrolyte complexes for colon delivery of vancomycin. *Carbohydrate Polymers, 143*, 124–130. https://doi.org/10.1016/j.carbpol.2016.02.020.

Chakraborty, S., Shukla, D., Mishra, B., & Singh, S. (2009). Lipid—An emerging platform for oral delivery of drugs with poor bioavailability. *European Journal of Pharmaceutics and Biopharmaceutics, 73*(1), 1–15. https://doi.org/10.1016/j.ejpb.2009.06.001.

Chang, C. H., Lin, Y. H., Yeh, C. L., Chen, Y. C., Chiou, S. F., Hsu, Y. M., et al. (2010). Nanoparticles incorporated in pH-sensitive hydrogels as amoxicillin delivery for eradication of *Helicobacter pylori*. *Biomacromolecules, 11*(1), 133–142. https://doi.org/10.1021/bm900985h.

Cheddadi, M., López-Cabarcos, E., Slowing, K., Barcia, E., & Fernández-Carballido, A. (2011). Cytotoxicity and biocompatibility evaluation of a poly(magnesium acrylate) hydrogel synthesized for drug delivery. *International Journal of Pharmaceutics, 413*(1–2), 126–133. https://doi.org/10.1016/j.ijpharm.2011.04.042.

Chen, Y., Lu, Y., Chen, J., Lai, J., Sun, J., Hu, F., et al. (2009). Enhanced bioavailability of the poorly water-soluble drug fenofibrate by using liposomes containing a bile salt. *International Journal of Pharmaceutics, 376*(1–2), 153–160. https://doi.org/10.1016/j.ijpharm.2009.04.022.

Choi, H. J., Kim, M. C., Kang, S. M., & Montemagno, C. D. (2014). The osmotic stress response of split influenza vaccine particles in an acidic environment. *Archives of Pharmacal Research, 37*(12), 1607–1616. https://doi.org/10.1007/s12272-013-0257-5.

Collnot, E. M., Baldes, C., Wempe, M. F., Hyatt, J., Navarro, L., Edgar, K. J., et al. (2006). Influence of vitamin E TPGS poly(ethylene glycol) chain length on apical efflux transporters in Caco-2 cell monolayers. *Journal of Controlled Release, 111*(1–2), 35–40. https://doi.org/10.1016/j.jconrel.2005.11.005.

Croitoru, C., Pop, M. A., Bedo, T., Cosnita, M., Roata, I. C., & Hulka, I. (2020). Physically cross-linked poly(vinyl alcohol)/kappa-carrageenan hydrogels: Structure and applications. *Polymers, 12*(3).

Crucho, C. I. C., & Barros, M. T. (2017). Polymeric nanoparticles: A study on the preparation variables and characterization methods. *Materials Science and Engineering C, 80*, 771–784. https://doi.org/10.1016/j.msec.2017.06.004.

Dan, N., Samanta, K., & Almoazen, H. (2020). An update on pharmaceutical strategies for oral delivery of therapeutic peptides and proteins in adults and pediatrics. *Children, 7*(12), 307. https://doi.org/10.3390/children7120307.

Dong, P., Sahle, F. F., Lohan, S. B., Saeidpour, S., Albrecht, S., Teutloff, C., et al. (2019). pH-sensitive Eudragit® L 100 nanoparticles promote cutaneous penetration and drug release on the skin. *Journal of Controlled Release, 295*, 214–222. https://doi.org/10.1016/j.jconrel.2018.12.045.

El-Maghawry, E., Tadros, M. I., Elkheshen, S. A., & Abd-Elbary, A. (2020). Eudragit®-S100 coated PLGA nanoparticles for colon targeting of Etoricoxib: Optimization and pharmacokinetic assessments in healthy human volunteers. *International Journal of Nanomedicine, 15*.

El-Say, K. M., & El-Sawy, H. S. (2017). Polymeric nanoparticles: Promising platform for drug delivery. *International Journal of Pharmaceutics, 675–691.* https://doi.org/10.1016/j.ijpharm.2017.06.052.

Ensign, L. M., Cone, R., & Hanes, J. (2012). Oral drug delivery with polymeric nanoparticles: The gastrointestinal mucus barriers. *Advanced Drug Delivery Reviews, 64*(6), 557–570. https://doi.org/10.1016/j.addr.2011.12.009.

Ensign, L. M., Henning, A., Schneider, C. S., Maisel, K., Wang, Y. Y., Porosoff, M. D., et al. (2013). Ex vivo characterization of particle transport in mucus secretions coating freshly excised mucosal tissues. *Molecular Pharmaceutics, 10*(6), 2176–2182. https://doi.org/10.1021/mp400087y.

Faralli, A., Shekarforoush, E., Ajalloueian, F., Mendes, A. C., & Chronakis, I. S. (2019). In vitro permeability enhancement of curcumin across Caco-2 cells monolayers using electrospun xanthan-chitosan nanofibers. *Carbohydrate Polymers, 206*, 38–47. https://doi.org/10.1016/j.carbpol.2018.10.073.

Francis, M. F., Cristea, M., & Winnik, F. M. (2005). Exploiting the vitamin B12 pathway to enhance oral drug delivery via polymeric micelles. *Biomacromolecules, 6*(5), 2462–2467. https://doi.org/10.1021/bm0503165.

Frank, L., Onzi, G., Morawski, A., Pohlmann, A., Guterres, S., & Contri, R. (2020). Chitosan as a coating material for nanoparticles intended for biomedical applications. *Reactive and Functional Polymers, 147.*

Franz-Montan, M., de Araújo, D. R., de Morais Ribeiro, L. N., de Melo, N. F. S., & de Paula, E. (2017). Nanostructured systems for transbuccal drug delivery. In *Nanostructures for oral medicine* (pp. 87–121). Elsevier Inc. https://doi.org/10.1016/B978-0-323-47720-8.00005-5.

Fricker, G., Kromp, T., Wendel, A., Blume, A., Zirkel, J., Rebmann, H., et al. (2010). Phospholipids and lipid-based formulations in oral drug delivery. *Pharmaceutical Research, 27*(8), 1469–1486. https://doi.org/10.1007/s11095-010-0130-x.

Gao, L., Liu, G., Ma, J., Wang, X., Zhou, L., Li, X., et al. (2013). Application of drug nanocrystal technologies on oral drug delivery of poorly soluble drugs. *Pharmaceutical Research, 30*(2), 307–324. https://doi.org/10.1007/s11095-012-0889-z.

Garg, T., Kumar, A., Rath, G., & Goyal, A. K. (2014). Gastroretentive drug delivery systems for therapeutic management of peptic ulcer. *Critical Reviews in Therapeutic Drug Carrier Systems, 31*(6), 531–557. https://doi.org/10.1615/CritRevTherDrugCarrierSyst.2014011104.

Goldberg, D. S., Vijayalakshmi, N., Swaan, P. W., & Ghandehari, H. (2011). G3.5 PAMAM dendrimers enhance transepithelial transport of SN38 while minimizing gastrointestinal toxicity. *Journal of Controlled Release, 150*(3), 318–325. https://doi.org/10.1016/j.jconrel.2010.11.022.

Gou, S., Huang, Y., Wan, Y., Ma, Y., Zhou, X., Tong, X., et al. (2019). Multi-bioresponsive silk fibroin-based nanoparticles with on-demand cytoplasmic drug release capacity for CD44-targeted alleviation of ulcerative colitis. *Biomaterials, 212*, 39–54. https://doi.org/10.1016/j.biomaterials.2019.05.012.

Goyal, P., Singh, M., Kumar, P., & Gupta, A. (2021). Chol-Dex nanomicelles: Synthesis, characterization and evaluation as efficient drug carriers for colon targeting. *Carbohydrate Research, 500*, 108255. https://doi.org/10.1016/j.carres.2021.108255.

Hamidi, M., Azadi, A., & Rafiei, P. (2008). Hydrogel nanoparticles in drug delivery. *Advanced Drug Delivery Reviews, 60*(15), 1638–1649. https://doi.org/10.1016/j.addr.2008.08.002.

Hamidi, M., Shahbazi, M. A., & Rostamizadeh, K. (2012). Copolymers: Efficient carriers for intelligent nanoparticulate drug targeting and gene therapy. *Macromolecular Bioscience, 12*(2), 144–164. https://doi.org/10.1002/mabi.201100193.

Han, S., Quach, T., Hu, L., Wahab, A., Charman, W. N., Stella, V. J., et al. (2014). Targeted delivery of a model immunomodulator to the lymphatic system: Comparison of alkyl ester versus triglyceride mimetic lipid prodrug strategies. *Journal of Controlled Release, 177*(1), 1–10. https://doi.org/10.1016/j.jconrel.2013.12.031.

Hofheinz, R.-D., Gnad-Vogt, S. U., Beyer, U., & Hochhaus, A. (2005). Liposomal encapsulated anti-cancer drugs. *Anti-Cancer Drugs*, *16*(7), 691–707. https://doi.org/10.1097/01.cad.0000167902.53039.5a.

Homayun, B., Lin, X., & Choi, H. J. (2019). Challenges and recent progress in oral drug delivery systems for biopharmaceuticals. *Pharmaceutics*, *11*(3). https://doi.org/10.3390/pharmaceutics11030129.

Hua, S., Marks, E., Schneider, J. J., & Keely, S. (2015). Advances in oral nano-delivery systems for colon targeted drug delivery in inflammatory bowel disease: Selective targeting to diseased versus healthy tissue. *Nanomedicine: Nanotechnology, Biology, and Medicine*, *11*(5), 1117–1132. https://doi.org/10.1016/j.nano.2015.02.018.

Jain, A., & Jain, S. K. (2008). In vitro and cell uptake studies for targeting of ligand anchored nanoparticles for colon tumors. *European Journal of Pharmaceutical Sciences*, *35*(5), 404–416. https://doi.org/10.1016/j.ejps.2008.08.008.

Janaszewska, A., Lazniewska, J., Trzepiński, P., Marcinkowska, M., & Klajnert-Maculewicz, B. (2019). Cytotoxicity of dendrimers. *Biomolecules*, *9*.

Kabanov, A. V., Batrakova, E. V., Melik-Nubarov, N. S., Fedoseev, N. A., Dorodnich, T. Y., Alakhov, V. Y., et al. (1992). A new class of drug carriers: Micelles of poly(oxyethylene)-poly (oxypropylene) block copolymers as microcontainers for drug targeting from blood in brain. *Journal of Controlled Release*, *22*(2), 141–157. https://doi.org/10.1016/0168-3659 (92)90199-2.

Kaur, I. P., Bhandari, R., Bhandari, S., & Kakkar, V. (2008). Potential of solid lipid nanoparticles in brain targeting. *Journal of Controlled Release*, *127*(2), 97–109. https://doi.org/10.1016/j.jconrel.2007.12.018.

Khalil, I. A., Kogure, K., Futaki, S., & Harashima, H. (2006). High density of octaarginine stimulates macropinocytosis leading to efficient intracellular trafficking for gene expression. *Journal of Biological Chemistry*, *281*(6), 3544–3551. https://doi.org/10.1074/jbc.M503202200.

Köllner, S., Dünnhaupt, S., Waldner, C., Hauptstein, S., Pereira De Sousa, I., & Bernkop-Schnürch, A. (2015). Mucus permeating thiomer nanoparticles. *European Journal of Pharmaceutics and Biopharmaceutics*, *97*, 265–272. https://doi.org/10.1016/j.ejpb.2015.01.004.

Kuzum, D., Yu, S., & Philip Wong, H.-S. (2013). Synaptic electronics: Materials, devices and applications. *Nanotechnology*, 382001. https://doi.org/10.1088/0957-4484/24/38/382001.

Lavrador, P., Esteves, M. R., Gaspar, V. M., & Mano, J. F. (2021). Stimuli-responsive nanocomposite hydrogels for biomedical applications. *Advanced Functional Materials*, *31*, 8.

Leal, J., Dong, T., Taylor, A., Siegrist, E., Gao, F., Smyth, H. D. C., et al. (2018). Mucus-penetrating phage-displayed peptides for improved transport across a mucus-like model. *International Journal of Pharmaceutics*, *553*(1–2), 57–64. https://doi.org/10.1016/j.ijpharm.2018.09.055.

Leo, D., Toro, D., Decorti, N., Malusà, G., Ventura, N., & Not, A. (2010). Fasting increases tobramycin oral absorption in mice. *Antimicrobial Agents and Chemotherapy*, *54*(4).

Lerma, T. A., Garcés, V., & Palencia, M. (2020). Novel multi- and bio-functional hybrid polymer hydrogels based on bentonite-poly(acrylic acid) composites and sorbitol polyesters: Structural and functional characterization. *European Polymer Journal*, *128*.

Liarou, E., Varlas, S., Skoulas, D., Tsimblouli, C., Sereti, E., Dimas, K., et al. (2018). Smart polymersomes and hydrogels from polypeptide-based polymer systems through α-amino acid N-carboxyanhydride ring-opening polymerization. From chemistry to biomedical applications. *Progress in Polymer Science*, *83*, 28–78. https://doi.org/10.1016/j.progpolymsci.2018.05.001.

Lilia Romero, E., & Morilla, M. J. (2011). Topical and mucosal liposomes for vaccine delivery. *Wiley Interdisciplinary Reviews: Nanomedicine and Nanobiotechnology*, *3*(4), 356–375. https://doi.org/10.1002/wnan.131.

Liu, Y., Yang, T., Wei, S., Zhou, C., Lan, Y., Cao, A., et al. (2018). Mucus adhesion- and penetration-enhanced liposomes for paclitaxel oral delivery. *International Journal of Pharmaceutics*, *537*(1–2), 245–256. https://doi.org/10.1016/j.ijpharm.2017.12.044.

Lombardo, D., Kiselev, M. A., & Caccamo, M. T. (2019). Smart nanoparticles for drug delivery application: Development of versatile nanocarrier platforms in biotechnology and nanomedicine. *Journal of Nanomaterials, 2019*. https://doi.org/10.1155/2019/3702518.

Lueschow, S. R., & McElroy, S. J. (2020). The paneth cell: The curator and defender of the immature small intestine. *Frontiers in Immunology, 11*. https://doi.org/10.3389/fimmu.2020.00587.

Luo, F., Wang, M., Huang, L., Wu, Z., Wang, W., Zafar, A., et al. (2020). Synthesis of zinc oxide eudragit FS30D nanohybrids: Structure, characterization, and their application as an intestinal drug delivery system. *ACS Omega, 5*(20), 11799–11808. https://doi.org/10.1021/acsomega.0c01216.

Madni, A., Rehman, S., Sultan, H., Khan, M. M., Ahmad, F., Raza, M. R., et al. (2021). Mechanistic approaches of internalization, subcellular trafficking, and cytotoxicity of nanoparticles for targeting the small intestine. *AAPS PharmSciTech, 22*(1). https://doi.org/10.1208/s12249-020-01873-z.

Maisel, K., Ensign, L., Reddy, M., Cone, R., & Hanes, J. (2015). Effect of surface chemistry on nanoparticle interaction with gastrointestinal mucus and distribution in the gastrointestinal tract following oral and rectal administration in the mouse. *Journal of Controlled Release, 197*, 48–57. https://doi.org/10.1016/j.jconrel.2014.10.026.

Managuli, R. S., Raut, S. Y., Reddy, M. S., & Mutalik, S. (2018). Targeting the intestinal lymphatic system: A versatile path for enhanced oral bioavailability of drugs. *Expert Opinion on Drug Delivery, 15*(8), 787–804. https://doi.org/10.1080/17425247.2018.1503249.

Mishra, B., Patel, B. B., & Tiwari, S. (2010). Colloidal nanocarriers: A review on formulation technology, types and applications toward targeted drug delivery. *Nanomedicine: Nanotechnology, Biology, and Medicine, 6*(1), 9–24. https://doi.org/10.1016/j.nano.2009.04.008.

Moghimipour, E., Rezaei, M., Kouchak, M., Fatahiasl, J., Angali, K. A., Ramezani, Z., et al. (2018). Effects of coating layer and release medium on release profile from coated capsules with Eudragit FS 30D: An in vitro and in vivo study. *Drug Development and Industrial Pharmacy, 44*(5), 861–867. https://doi.org/10.1080/03639045.2017.1415927.

More, S., Gavali, K., Doke, O., & Kasgawade, P. (2018). Gastroretentive drug delivery system. *Journal of Drug Delivery and Therapeutics*. https://doi.org/10.22270/jddt.v8i4.1788.

Moroz, E., Matoori, S., & Leroux, J. C. (2016). Oral delivery of macromolecular drugs: Where we are after almost 100 years of attempts. *Advanced Drug Delivery Reviews, 101*, 108–121. https://doi.org/10.1016/j.addr.2016.01.010.

Moulari, B., Béduneau, A., Pellequer, Y., & Lamprecht, A. (2014). Lectin-decorated nanoparticles enhance binding to the inflamed tissue in experimental colitis. *Journal of Controlled Release, 188*, 9–17. https://doi.org/10.1016/j.jconrel.2014.05.046.

Müller, C., Perera, G., König, V., & Bernkop-Schnürch, A. (2014). Development and in vivo evaluation of papain-functionalized nanoparticles. *European Journal of Pharmaceutics and Biopharmaceutics, 87*(1), 125–131. https://doi.org/10.1016/j.ejpb.2013.12.012.

Nabi, B., Rehman, S., Baboota, S., & Ali, J. (2019). Insights on oral drug delivery of lipid nanocarriers: A win-win solution for augmenting bioavailability of antiretroviral drugs. *AAPS PharmSciTech, 20*(2). https://doi.org/10.1208/s12249-018-1284-9.

Najlah, M., Freeman, S., Attwood, D., & D'Emanuele, A. (2007). In vitro evaluation of dendrimer prodrugs for oral drug delivery. *International Journal of Pharmaceutics, 336*(1), 183–190. https://doi.org/10.1016/j.ijpharm.2006.11.047.

O'Neill, M. J., Bourre, L., Melgar, S., & O'Driscoll, C. M. (2011). Intestinal delivery of non-viral gene therapeutics: Physiological barriers and preclinical models. *Drug Discovery Today, 16*(5–6), 203–218. https://doi.org/10.1016/j.drudis.2011.01.003.

Ofridam, F., Lebaz, N., Gagnière, É., Mangin, D., & Elaissari, A. (2020). Effect of secondary polymer on self-precipitation of pH-sensitive polymethylmethacrylate derivatives Eudragit E100 and

Eudragit L100. *Polymers for Advanced Technologies, 31*(6), 1270–1279. https://doi.org/10.1002/pat.4856.

Ofridam, F., Tarhini, M., Lebaz, N., Gagnière, É., Mangin, D., & Elaissari, A. (2021). pH-sensitive polymers: Classification and some fine potential applications. *Polymers for Advanced Technologies, 32*(4), 1455–1484. https://doi.org/10.1002/pat.5230.

Pandey, R., Sharma, S., & Khuller, G. K. (2005). Oral solid lipid nanoparticle-based antitubercular chemotherapy. *Tuberculosis, 85*(5–6), 415–420. https://doi.org/10.1016/j.tube.2005.08.009.

Pandya, P., Giram, P., Bhole, R. P., Chang, H. I., & Raut, S. Y. (2021). Nanocarriers based oral lymphatic drug targeting: Strategic bioavailability enhancement approaches. *Journal of Drug Delivery Science and Technology, 64*. https://doi.org/10.1016/j.jddst.2021.102585.

Pardeshi, C., Rajput, P., Belgamwar, V., Tekade, A., Patil, G., Chaudhary, K., et al. (2012). Solid lipid based nanocarriers: An overview. *Acta Pharmaceutica, 62*(4), 433–472. https://doi.org/10.2478/v10007-012-0040-z.

Patel, V. R., & Amiji, M. M. (1996). Preparation and characterization of freeze-dried chitosan-poly(-ethylene oxide) hydrogels for site-specific antibiotic delivery in the stomach. *Pharmaceutical Research, 13*(4), 588–593. https://doi.org/10.1023/A:1016054306763.

Patil, S. B., Inamdar, S. Z., Reddy, K. R., Raghu, A. V., Soni, S. K., & Kulkarni, R. V. (2019). Novel biocompatible poly(acrylamide)-grafted-dextran hydrogels: Synthesis, characterization and biomedical applications. *Journal of Microbiological Methods, 159*, 200–210. https://doi.org/10.1016/j.mimet.2019.03.009.

Pawar, V. K., Meher, J. G., Singh, Y., Chaurasia, M., Surendar Reddy, B., & Chourasia, M. K. (2014). Targeting of gastrointestinal tract for amended delivery of protein/peptide therapeutics: Strategies and industrial perspectives. *Journal of Controlled Release, 196*, 168–183. https://doi.org/10.1016/j.jconrel.2014.09.031.

Pereira De Sousa, I., Cattoz, B., Wilcox, M. D., Griffiths, P. C., Dalgliesh, R., Rogers, S., et al. (2015). Nanoparticles decorated with proteolytic enzymes, a promising strategy to overcome the mucus barrier. *European Journal of Pharmaceutics and Biopharmaceutics, 97*, 257–264. https://doi.org/10.1016/j.ejpb.2015.01.008.

Petri, B., Bootz, A., Khalansky, A., Hekmatara, T., Müller, R., Uhl, R., et al. (2007). Chemotherapy of brain tumour using doxorubicin bound to surfactant-coated poly(butyl cyanoacrylate) nanoparticles: Revisiting the role of surfactants. *Journal of Controlled Release, 117*(1), 51–58. https://doi.org/10.1016/j.jconrel.2006.10.015.

Pettinelli, N., Rodríguez-Llamazares, S., Farrag, Y., Bouza, R., Barral, L., Feijoo-Bandín, S., et al. (2020). Poly(hydroxybutyrate-co-hydroxyvalerate) microparticles embedded in κ-carrageenan/locust bean gum hydrogel as a dual drug delivery carrier. *International Journal of Biological Macromolecules, 146*, 110–118. https://doi.org/10.1016/j.ijbiomac.2019.12.193.

Plaza-Oliver, M., Santander-Ortega, M. J., & Lozano, M. V. (2021). Current approaches in lipid-based nanocarriers for oral drug delivery. *Drug Delivery and Translational Research, 11*, 471–497. https://doi.org/10.1007/s13346-021-00908-7.

Porter, C. J. H., & Charman, W. N. (2001). Intestinal lymphatic drug transport: An update. *Advanced Drug Delivery Reviews, 50*(1–2), 61–80. https://doi.org/10.1016/S0169-409X(01)00151-X.

Prajapati, V. D., Jani, G. K., Khutliwala, T. A., & Zala, B. S. (2013). Raft forming system—An upcoming approach of gastroretentive drug delivery system. *Journal of Controlled Release, 168*(2), 151–165. https://doi.org/10.1016/j.jconrel.2013.02.028.

Pridgen, E. M., Alexis, F., & Farokhzad, O. C. (2015). Polymeric nanoparticle drug delivery technologies for oral delivery applications. *Expert Opinion on Drug Delivery, 12*(9), 1459–1473. https://doi.org/10.1517/17425247.2015.1018175.

Qi, D., Shi, W., Black, A. R., Kuss, M. A., Pang, X., He, Y., et al. (2020). Repair and regeneration of small intestine: A review of current engineering approaches. *Biomaterials, 240*.

Qi, X., Wang, L., Zhu, J., Hu, Z., & Zhang, J. (2011). Self-double-emulsifying drug delivery system (SDEDDS): A new way for oral delivery of drugs with high solubility and low permeability. *International Journal of Pharmaceutics*, *409*(1–2), 245–251. https://doi.org/10.1016/j.ijpharm.2011.02.047.

Riehemann, K., Schneider, S. W., Luger, T. A., Godin, B., Ferrari, M., & Fuchs, H. (2009). Nanomedicine—Challenge and perspectives. *Angewandte Chemie International Edition*, *48*(5), 872–897. https://doi.org/10.1002/anie.200802585.

Rieux, A. D., Ragnarsson, E. G. E., Gullberg, E., Préat, V., Schneider, Y. J., & Artursson, P. (2005). Transport of nanoparticles across an in vitro model of the human intestinal follicle associated epithelium. *European Journal of Pharmaceutical Sciences*, *25*(4–5), 455–465. https://doi.org/10.1016/j.ejps.2005.04.015.

Russell-Jones, G. J. (1996). The potential use of receptor-mediated endocytosis for oral drug delivery. *Advanced Drug Delivery Reviews*, *20*(1), 83–97. https://doi.org/10.1016/0169-409X(95)00131-P.

Sabir, F., Qindeel, M., Rehman, A. U., Ahmad, N. M., Khan, G. M., Csoka, I., et al. (2021). An efficient approach for development and optimisation of curcumin-loaded solid lipid nanoparticles' patch for transdermal delivery. *Journal of Microencapsulation*, *38*(4), 233–248. https://doi.org/10.1080/02652048.2021.1899321.

Salah, E., Abouelfetouh, M. M., Pan, Y., Chen, D., & Xie, S. (2020). Solid lipid nanoparticles for enhanced oral absorption: A review. *Colloids and Surfaces B: Biointerfaces*, *196*, 111305. https://doi.org/10.1016/j.colsurfb.2020.111305.

Sarangi, M. K., Rao, M. E. B., Parcha, V., & Upadhyay, A. (2020). Tailoring of colon targeting with sodium alginate-Assam bora rice starch based multi particulate system containing naproxen. *Starch-Stärke*, *72*(7–8), 1900307. https://doi.org/10.1002/star.201900307.

Schaffazick, S. R., Pohlmann, A. R., Dalla-Costa, T., & Guterres, S. S. (2003). Freeze-drying polymeric colloidal suspensions: Nanocapsules, nanospheres and nanodispersion. A comparative study. *European Journal of Pharmaceutics and Biopharmaceutics*, *56*(3), 501–505. https://doi.org/10.1016/S0939-6411(03)00139-5.

Shan, W., Zhu, X., Liu, M., Li, L., Zhong, J., Sun, W., et al. (2015). Overcoming the diffusion barrier of mucus and absorption barrier of epithelium by self-assembled nanoparticles for oral delivery of insulin. *ACS Nano*, *9*(3), 2345–2356. https://doi.org/10.1021/acsnano.5b00028.

Shen, M. Y., Liu, T. I., Yu, T. W., Kv, R., Chiang, W. H., Tsai, Y. C., et al. (2019). Hierarchically targetable polysaccharide-coated solid lipid nanoparticles as an oral chemo/thermotherapy delivery system for local treatment of colon cancer. *Biomaterials*, *197*, 86–100. https://doi.org/10.1016/j.biomaterials.2019.01.019.

Shi, S., Zhang, L., Zhu, M., Wan, G., Li, C., Zhang, J., et al. (2018). Reactive oxygen species-responsive nanoparticles based on PEGlated prodrug for targeted treatment of oral tongue squamous cell carcinoma by combining photodynamic therapy and chemotherapy. *ACS Applied Materials and Interfaces*, *10*(35), 29260–29272. https://doi.org/10.1021/acsami.8b08269.

Sladek, S., McCartney, F., Eskander, M., Dunne, D. J., Santos-Martinez, M. J., Benetti, F., et al. (2020). An enteric-coated polyelectrolyte nanocomplex delivers insulin in rat intestinal instillations when combined with a permeation enhancer. *Pharmaceutics*, *12*(3). https://doi.org/10.3390/pharmaceutics12030259.

Sood, A., Dev, A., Mohanbhai, S. J., Shrimali, N., Kapasiya, M., Kushwaha, A. C., et al. (2019). Disulfide-bridged chitosan-Eudragit S-100 nanoparticles for colorectal cancer. *ACS Applied Nano Materials*, *2*(10), 6409–6417. https://doi.org/10.1021/acsanm.9b01377.

Sorroza-Martínez, K., Ruiu, A., González-Méndez, I., & Rivera, E. (2021). *Design and properties of dendrimers for pharmaceutical applications* (pp. 15–31). Elsevier BV. https://doi.org/10.1016/b978-0-12-821250-9.00002-0.

Storms, W. W., & Miller, J. E. (2018). Daily use of guaifenesin (Mucinex) in a patient with chronic bronchitis and pathologic mucus hypersecretion: A case report. *Respiratory Medicine Case Reports*, *23*, 156–157. https://doi.org/10.1016/j.rmcr.2018.02.009.

Su, R., Jin, X., Li, H., Huang, L., & Li, Z. (2020). The mechanisms of PM2. 5 and its main components penetrate into HUVEC cells and effects on cell organelles. *Chemosphere*, *241*.

Szczęch, M., & Szczepanowicz, K. (2020). Polymeric core-shell nanoparticles prepared by spontaneous emulsification solvent evaporation and functionalized by the layer-by-layer method. *Nanomaterials*, *10*(3), 496. https://doi.org/10.3390/nano10030496.

Taghipour-Sabzevar, V., Sharifi, T., & Moghaddam, M. M. (2019). Polymeric nanoparticles as carrier for targeted and controlled delivery of anticancer agents. *Therapeutic Delivery*, *10*(8), 527–550. https://doi.org/10.4155/tde-2019-0044.

Trevaskis, N. L., Charman, W. N., & Porter, C. J. H. (2008). Lipid-based delivery systems and intestinal lymphatic drug transport: A mechanistic update. *Advanced Drug Delivery Reviews*, *60*(6), 702–716. https://doi.org/10.1016/j.addr.2007.09.007.

Trevaskis, N. L., Kaminskas, L. M., & Porter, C. J. H. (2015). From sewer to saviour-targeting the lymphatic system to promote drug exposure and activity. *Nature Reviews Drug Discovery*, *14*(11), 781–803. https://doi.org/10.1038/nrd4608.

Trivedi, R., & Kompella, U. B. (2010). Nanomicellar formulations for sustained drug delivery: Strategies and underlying principles. *Nanomedicine*, *5*(3), 485–505. https://doi.org/10.2217/nnm.10.10.

Vahdatpour, S., Mamaghani, A., Goloujeh, M., Maheri-Sis, N., & Mahmoodpour, H. (2016). the systematic review of proteins digestion and new strategies for delivery of small peptides. *Electronic Journal of Biology*, *12*(3), 265–275.

van Hoogevest, P., Liu, X., & Fahr, A. (2011). Drug delivery strategies for poorly water-soluble drugs: The industrial perspective. *Expert Opinion on Drug Delivery*, *8*(11), 1481–1500. https://doi.org/10.1517/17425247.2011.614228.

Vinarov, Z., Abdallah, M., Agundez, J. A. G., Allegaert, K., Basit, A. W., Braeckmans, M., et al. (2021). Impact of gastrointestinal tract variability on oral drug absorption and pharmacokinetics: An UNGAP review. *European Journal of Pharmaceutical Sciences*, *162*, 105812. https://doi.org/10.1016/j.ejps.2021.105812.

Vllasaliu, D., Fowler, R., Garnett, M., Eaton, M., & Stolnik, S. (2011). Barrier characteristics of epithelial cultures modelling the airway and intestinal mucosa: A comparison. *Biochemical and Biophysical Research Communications*, *415*(4), 579–585. https://doi.org/10.1016/j.bbrc.2011.10.108.

Wilcox, M. D., Van Rooij, L. K., Chater, P. I., Pereira De Sousa, I., & Pearson, J. P. (2015). The effect of nanoparticle permeation on the bulk rheological properties of mucus from the small intestine. *European Journal of Pharmaceutics and Biopharmaceutics*, *96*, 484–487. https://doi.org/10.1016/j.ejpb.2015.02.029.

Wolfram, J., Zhu, M., Yang, Y., Shen, J., Gentile, E., Paolino, D., et al. (2015). Safety of nanoparticles in medicine. *Current Drug Targets*, *16*(14), 1671–1681. https://doi.org/10.2174/1389450115666140804124808.

Xiao, B., Chen, Q., Zhang, Z., Wang, L., Kang, Y., Denning, T., et al. (2018). TNFα gene silencing mediated by orally targeted nanoparticles combined with interleukin-22 for synergistic combination therapy of ulcerative colitis. *Journal of Controlled Release*, 235–246. https://doi.org/10.1016/j.jconrel.2018.08.021.

Xu, X., Liu, K., Jiao, B., Luo, K., Ren, J., Zhang, G., et al. (2020). Mucoadhesive nanoparticles based on ROS activated gambogic acid prodrug for safe and efficient intravesical instillation chemotherapy of bladder cancer. *Journal of Controlled Release*, *324*, 493–504. https://doi.org/10.1016/j.jconrel.2020.03.028.

Xu, Y., Shrestha, N., Préat, V., & Beloqui, A. (2020). Overcoming the intestinal barrier: A look into targeting approaches for improved oral drug delivery systems. *Journal of Controlled Release, 322*, 486–508. https://doi.org/10.1016/j.jconrel.2020.04.006.

Xu, Y., Shrestha, N., Préat, V., & Beloqui, A. (2021). An overview of in vitro, ex vivo and in vivo models for studying the transport of drugs across intestinal barriers. *Advanced Drug Delivery Reviews*. https://doi.org/10.1016/j.addr.2021.05.005.

Yamagata, T., Morishita, M., Kavimandan, N. J., Nakamura, K., Fukuoka, Y., Takayama, K., et al. (2006). Characterization of insulin protection properties of complexation hydrogels in gastric and intestinal enzyme fluids. *Journal of Controlled Release, 112*(3), 343–349. https://doi.org/10.1016/j.jconrel.2006.03.005.

Yan, Y., Sun, Y., Wang, P., Zhang, R., Huo, C., Gao, T., et al. (2020). Mucoadhesive nanoparticles-based oral drug delivery systems enhance ameliorative effects of low molecular. *Carbohydrate Polymers, 246*, 1–12. https://doi.org/10.1016/j.carbpol.2020.116660.

Yang, C., Gao, S., Dagnæs-Hansen, F., Jakobsen, M., & Kjems, J. (2017). Impact of PEG chain length on the physical properties and bioactivity of PEGylated chitosan/siRNA nanoparticles in vitro and in vivo. *ACS Applied Materials and Interfaces, 9*(14), 12203–12216. https://doi.org/10.1021/acsami.6b16556.

Yeh, T. H., Hsu, L. W., Tseng, M. T., Lee, P. L., Sonjae, K., Ho, Y. C., et al. (2011). Mechanism and consequence of chitosan-mediated reversible epithelial tight junction opening. *Biomaterials, 32*(26), 6164–6173. https://doi.org/10.1016/j.biomaterials.2011.03.056.

Yildiz, H. M., McKelvey, C. A., Marsac, P. J., & Carrier, R. L. (2015). Size selectivity of intestinal mucus to diffusing particulates is dependent on surface chemistry and exposure to lipids. *Journal of Drug Targeting, 23*(7–8), 768–774. https://doi.org/10.3109/1061186X.2015.1086359.

Zeeshan, M., Ali, H., Khan, S., Khan, S. A., & Weigmann, B. (2019). Advances in orally-delivered pH-sensitive nanocarrier systems; an optimistic approach for the treatment of inflammatory bowel disease. *International Journal of Pharmaceutics, 558*, 201–214. https://doi.org/10.1016/j.ijpharm.2018.12.074.

Zhaeentan, S., Amjadi, F. S., Zandie, Z., Joghataei, M. T., Bakhtiyari, M., & Aflatoonian, R. (2018). The effects of hydrocortisone on tight junction genes in an in vitro model of the human fallopian epithelial cells. *European Journal of Obstetrics, Gynecology, and Reproductive Biology, 229*, 127–131. https://doi.org/10.1016/j.ejogrb.2018.05.034.

Zhang, X., Cheng, H., Dong, W., Zhang, M., Liu, Q., Wang, X., et al. (2018). Design and intestinal mucus penetration mechanism of core-shell nanocomplex. *Journal of Controlled Release, 272*, 29–38. https://doi.org/10.1016/j.jconrel.2017.12.034.

Zhang, F., Trent Magruder, J., Lin, Y. A., Crawford, T. C., Grimm, J. C., Sciortino, C. M., et al. (2017). Generation-6 hydroxyl PAMAM dendrimers improve CNS penetration from intravenous administration in a large animal brain injury model. *Journal of Controlled Release, 249*, 173–182. https://doi.org/10.1016/j.jconrel.2017.01.032.

Zhang, Y., Xiong, M., Ni, X., Wang, J., Rong, H., Su, Y., et al. (2021). Virus-mimicking mesoporous silica nanoparticles with an electrically neutral and hydrophilic surface to improve the oral absorption of insulin by breaking through dual barriers of the mucus layer and the intestinal epithelium. *ACS Applied Materials and Interfaces, 13*(15), 18077–18088. https://doi.org/10.1021/acsami.1c00580.

Zhao, S., Dai, W., He, B., Wang, J., He, Z., Zhang, X., et al. (2012). Monitoring the transport of polymeric micelles across MDCK cell monolayer and exploring related mechanisms. *Journal of Controlled Release, 158*(3), 413–423. https://doi.org/10.1016/j.jconrel.2011.12.018.

Zhao, L., Zhu, J., Cheng, Y., Xiong, Z., Tang, Y., Guo, L., et al. (2015). Chlorotoxin-conjugated multifunctional dendrimers labeled with radionuclide 131I for single photon emission computed tomography imaging and radiotherapy of gliomas. *ACS Applied Materials and Interfaces, 7*(35), 19798–19808. https://doi.org/10.1021/acsami.5b05836.

CHAPTER 9

Nanomaterials as promising therapeutic platform for bone tissues engineering

Introduction

Bone fractures are one of the most prevalent incidents of organ injuries worldwide, which may occur as a result of high-energy trauma like motorcycle and vehicle accidents or sports injuries (e.g., paragliding, rugby, mountain bike, etc.). Also, work accidents are a common cause of fractures in developing countries, owing to the load in economic activity and the related working environment (Ho-Shui-Ling et al., 2018). Bone defects are usually divided into three subfields based on where they occur such as long bones and spine, craniofacial and maxillofacial. Fig. 9.1 depicts the most common bone fracture sites: ankle (distal tibia/fibula fractures, over the joint), femur, wrist (ulna/radius), maxillo-and craniofacial (calvaria, jawbone), shoulder (mainly humerus), vertebral, hip (femoral neck) as well as tibia (distal third) fractures. Although, the combined action of osteogenic cells, immune system cells, and mesenchymal cells allows the bone to self-repair tiny flaws (Marsell & Einhorn, 2011). The physicochemical and mechanical properties of self-repaired bone are frequently associated with a recapitulation of the replaced bone (Dimitriou, Jones, McGonagle, & Giannoudis, 2011). Larger defects, on the other hand, are unable to heal on their own, and regenerative therapies are critical to redressing such clinical issues (Ho-Shui-Ling et al., 2018). In general, bone grafts are considered one of the most common tissue transplantations around the world, with approximately more than 2.2 million performed each year (Giannoudis, Dinopoulos, & Tsiridis, 2005). Despite the significant limitations, allograft and autologous bone are usually attributed as the gold standard in clinical care for bone restoration (Gupta et al., 2015). However, the usefulness of these procedures is extensively hindered by the donor site morbidity, high rates of failure (up to 50%) (Bajaj, Wongworawat, & Punjabi, 2003; Clavero & Lundgren, 2003). In some cases, immunogenicity (Bauer & Muschler, 2000; Stevenson & Horowitz, 1992) and the issues related to the quality and supply of host bone (Angevine, Zivin, & McCormick, 2005). Because of these limitations, the search for synthetic biomaterials to replace bone tissue has been continued. These

Nanocarriers for Organ-Specific and Localized Drug Delivery. https://doi.org/10.1016/B978-0-12-821093-2.00007-4

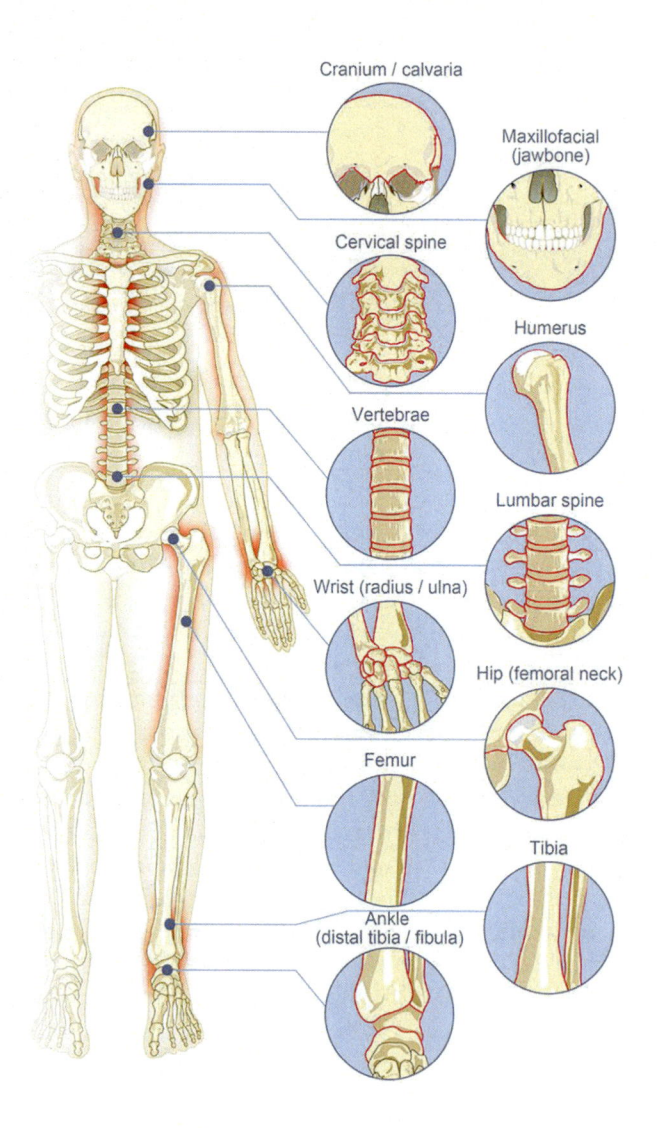

FIG. 9.1

The key fracture locations with in body where synthetic bone graft substitutes, bioactive compounds, and/or stem cells are necessary to restore bones in challenging clinical settings. *Reproduced with permission from Ho-Shui-Ling, A., Bolander, J., Rustom, L. E., Johnson, A. W., Luyten, F. P., & Picart, C. (2018). Bone regeneration strategies: Engineered scaffolds, bioactive molecules and stem cells current stage and future perspectives.* Biomaterials, *143–162. https://doi.org/10.1016/j.biomaterials.2018.07.017.*

synthetic materials, on the other hand, are hampered/limited by the possible risk of foreign-body responses as well as infection. More recently, novel approaches like porous 3D scaffolds and nano-engineered particles that promote the growth of new bone, have received a lot of interest in the field of bone tissue engineering. However, various important aspects of bone regeneration must be critically considered to achieve all its desired outcomes. For instance, natural bone is made up of 70% inorganic calcium phosphate crystals and 30% w/v organic collagen fibrils. On a macro- and nanoscale level, this composition has been used as a template to simulate bone structure (Ryu, Kim, Kang, & Park, 2010). The use of polymeric matrices connecting calcium phosphates and components like chitosan to treat various bone deformities has been evident (Liu et al., 2006). Nanotherapeutic advances, on the other hand, have enabled further modification of the extracellular matrix to offer more suitable surface chemistry and interlinked porosity for cell growth and angiogenesis. Another key element is the necessity for signaling molecules to be delivered in a controlled spatial and temporal manner to govern cellular differentiation and survival. After all, biocompatibility is critical, as synthetic nanomaterials must be inert or, ideally, resorb in a controlled and predictable manner to allow for reconstruction.

Bone tissue defects are currently causing an increased burden on healthcare systems around the world. Current bone tissue engineering methods face challenges such as ineffective cell growth, a lack of appropriate biomaterials, and techniques for capturing appropriate physiological architectures, as well as insufficient and unstable growth factor production to stimulate cell communication and proper response. Nanomaterials at the cutting edge of nanotechnology, with their unique size-dependent characteristics, have shown promise in overcoming many of the challenges that bone tissue engineering faces today. The previous and current advances in bone tissue therapy based on nanotechnological techniques are discussed in this chapter. Nonotherapeutic strategies for delivering drugs and growth factors that promote bone development, as well as gene therapy materials, are among the most notable. Nanomaterials have also made significant progress in stem cell targeting and imaging, which will be described. Finally, current developments in nano-composite construction and scaffold adaptations will be discussed to increase our understanding of the biocompatibility, cellular survival, and mechanical stability of implanted constructs.

Nanoparticles are small particles with a particle size of less than 100 nm that can be used as a tool for the delivery of various drugs, genetic materials, and growth factors as well (Wahajuddin, 2012). The size of nanoparticles greatly influences their distribution and half-life. The kidney readily clears particles smaller than 10 nm, but the spleen phagocytoses and eliminates particles larger than 200 nm (Fernández-Urrusuno et al., 1996). The majority of therapeutic nanoparticles are therefore ranging in size from 10 to 100 nm, allowing them to distribute throughout the systemic circulation and to pass through microscopic capillaries (Rolland, Verge, Collet, & Toujas, 1989). Also, surface characteristics have been demonstrated to have a substantial impact on nanoparticle stability, penetration, and localization into diverse target cells, and the charge has been proven to be a major factor in this context

(Wahajuddin, 2012). For instance, under the influence of a magnetic field, superparamagnetic iron oxide nanoparticles (SPIONs) have been used to deliver drugs or genetic material to desired tissues/cells in the body. Surface-modified hydrophilic polymers such as polyethylene glycol having amino or hydroxyl functional groups enable the outflux of nanoparticles from reticulo-endothelial cells, whereas hydrophobic surfaces enhance entrapment through circulating macrophages (Veiseh, Gunn, & Zhang, 2010). Notably, the physical features of nanocarriers must allow for loading without affecting the package's effectiveness, distribution to target sites, and ultimate release at a predetermined rate.

In this chapter, we discuss previous and present achievements in bone tissues therapies, based on nanotechnological approaches. Among them, nanotherapeutic techniques for delivering drugs and growth factors that promote bone development and gene therapy materials such as plasmid DNA or siRNAs are noteworthy. Nanomaterials have also enabled substantial advancements in stem cell targeting and imaging, which will be discussed. Finally, recent advances in nano-composite construction and scaffold adaptations will be elaborated, to improve implanted construct's biocompatibility, cellular survival, and mechanical stability.

Nanoparticle-based delivery in bone tissues engineering

Generally, nanoparticles could be used locally in bone tissue engineering (BTE) to promote tissue regeneration, improve implant osseointegration, and stop infections (Tautzenberger, Kovtun, & Ignatius, 2012). Given the poor results of several currently available biomaterials when used alone for bone replacement, researchers are increasingly interested in using biologically active molecules to encourage bone growth. The inherent constraints of small molecules, such as poor cell membrane permeability, limited physiological stability, and non-specific targeting, make the direct administration of therapeutic materials more challenging (Zhuang et al., 2014). In many instances, supraphysiological doses are frequently required to overcome the pharmacokinetics-related issues of these compounds, hence raising the risk of adverse effects (Faraji & Wipf, 2009). Nanocarriers based techniques can defeat such challenges by increasing the stability of bioactive molecules via entrapment or surface attachment, thereby promoting cellular penetration, directing cellular delivery, and enabling regulated drug release at the chosen target (Faraji & Wipf, 2009; Trewyn, Slowing, Giri, Chen, & Lin, 2007) (Fig. 9.2).

Nanospheres are extensively acknowledged tools and a viable method for controlled drug administration, because of their intrinsic small size and large specific surface area, higher responsiveness toward neighboring tissues in vivo, high drug-carrying efficiency, and ease of penetration of entrapped drug

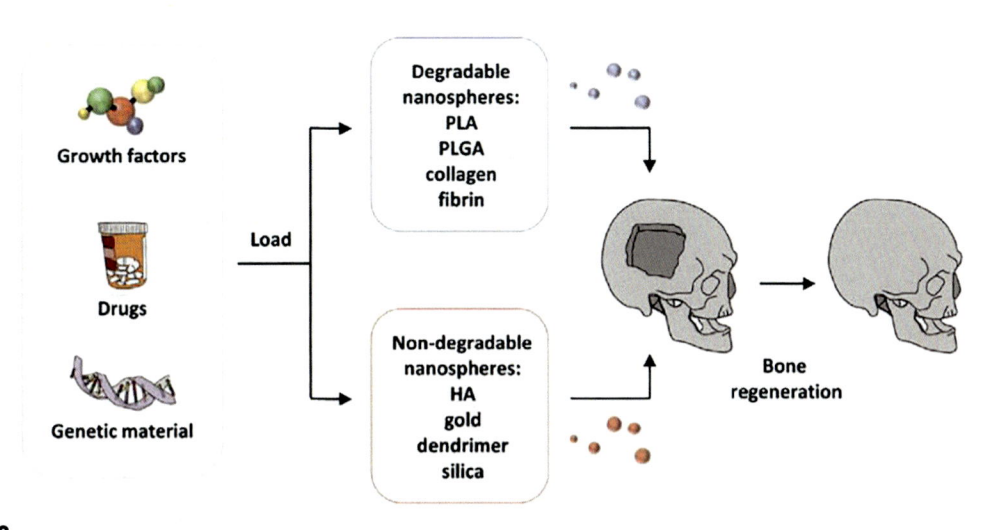

FIG. 9.2

Bone regeneration techniques based on nanoparticles. Degradable (poly(L-lactide) (PLA), poly(L-lactide-*co*-glycolic) (PLGA), collagen, fibrin), and non-degradable (hydroxyapatite (HA), gold, dendrimer, silica) nanoparticles are available, each with their own set of advantages and disadvantages. To improve bone production at disease and fracture sites, medicines, growth agents, or genes can be loaded. Reproduced with permission from *Walmsley, G. G., McArdle, A., Tevlin, R., Momeni, A., Atashroo, D., Hu, M. S., et al. (2015). Nanotechnology in bone tissue engineering.* Nanomedicine: Nanotechnology, Biology, and Medicine, 11*(5), 1253–1263. https://doi.org/10.1016/j.nano.2015.02.013.*

particles (Wang, Leeuwenburgh, Li, & Jansen, 2012). The targeted delivery of medications is one of the goals of current clinical therapies. So far, nanospheres' small size enables them to respond actively to stimuli from the surroundings such as pH, ultrasounds, irradiation, and magnetic fields, hence, these spheres can be used to administer chemically or biologically active compounds in a stimulus-driven manner, establishing triggered release as a result of external stimulation (Fundueanu, Constantin, Stanciu, Theodoridis, & Ascenzi, 2009). Following drug loading in nanospheres, either of non-degradable or degradable properties, drugs, genetic materials, or growth factors could be easily delivered. Gold, silica, hydroxyapatite, and dendrimers are the well-known examples of non-degradable nanoparticles (Fanord et al., 2010; Jensen et al., 2011; Oliveira et al., 2009), whereas poly(L-lactide-*co*-glycolic) (PLGA) or poly(L-lactide) (PLA) nanoparticles are the examples of degradable nanoparticles (Chen, Liu, Li, Tan, & Zhang, 2011; Mercado, Ma, He, & Jabbari, 2009).

Desired features of nanomaterials for bone tissues engineering

The core biomaterial for nanosphere synthesis is determined by the ultimate application criteria. It is determined by several factors, including (i) the desired nanoparticle size, (ii) the properties of materials to be loaded in the polymer (e.g., stability, water-solubility, etc.), (iii) surface properties and reactivity, (iv) the extent of biocompatibility and biodegradability, and (v) end product drug release profile (Mahapatro & Singh, 2011). Nanoparticles are frequently coupled with scaffolds like biodegradable polymeric matrices or proteinaceous hydrogels to make their application easier in bone treatments. The interaction between osteoclasts and osteoblasts is complex, and their activity is critical for bone homeostasis. Nanoparticle-based drug/growth factor (GF) administration can boost osteoblasts, while nanoparticles that release particular inhibitors locally can control osteoclasts (Tautzenberger et al., 2012).

Natural polymers based nanomaterials for bone tissues engineering

Natural polymers such as polysaccharides and proteins, as well as synthetic polymers, can be used to construct biodegradable nanospheres. Unlike injected proteins that are normally removed from the body quickly, locally deposited proteins are liberated through diffusion or desorption or and can thus be preserved for longer (Kofron, Li, & Laurencin, 2004). For this reason, nanospheres have been investigated as spectacularly modifiable drug delivery devices in terms of drug release site and period. Local drug administration is preferable compared to systemic drug delivery in virtue of minimum adverse effects. Furthermore, properly tailored nanoparticles enable a time-controlled, continuous delivery that meets the criteria. Additionally, the inadequate delivery of certain bioactive molecules including drugs, growth factors, and genes is usually caused by their unbalanced biological activity (Babensee, McIntire,& Mikos, 2000). On the implanted scaffold's surface, a carrier delivery system delivers controlled, long-term release with acceptable efficacy when compared to direct adsorption of a bioactive molecule (Orban, Marra, & Hollinger, 2002). The construction of delivery carriers requires materials with certain properties such as biodegradable, biocompatible, and best suited for the encapsulation of bioactive materials. Particularly, the entrapped growth factors could be released when the polymer degrades according to a preset and controlled profile, which is an important aspect of biodegradable nanosphere design. Hence, nanospheres are extensively being regarded as remarkably adaptable drug delivery systems that can control the position and duration of drug release, at the same time, also safeguarding the therapeutic agent from the biological environment.

Natural polymers in bone development

Natural polymers such as chitosan, fibrin, alginate, collagen, fibrin, and gelatin have garnered great interest because of their inherent biodegradability and biocompatibility properties. Collagen, for example, has a significant benefit in BTE because the extracellular matrix (ECM) of bone is primarily constituted of collagen in its organic stage, giving it excellent biodegradability, biocompatibility, and immunogenicity properties. To enhance the biological efficiency of thin bioactive coatings on metal implants, which are employed in BTE, the encapsulation of recombinant human morphogens and preventing its burst release, are significant problems. To address this problem, a study has recently demonstrated that nano-based colloidal gelatin gels possess significant potentials for the prolonged delivery of therapeutic osteogenic proteins such as alkaline phosphatase and bone morphogenetic protein-2 (BMP-2) (Wang, Boerman, et al., 2012).

Natural polymers contained numerous side groups on the molecular chain that enable more modification as well as functionalization (Wang, Boerman, et al., 2012). For this reason, natural polymers like gelatin and collagen have motifs like arginine-glycine-aspartic acid (RGD) sequences that can alter cell binding, promoting cellular activities over polymers that don't have sites for cell-recognition. Chitosan, which is a linear polysaccharide derived from crustacean shells, has significant advantages in nanosphere design since it is soluble in aqueous environments, eliminates the need for organic solvents, and eliminates the need for further nanoparticle purification (Hosseinzadeh, Atyabi, Dinarvand, & Ostad, 2012).

Generally, chitosan exhibits cationic character due to the presence of free amine groups in its structure, and it can interact with numerous cross-linkers to create nanoparticles. However, glutaraldehyde like crosslinkers may be harmful to the biological systems, chitosan can be ionically crosslinked using multivalent anions such as tripolyphosphate. Ionic gelation, a gentle method that produces nanoparticles with diameters less than 200 nm and has been shown to incorporate a variety of biological and active chemicals, has some advantages (Calvo, Remuñán-López, Vila-Jato, & Alonso, 1997; Shah, Pal, Kaushik, & Devi, 2009). Based on the large surface area of the nanosphere and the high affinity of chitosan for proteins, Kong and his associates (Kong et al., 2014) have recently applied the inclusion of nanospheres into thin coatings of collagen (i.e., mineralized), to maximize the rhBMP-2 encapsulation and increase morphogen release kinetics. Other natural polymers have the disadvantage of losing biological activity during processing, which can lead to the development of an immunological response. However, most of BTE research is focused on formulating the recombinant collagen, usually a safe alternative that could be chemically tailored to achieve clinical requirements (e.g., several chemical preparations enhance the preservation of a spherical shape), synthetic polymers may serve as a feasible alternative that offers additional benefits to the BTE platform.

Synthetic biodegradable polymers based nanomaterials for tissue bone regeneration

Synthetic biodegradable polymers, such as poly(L-lactide) (PLA) or poly(L-lactide-*co*-glycolic) (PLGA), have typical benefits such as intrinsic ease of preparation and adaptation, as well as cost-effectiveness. Also, the degradation profiles of polymer can be customized to provide the best bioactive material release profile. Poly (gly*co*lic acid) (PGA), PLA, and their co-polymers like poly(lactide-*co*-glycolide) (PLGA) have been the most extensively utilized polymers for nanoparticle and nanosphere production (PLGA). These polymers are notable for their biocompatibility as well as their natural resorbability. The enclosing of molecules in a phospholipid bilayer shell can improve site-specific targeting, degradation resistance, and the transportation of a large amount of medicine (Skouras et al., 2011; Wahajuddin, 2012).

Bisphosphonates, like alendronate, are generally employed for the management of osteoporosis. Because of their limited absorption, high doses are usually necessary to be clinically successful, resulting in systemic toxicity. As a result, a localized, long-lasting route of alendronate delivery is preferred, and outstanding outcomes have been observed with alendronate encapsulation in the PLGA nanoparticles. This technique has been appeared to be more successful than unbound drugs alone at producing osteoclast apoptosis and damaging osteoclast function (Cohen-Sela, Chorny, Koroukhov, Danenberg, & Golomb, 2009; Zhu et al., 2014). During in vitro research, employing poly (L-lysine) (PLL) nanoparticles to administer BMP-2 embedded in fibrin hydrogel displayed an improved osteogenic differentiation of mesenchymal cells derived from bone marrows (Park, Kim, Moon, & Na, 2009). Likewise, using this method, BMP-2-coated PLGA nanoparticles incorporated in fibrin hydrogel complex were found to greatly enhance bone healing in a critical-sized rat calvarial lesion during in vivo investigations (Qiao et al., 2013). BMP-7 was loaded in PLGA nanospheres, similar to BMP-2, which display a controlled release ectopic bone healing in rats after subcutaneous implantation on nano-fibrous PLA scaffolds (Wei, Jin, Giannobile, & Ma, 2007). These findings reveal that nanoparticles can be used to deliver growth factors in innovative bone regenerating techniques.

Non-biodegradable nanoparticles are primarily made up of ceramic nanoparticles (e.g., alumina, silica), metal sulfides, metal oxides, and certain other metals, which can be combined to create a wide range of nanostructures.

Inorganic nanoparticles for bone tissues engineering

Inorganic nanoparticles of various sizes and surface compositions can be created to avoid the reticuloendothelial system in general. Bioactive glasses are being described more and more for utilization in BTE. Bioactive glasses are a class of surface-responsive materials that can attach to bone in a physiological setting. They were first formed in 1969 (Hench, 2006). The most frequent bioactive glasses employed in BTE are silicate networks including phosphorous, calcium, and sodium but variants containing zinc, silver, magnesium, potassium, iron, boron, strontium, or fluorine have also been reported in the literature (Boccaccini et al., 2010; Oki, Parveen, Hossain, Adeniji, & Donahue, 2004; Sepulveda, Jones, & Hench, 2001). About its optimal biocompatibility, chemical characteristics, and thermal stability mesoporous silica nanoparticles are frequently employed as a delivery reagent. Due to their ease of manufacturing and modification, as well as their ability to sustain the function of bioactive materials, silica nanoparticles (i.e., sol-gel derivatives) in soft circumstances are currently a hot topic. Silica's unique mesoporous construction allows for effective drug loading, which is followed by regulated release. Mesopore characteristics, such as porosity and pore size, along with surface properties, could be changed, which mostly depends on the additives chosen to make mesoporous silica nanoparticles. Nano-shells, which are calcium- and phosphate-based hollow nanoparticles, have surface pores that lead to a core reservoir (Boccaccini et al., 2010). The distinctive surface of silica allows for surface modification and medicinal molecule linkage through functionalization. Hence, the previously described pH-controlled smart-release device shows potential for tailored drug/morphogen delivery in diffuse settings like BTE (Zhang, Wu, & Kong, 2014). Because of the prolonged production and secretion of growth factors (GFs) obtained by gene transfection, administration of growth factor (GF) genes may be more effective than delivery of GFs alone (Kim & Fisher, 2007). Downregulation of undesired genes or overexpression of pro-osteogenic genes are two techniques that could be used, but delivering constructs efficiently while retaining integrity and stability is still a key challenge. Downregulation of undesired genes or overexpression of pro-osteogenic genes are two techniques that could be used, but delivering constructs proficiently while retaining stability and integrity is still a key challenge. In a sequential approach, transferring genetic material creates unique obstacles. Several processes must be followed to properly deliver exogenous DNA encoding GF. For instance, the following steps should be accomplished perfectly: (i) diffusion of the DNA-nanoparticle complex across the cell membrane, (ii) absorption of intracellular endosomes, (iii) discharge into the cytoplasm, (iv) the complex's nuclear uptake, (v) detachment from the vector, (vi) protein expression, and (vii) GF protein secretion (Kim & Fisher, 2007; Mansouri et al., 2004). The carrier vector must be tiny enough to be imported into the cell and able to elude detection by endosome lysosome processes to safeguard the DNA until it reaches the desired cell for effective gene transfer. As a result, the involvement of nanoparticle-based genetic material transport may be easily identified. Nanoparticles of sub-cellular size can quickly infiltrate specific cells and tissues and provide treatment via endocytosis, eliminating the intrinsic drawbacks of viral-based vector systems (Yamamoto & Tabata, 2006). Nanoparticles coated with cationic polymers, which react with negatively charged DNA have been produced, and they can assist to stabilize and transport constructions to their intended locations (Lungwitz, Breunig, Blunk, & Göpferich, 2005). Liposomes, DNA PEGylation, micelles, nanoscale inorganic substances approach, polymeric nanoparticle DNA entrapment, and dendrimers systems have all been explored as nanoparticle gene delivery methods (Kim & Fisher, 2007). Polyethylenimine (PEI), a cationic polymer that may interact with negatively charged DNA, is one of the most often used methods. Dextran-coated nanostructures having a negatively charged functional group can also pair with peptide oligomer having positively charged in a similar manner (Bankura et al., 2012; Petri-Fink, Steitz, Finka, Salaklang, & Hofmann, 2008). The delivery of microRNAs and siRNAs to human MC3T3-E1 preosteoblasts and mesenchymal cells (hMSCs), using PEI-coated superparamagnetic iron oxide nanoparticles (SPIONs) and nanostructured polymers have been reported to be successful with low to no harmful effects (Hsu et al., 2014).

Nanostructures transmission of MSCs has also been shown to be as effective as or more effective than lipofectamine in studies (Ding, Zheng, Yang, Zhou, & Li, 2014; Wang et al., 2011). Positively charged nanoparticles can promote DNA intake via endocytosis and avoid the lysosome for transport to the nucleus thanks to the proton-sponge effect (Wahajuddin, 2012). Moreover, increased translocation can be achieved by using magnetic fields to direct SPION-DNA complexes to specified locations across the body, reducing the random diffusion of such particles once they arrive (Petri-Fink & Hofmann, 2007). Nanoparticles with PEI coatings have been used to distribute plasmids expressing the BMP-2 gene to improve bone growth. Hence, using this technique for the transfection of MSCs, showed enhanced expression of osteogenic indicators, and also alizarin alkaline phosphatase red staining, in vitro investigations. In immunocompromised animals, the subcutaneous implantation of such cells on a calcium phosphate cement platform resulted in much more ectopic new bone development (Lü et al., 2012; Lü et al., 2012). Similarly, the BMP-4 gene has also been delivered by nanoparticles, the employment of PLGA nanoparticles to improve the transfection of adipose-derived stromal cells (ASCs) has been shown to improve osteo-chondrogenic variations (Shi et al., 2001). Ultimately, rabbits were given a complex of gold nanoparticles conjugated with PEI and plasmids that expressed the BMP-7 gene to affect fibrosis and wound healing (Tandon et al., 2013). Nanoparticles have been discovered to be extremely versatile when it comes to delivering medications, genetic materials, and growth factors. Nanoparticles could react with their cargo due to their physical features, allowing them not just to improve stability but also effectively transport complex materials to specific cells/sites all through the body. Thus, modifications to proliferation, cell survival, and differentiation can be accomplished using a nanotherapeutic method to improve bone repair.

Conclusion

Ineffective cell growth, a lack of appropriate biomaterials and techniques for capturing appropriate physiological architectures, as well as insufficient and unstable growth factor production to stimulate cell communication and proper response, are all challenges that current bone tissue engineering methods face. Thanks to their unique size-dependent properties, nanomaterials at the cutting edge of nanotechnology have shown promise in overcoming many of the obstacles that bone tissue engineering. Drugs, growth factors, and gene therapy materials have all been delivered via nanomaterials such as metal nanoparticles and synthetic and natural biodegradable polymer-based nanocarriers. Nanomaterials have also made substantial advances in the targeting and imaging of stem cells for correcting bone tissues defects. The application of nanomaterials in bone tissues engineering is still in its early stages, and more study is needed to design and develop innovative nanomaterials for real-time applications in the field.

References

Angevine, P. D., Zivin, J. G., & McCormick, P. C. (2005). Cost-effectiveness of single-level anterior cervical discectomy and fusion for cervical spondylosis. *Spine, 30*(17), 1989–1997. https://doi.org/10.1097/01.brs.0000176332.67849.ea.

Babensee, J. E., McIntire, L. V., & Mikos, A. G. (2000). Growth factor delivery for tissue engineering. *Pharmaceutical Research, 17*(5), 497–504. https://doi.org/10.1023/A:1007502828372.

Bajaj, A. K., Wongworawat, A. A., & Punjabi, A. (2003). Management of alveolar clefts. *The Journal of Craniofacial Surgery, 14*(6), 840–846. https://doi.org/10.1097/00001665-200311000-00005.

Bankura, K. P., Maity, D., Mollick, M. M. R., Mondal, D., Bhowmick, B., Bain, M. K., et al. (2012). Synthesis, characterization and antimicrobial activity of dextran stabilized silver nanoparticles in aqueous medium. *Carbohydrate Polymers, 89*(4), 1159–1165. https://doi.org/10.1016/j.carbpol.2012.03.089.

Bauer, T. W., & Muschler, G. F. (2000). Bone graft materials: An overview of the basic science. *Clinical Orthopaedics and Related Research, 371*, 10–27. Lippincott Williams and Wilkins https://doi.org/10.1097/00003086-200002000-00003.

Boccaccini, A. R., Erol, M., Stark, W. J., Mohn, D., Hong, Z., & Mano, J. F. (2010). Polymer/bioactive glass nanocomposites for biomedical applications: A review. *Composites Science and Technology, 70*(13), 1764–1776. https://doi.org/10.1016/j.compscitech.2010.06.002.

Calvo, P., Remuñán-López, C., Vila-Jato, J. L., & Alonso, M. J. (1997). Novel hydrophilic chitosan-polyethylene oxide nanoparticles as protein carriers. *Journal of Applied Polymer Science, 63*(1), 125–132. https://doi.org/10.1002/(SICI)1097-4628(19970103)63:1<125::AID-APP13>3.0.CO;2-4.

Chen, L., Liu, L., Li, C., Tan, Y., & Zhang, G. (2011). A new growth factor controlled drug release system to promote healing of bone fractures: Nanospheres of recombinant human bone morphogenetic-2 and polylactic acid. *Journal of Nanoscience and Nanotechnology*, *11*(4), 3107–3114. https://doi.org/10.1166/jnn.2011.3820.

Clavero, J., & Lundgren, S. (2003). Ramus or chin grafts for maxillary sinus inlay and local onlay augmentation: Comparison of donor site morbidity and complications. *Clinical Implant Dentistry and Related Research*, *5*(3), 154–160. https://doi.org/10.1111/j.1708-8208.2003.tb00197.x.

Cohen-Sela, E., Chorny, M., Koroukhov, N., Danenberg, H. D., & Golomb, G. (2009). A new double emulsion solvent diffusion technique for encapsulating hydrophilic molecules in PLGA nanoparticles. *Journal of Controlled Release*, *133*(2), 90–95. https://doi.org/10.1016/j.jconrel.2008.09.073.

Dimitriou, R., Jones, E., McGonagle, D., & Giannoudis, P. V. (2011). Bone regeneration: Current concepts and future directions. *BMC Medicine*. https://doi.org/10.1186/1741-7015-9-66.

Ding, L. F., Zheng, G., Yang, J., Zhou, Z. D., & Li, J. J. (2014). Study on bone mesenchymal stem cells transfected by polyethylene glycol/bone morphogenetic protein-2. *Zhongguo gu shang = China Journal of Orthopaedics and Traumatology*, *27*(1), 48–53.

Fanord, F., Fairbairn, K., Kim, H., Garces, A., Bhethanabotla, V., & Gupta, V. K. (2010). Bisphosphonate-modified gold nanoparticles: A useful vehicle to study the treatment of osteonecrosis of the femoral head. *Nanotechnology*, *22*(3).

Faraji, A. H., & Wipf, P. (2009). Nanoparticles in cellular drug delivery. *Bioorganic and Medicinal Chemistry*, *17*(8), 2950–2962. https://doi.org/10.1016/j.bmc.2009.02.043.

Fernández-Urrusuno, R., Fattal, E., Rodrigues, J. M., Féger, J., Bedossa, P., & Couvreur, P. (1996). Effect of polymeric nanoparticle administration on the clearance activity of the mononuclear phagocyte system in mice. *Journal of Biomedical Materials Research*, *31*(3), 401–408. https://doi.org/10.1002/(SICI)1097-4636(199607)31:3<401::AID-JBM15>3.0.CO;2-L.

Fundueanu, G., Constantin, M., Stanciu, C., Theodoridis, G., & Ascenzi, P. (2009). PH- and temperature-sensitive polymeric microspheres for drug delivery: The dissolution of copolymers modulates drug release. *Journal of Materials Science: Materials in Medicine*, *20*(12), 2465–2475. https://doi.org/10.1007/s10856-009-3807-0.

Giannoudis, P. V., Dinopoulos, H., & Tsiridis, E. (2005). Bone substitutes: An update. *Injury*, *36*(3), S20–S27. https://doi.org/10.1016/j.injury.2005.07.029.

Gupta, A., Kukkar, N., Sharif, K., Main, B. J., Albers, C. E., & El-Amin, S. F., III. (2015). Bone graft substitutes for spine fusion: A brief review. *World Journal of Orthopedics*, 6.

Hench, L. L. (2006). The story of Bioglass®. *Journal of Materials Science: Materials in Medicine*, *17*(11), 967–978. https://doi.org/10.1007/s10856-006-0432-z.

Ho-Shui-Ling, A., Bolander, J., Rustom, L. E., Johnson, A. W., Luyten, F. P., & Picart, C. (2018). Bone regeneration strategies: Engineered scaffolds, bioactive molecules and stem cells current stage and future perspectives. *Biomaterials*, 143–162. https://doi.org/10.1016/j.biomaterials.2018.07.017.

Hosseinzadeh, H., Atyabi, F., Dinarvand, R., & Ostad, S. N. (2012). Chitosan–Pluronic nanoparticles as oral delivery of anticancer gemcitabine: Preparation and in vitro study. *International Journal of Nanomedicine*, 7.

Hsu, E. W., Liu, S., Shrivats, A. R., Watt, A. C. S., McBride, S., Averick, S. E., et al. (2014). Cationic nanostructured polymers for siRNA delivery in murine calvarial pre-osteoblasts. *Journal of Biomedical Nanotechnology*, *10*(6), 1130–1136. https://doi.org/10.1166/jbn.2014.1823.

Jensen, T., Baas, J., Dolatshahi-Pirouz, A., Jacobsen, T., Singh, G., Nygaard, J. V., et al. (2011). Osteopontin functionalization of hydroxyapatite nanoparticles in a PDLLA matrix promotes bone formation. *Journal of Biomedical Materials Research. Part A*, *99*(1), 94–101. https://doi.org/10.1002/jbm.a.33166.

Kim, K., & Fisher, J. P. (2007). Nanoparticle technology in bone tissue engineering. *Journal of Drug Targeting, 15*(4), 241–252. https://doi.org/10.1080/10611860701289818.

Kofron, M. D., Li, X., & Laurencin, C. T. (2004). Protein- and gene-based tissue engineering in bone repair. *Current Opinion in Biotechnology, 15*(5), 399–405. https://doi.org/10.1016/j.copbio.2004.07.004.

Kong, Z., Lin, J., Yu, M., Yu, L., Li, J., Weng, W., et al. (2014). Enhanced loading and controlled release of rhBMP-2 in thin mineralized collagen coatings with the aid of chitosan nanospheres and its biological evaluations. *Journal of Materials Chemistry B, 2*(28), 4572–4582. https://doi.org/10.1039/c4tb00404c.

Liu, H., Li, H., Cheng, W., Yang, Y., Zhu, M., & Zhou, C. (2006). Novel injectable calcium phosphate/chitosan composites for bone substitute materials. *Acta Biomaterialia, 2*(5), 557–565. https://doi.org/10.1016/j.actbio.2006.03.007.

Lü, K., Xu, L., Xia, L., Zhang, Y., Zhang, X., Kaplan, D. L., et al. (2012). An ectopic study of apatite-coated silk fibroin scaffolds seeded with AdBMP-2-modified canine bMSCs. *Journal of Biomaterials Science, Polymer Edition, 23*(1–4), 509–526. https://doi.org/10.1163/092050610X552861.

Lü, K., Zeng, D., Zhang, W., Xia, L., Xu, L., Jiang, X., et al. (2012). Ectopic study of calcium phosphate cement seeded with pBMP-2 modified canine bMSCs mediated by a non-viral PEI derivative. *Cell Biology International, 36*(2), 119–128. https://doi.org/10.1042/CBI20100848.

Lungwitz, U., Breunig, M., Blunk, T., & Göpferich, A. (2005). Polyethylenimine-based non-viral gene delivery systems. *European Journal of Pharmaceutics and Biopharmaceutics, 60*(2), 247–266. https://doi.org/10.1016/j.ejpb.2004.11.011.

Mahapatro, A., & Singh, D. K. (2011). Biodegradable nanoparticles are excellent vehicle for site directed in-vivo delivery of drugs and vaccines. *Journal of Nanobiotechnology, 55*. https://doi.org/10.1186/1477-3155-9-55.

Mansouri, S., Lavigne, P., Corsi, K., Benderdour, M., Beaumont, E., & Fernandes, J. C. (2004). Chitosan-DNA nanoparticles as non-viral vectors in gene therapy: Strategies to improve transfection efficacy. *European Journal of Pharmaceutics and Biopharmaceutics, 57*(1), 1–8. https://doi.org/10.1016/S0939-6411(03)00155-3.

Marsell, R., & Einhorn, T. A. (2011). The biology of fracture healing. *Injury, 42*(6), 551–555. https://doi.org/10.1016/j.injury.2011.03.031.

Mercado, A. E., Ma, J., He, X., & Jabbari, E. (2009). Release characteristics and osteogenic activity of recombinant human bone morphogenetic protein-2 grafted to novel self-assembled poly(lactide-co-glycolide fumarate) nanoparticles. *Journal of Controlled Release, 140*(2), 148–156. https://doi.org/10.1016/j.jconrel.2009.08.009.

Oki, A., Parveen, B., Hossain, S., Adeniji, S., & Donahue, H. (2004). Preparation and in vitro bioactivity of zinc containing sol-gel-derived bioglass materials. *Journal of Biomedical Materials Research. Part A, 69*(2), 216–221. https://doi.org/10.1002/jbm.a.20070.

Oliveira, J. M., Sousa, R. A., Kotobuki, N., Tadokoro, M., Hirose, M., Mano, J. F., et al. (2009). The osteogenic differentiation of rat bone marrow stromal cells cultured with dexamethasone-loaded carboxymethylchitosan/poly(amidoamine) dendrimer nanoparticles. *Biomaterials, 30*(5), 804–813. https://doi.org/10.1016/j.biomaterials.2008.10.024.

Orban, J. M., Marra, K. G., & Hollinger, J. O. (2002). Composition options for tissue-engineered bone. *Tissue Engineering, 8*(4), 529–539. https://doi.org/10.1089/107632702760240454.

Park, K. H., Kim, H., Moon, S., & Na, K. (2009). Bone morphogenic protein-2 (BMP-2) loaded nanoparticles mixed with human mesenchymal stem cell in fibrin hydrogel for bone tissue engineering. *Journal of Bioscience and Bioengineering, 108*(6), 530–537. https://doi.org/10.1016/j.jbiosc.2009.05.021.

Petri-Fink, A., & Hofmann, H. (2007). Superparamagnetic iron oxide nanoparticles (SPIONs): From synthesis to in vivo studies—A summary of the synthesis, characterization, in vitro, and in vivo investigations of SPIONs with particular focus on surface and colloidal properties. *IEEE Transactions on Nanobioscience, 6*(4), 289–297. https://doi.org/10.1109/TNB.2007.908987.

Petri-Fink, A., Steitz, B., Finka, A., Salaklang, J., & Hofmann, H. (2008). Effect of cell media on polymer coated superparamagnetic iron oxide nanoparticles (SPIONs): Colloidal stability, cytotoxicity, and cellular uptake studies. *European Journal of Pharmaceutics and Biopharmaceutics, 68*(1), 129–137. https://doi.org/10.1016/j.ejpb.2007.02.024.

Qiao, C., Zhang, K., Jin, H., Miao, L., Shi, C., Liu, X., et al. (2013). Using poly (lactic-co-glycolic acid) microspheres to encapsulate plasmid of bone morphogenetic protein 2/polyethylenimine nanoparticles to promote bone formation in vitro and in vivo. *International Journal of Nanomedicine, 8*.

Rolland, A., Verge, R. L., Collet, B., & Toujas, L. (1989). Blood clearance and organ distribution of intravenously administered polymethacrylic nanoparticles in mice. *Journal of Pharmaceutical Sciences, 78*(6), 481–484. https://doi.org/10.1002/jps.2600780613.

Ryu, J., Kim, S. W., Kang, K., & Park, C. B. (2010). Mineralization of self-assembled peptide nanofibers for rechargeable lithium ion batteries. *Advanced Materials, 22*(48), 5537–5541. https://doi.org/10.1002/adma.201000669.

Sepulveda, P., Jones, J. R., & Hench, L. L. (2001). Characterization of melt-derived 45S5 and sol-gel-derived 58S bioactive glasses. *Journal of Biomedical Materials Research, 58*(6), 734–740. https://doi.org/10.1002/jbm.10026.

Shah, S., Pal, A., Kaushik, V. K., & Devi, S. (2009). Preparation and characterization of venlafaxine hydrochloride-loaded chitosan nanoparticles and in vitro release of drug. *Journal of Applied Polymer Science, 112*(5), 2876–2887. https://doi.org/10.1002/app.29807.

Shi, J., Zhang, X., Zhu, J., Pi, Y., Hu, X., Zhou, C., et al. (2001). Nanoparticle delivery of the bone morphogenetic protein 4 gene to adipose-derived stem cells promotes articular cartilage repair in vitro and in vivo. *Arthroscopy: The Journal of Arthroscopic and Related Surgery, 29*(12).

Skouras, A., Mourtas, S., Markoutsa, E., De Goltstein, M. C., Wallon, C., Catoen, S., et al. (2011). Magnetoliposomes with high USPIO entrapping efficiency, stability and magnetic properties. *Nanomedicine: Nanotechnology, Biology, and Medicine, 7*(5), 572–579. https://doi.org/10.1016/j.nano.2011.06.010.

Stevenson, S., & Horowitz, M. (1992). The response to bone allografts. *Journal of Bone and Joint Surgery, 74*(6), 939–950. https://doi.org/10.2106/00004623-199274060-00017.

Tandon, A., Sharma, A., Rodier, J. T., Klibanov, A. M., Rieger, F. G., & Mohan, R. R. (2013). BMP7 gene transfer via gold nanoparticles into stroma inhibits corneal fibrosis in vivo. *PLoS One, 8*(6).

Tautzenberger, A., Kovtun, A., & Ignatius, A. (2012). Nanoparticles and their potential for application in bone. *International Journal of Nanomedicine, 7*, 4545–4557. https://doi.org/10.2147/IJN.S34127.

Trewyn, B. G., Slowing, I. I., Giri, S., Chen, H. T., & Lin, V. S. Y. (2007). Synthesis and functionalization of a mesoporous silica nanoparticle based on the sol-gel process and applications in controlled release. *Accounts of Chemical Research, 40*(9), 846–853. https://doi.org/10.1021/ar600032u.

Veiseh, O., Gunn, J. W., & Zhang, M. (2010). Design and fabrication of magnetic nanoparticles for targeted drug delivery and imaging. *Advanced Drug Delivery Reviews, 62*(3), 284–304. https://doi.org/10.1016/j.addr.2009.11.002.

Wahajuddin, S. A. (2012). Superparamagnetic iron oxide nanoparticles: Magnetic nanoplatforms as drug carriers. *International Journal of Nanomedicine, 7*, 3445–3471. https://doi.org/10.2147/IJN.S30320.

Wang, H., Boerman, O. C., Sariibrahimoglu, K., Li, Y., Jansen, J. A., & Leeuwenburgh, S. C. G. (2012). Comparison of micro- vs. nanostructured colloidal gelatin gels for sustained delivery of osteogenic proteins: Bone morphogenetic protein-2 and alkaline phosphatase. *Biomaterials*, *33*(33), 8695–8703. https://doi.org/10.1016/j.biomaterials.2012.08.024.

Wang, H., Leeuwenburgh, S. C. G., Li, Y., & Jansen, J. A. (2012). The use of micro-and nanospheres as functional components for bone tissue regeneration. *Tissue Engineering, Part B: Reviews*, *18*(1), 24–39. https://doi.org/10.1089/ten.teb.2011.0184.

Wang, W., Li, W., Ou, L., Flick, E., Mark, P., Nesselmann, C., et al. (2011). Polyethylenimine-mediated gene delivery into human bone marrow mesenchymal stem cells from patients. *Journal of Cellular and Molecular Medicine*, *15*(9), 1989–1998. https://doi.org/10.1111/j.1582-4934.2010.01130.x.

Wei, G., Jin, Q., Giannobile, W. V., & Ma, P. X. (2007). The enhancement of osteogenesis by nano-fibrous scaffolds incorporating rhBMP-7 nanospheres. *Biomaterials*, *28*(12), 2087–2096. https://doi.org/10.1016/j.biomaterials.2006.12.028.

Yamamoto, M., & Tabata, Y. (2006). Tissue engineering by modulated gene delivery. *Advanced Drug Delivery Reviews*, *58*(4), 535–554. https://doi.org/10.1016/j.addr.2006.03.003.

Zhang, P., Wu, T., & Kong, J. L. (2014). In situ monitoring of intracellular controlled drug release from mesoporous silica nanoparticles coated with pH-responsive charge-reversal polymer. *ACS Applied Materials and Interfaces*, *6*(20), 17446–17453. https://doi.org/10.1021/am5059519.

Zhu, X., Gu, J., Wang, Y., Li, B., Li, Y., Zhao, W., et al. (2014). Inherent anchorages in UiO-66 nano-particles for efficient capture of alendronate and its mediated release. *Chemical Communications*, *50*(63), 8779–8782. https://doi.org/10.1039/c4cc02570a.

Zhuang, J., Kuo, C. H., Chou, L. Y., Liu, D. Y., Weerapana, E., & Tsung, C. K. (2014). Optimized metal-organic-framework nanospheres for drug delivery: Evaluation of small-molecule encapsulation. *ACS Nano*, *8*(3), 2812–2819. https://doi.org/10.1021/nn406590q.

CHAPTER 10

Overcoming ocular barriers through nanocarrier-based drug delivery systems

Introduction

Ophthalmic drug delivery is one of the most attractive and challenging fields for pharmaceutical scientists. Eyes have unique anatomy which consists of anterior and posterior segments (Fig. 10.1) (Patel, 2013). The main parts of the anterior segment include the cornea, ciliary body, conjunctiva, iris, aqueous humor, and lens while the posterior part consists of the sclera, retinal pigment epithelium, choroid, neural retina, vitreous humor, and optic nerve. Eyes can be affected by ocular pathological diseases which are usually referred to as anterior and posterior segment disorders. Most common diseases which affect the anterior segment include glaucoma, anterior uveitis, cataract, and allergic conjunctivitis (Molokhia, Thomas, Garff, Mandell, & Wirostko, 2013) while diabetic retinopathy and age-related macular degeneration (AMD) are the most prevalent disorders of the posterior segment of the eye (Varela-Fernández et al., 2020). There are different physiological barriers between anterior and posterior segments of the eye (Fig. 10.2) (Fan et al., 2020). Ocular barriers serve as both a protective barrier and a barrier to medication entry into the eye. Several ways have been tested to overcome these obstacles and successfully transport therapeutically active chemicals to both the anterior and posterior parts of the eye (Weng et al., 2017). For the treatment of ocular anterior segment disorders, usually non-invasive conventional methods are used which are based on topical drop installation over the anterior segment. Ninety percent of ophthalmic formulations being marketed consist of eye drops due to the ease of administration, patient compliance, and the non-invasive nature of drug dosage for anterior segment diseases. Ointments, emulsion, aqueous gels, and suspensions are the most extensively used conventional formulations for ocular drug administration (Patel, 2013). The major problems associated with topical drop administration include low bioavailability, rapid clearance, and extensive loss of drug from the eye due to the lachrymal drainage and dilution by tears. Due to these drawbacks, a small amount of administered ophthalmic drug penetrates the corneal layer and reaches the internal tissues of the eye which results

225

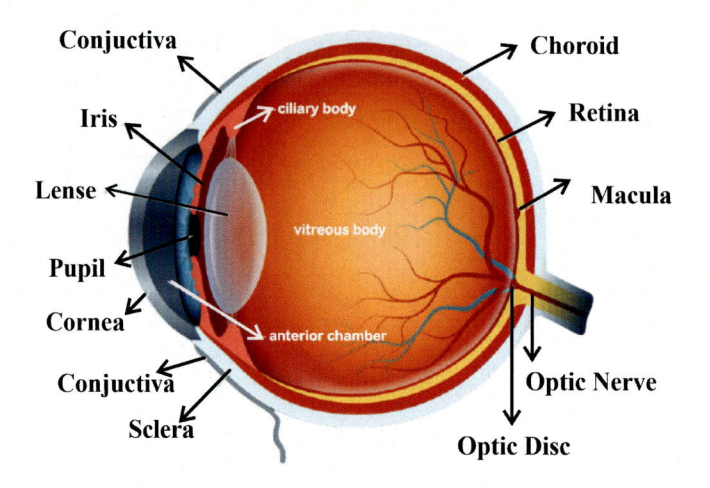

FIG. 10.1

Anatomy of the eye. *From Bahamonde-Norambuena, D., Molina-Pereira, A., Cantin, M., Muñoz, M., Zepeda, K., & Vilos, C. (2015). Nanopartículas poliméricas en dermocosmética.* International Journal of Morphology, 33*(4), 1563–1568. https://doi.org/10.4067/S0717-95022015000400061.*

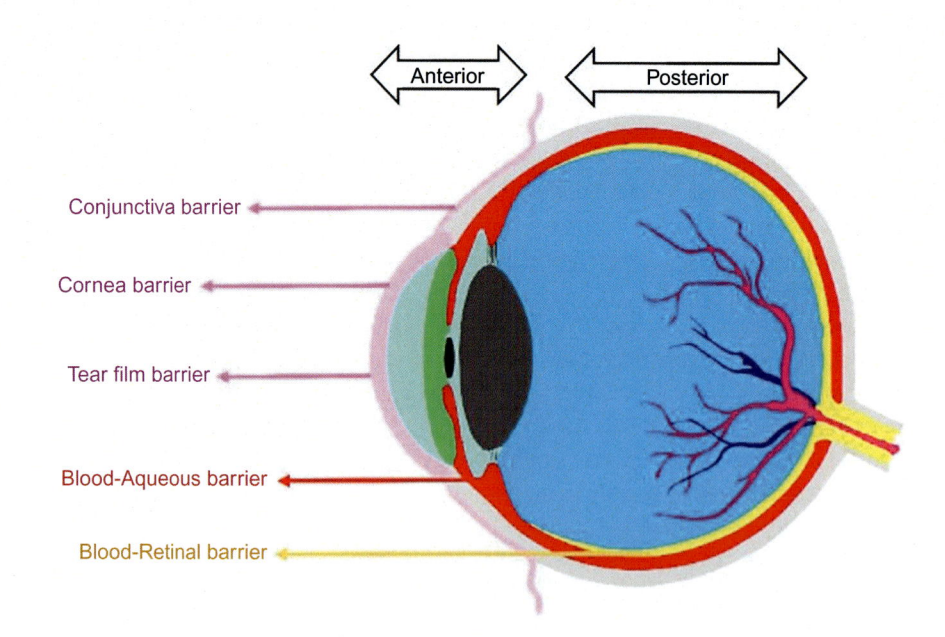

Ocular Barriers

FIG. 10.2

Physiological barriers of the eye. *Reproduced with permission from Fan, X., Torres-Luna, C., Azadi, M., Domszy, R., Hu, N., Yang, A., et al. (2020). Evaluation of commercial soft contact lenses for ocular drug delivery: A review.* Acta Biomaterialia, 115, *60–74. https://doi.org/10.1016/j.actbio.2020.08.025.*

in reduced bioavailability and unwanted side effects. In the light of the above-mentioned issues associated with the conventional ocular drug administration methods, novel drug delivery systems have been developed by researchers over the last two decades to overcome the major barriers of ocular drug delivery. The main purpose for the design and development of an ocular drug delivery system is to provide controlled and sustained release of drug to achieve enhanced and prolonged corneal penetration, simple and efficient route of administration, non-irritative and biocompatible nature with minimum side effects (Xu, Kambhampati, & Kannan, 2013).

Nanocarrier based ocular drug delivery system

Nanotechnology-based drug delivery systems have received great attention for the treatment of ophthalmic disorders. The development of nanomedicines for ocular delivery is a unique and effective approach compared to conventional medications since they offer great advantages including superior efficacy, improved bioavailability, and enhanced transport through the biological barriers which provide greater opportunities for disease targeting, prolonged circulation, and reduced toxicity. The utilization of nanocarriers has been increased in the biomedical field as compared to traditional conventional methods. Nanocarriers-based therapeutics for ocular drug delivery has proved to be highly effective for the efficient treatment and targeted biomolecular interaction with lower side effects (Wadhwa, Paliwal, & Vyas, 2009). The nanocarriers are designed to be highly biocompatible, safe, and acquire the ability for sustained release of the drug. These characteristics are developed by modifying the physiochemical properties including shape, size, and morphology of the nanocarriers which has a high impact on their physiological behaviors in biological systems. Nanoparticles, nanoemulsions, nanomicelles, liposomes, and dendrimers are the most commonly employed nanocarriers for ocular drug delivery (Fig. 10.3) (Weng et al., 2017). These nanocarriers have lower irritation and have better bioavailability of drug with the unique ability for targeted site action which enhances the interaction of drug with ocular tissues. Here we discuss the advanced nanocarriers which are being developed for the treatment of anterior and posterior segments of the eye.

Anterior ocular drug delivery

The anterior segment of the eye remains the most affected area due to exposure to the external environment which results in inflammation due to dust and microbes. Ophthalmic formulations in the form of eye drops remain the most employed method for drug delivery to ocular tissues. Recent advancement in the development of ophthalmic formulations is focused on finding new strategies for the delivery of therapeutic agents to the anterior segment of eyes that

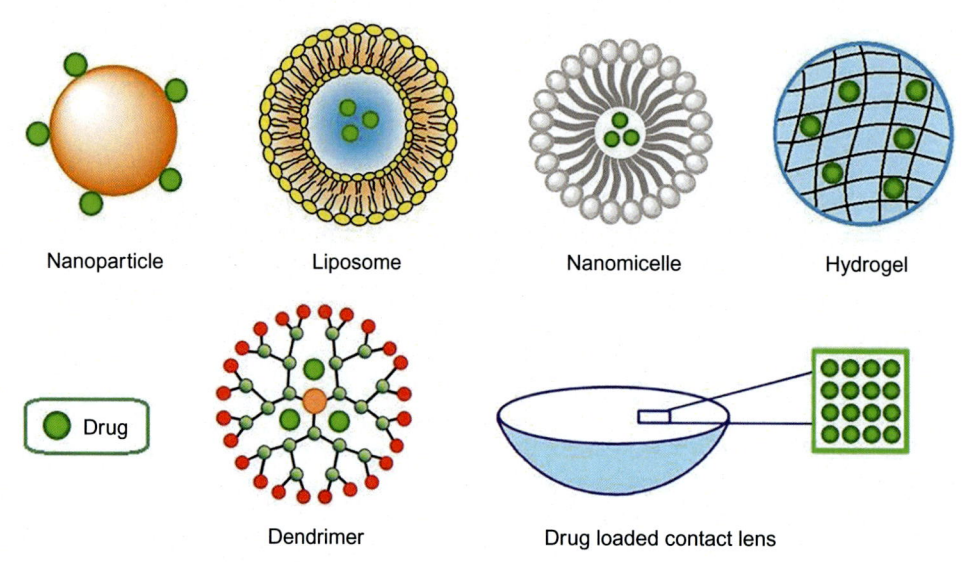

FIG. 10.3

Types of nanocarriers for ocular drug delivery. *No Permission Required.*

can efficiently target the specific ocular tissues, maintain high therapeutic drug levels, have prolonged and sustained release, achieve longer precorneal residence time with minimum side effects (Cholkar, Patel, Vadlapudi, & Mitra, 2013). Most frequently used conventional ophthalmic formulations are in the form of the oily and aqueous solution, emulsions, ointments, and suspensions which were developed to achieve enhanced drug solubility, greater precorneal retention time with increased bioavailability as compared to eye drops (Behar-Cohen, 2012). Nanotechnology-based ophthalmic formulations have been a developing area in the field of research for the generation of nanosystems that have adequate particle size to achieve enhanced bioavailability and lower irritation to the ocular tissues. Nanotechnology-based formulations are an attractive alternative to overcome the drawbacks of conventional ophthalmic formulations. The advancement in nanotechnology has led to the modification of topical formulations and the development of nanoformulations with permeation and viscosity enhancers bac (Bachu, Chowdhury, Al-Saedi, Karla, & Boddu, 2018). Colloidal dispersions of nanosized particles having a high capacity to dissolve both lipophilic and hydrophilic drugs with greater stability and bioavailability have been introduced to enhance drug permeability across the cornea. Various ligands including charged polymeric compounds having greater therapeutic efficacy and affinity toward targeted ocular tissues have been added in the nanoformulations accompanied with surfactants to increase corneal penetration (Tsai et al., 2018).

Emulsions

Ophthalmic drug delivery through conventional eye drop remains the most common approach for ocular drug administration despite the lower bioavailability of less than 5%. Various drug delivery approaches have been used to overcome the disadvantages of eye drops. Microemulsion-based drug delivery systems are an attractive approach for the delivery of less water-soluble drugs that comprise dispersions of oil and water mixed with a surfactant (Vandamme, 2002). Emulsions deliver hydrophobic drugs to the cornea due to the possibility of greater drug loading in the oil particles. Microemulsions have improved stability, greater solubility, and improved corneal penetration. The design of these emulsions provides greater opportunities that enhance the possibility of increasing the residence time of drugs due to the binding between the formulation and cornea (Gautam & Kesavan, 2017). Microemulsion dispersed in contact lenses is also an advanced approach for the increment in residence time. The parameters which have a greater impact on the stability of microemulsions are the aqueous and organic phase as well as the surfactant used for the stabilization of the formulations (Ameeduzzafar et al., 2016). Despite the attractive features of emulsions, there are also some drawbacks associated with these kinds of formulations which are toxicity and stability of emulsions. Furthermore, ocular physiology including rapid tear turnover and tear biochemistry also limits the effectiveness of using microemulsions. Drugs such as sirolimus, timolol, and chloramphenicol were incorporated in microemulsions with improved solubility, stability, and enhanced bioavailability (Buech, Bertelmann, Pleyer, Siebenbrodt, & Borchert, 2007). Nanotechnology-based recent approaches have led to the development of nanoemulsions that are the most efficient and appropriate for the delivery of bioactive molecules to ocular tissues. Nanosuspensions are submicron colloidal systems consisting of inert polymeric resins having drugs suspended in a suitable dispersion medium. Cationic nanoemulsions are nanosystems having bioadhesive properties and provide greater solubility of the drug, enhanced bioavailability with less irritation in the eye. The strategy for developing cationic nanoemulsions is to prolong residence time due to the development of electrostatic interaction between the emulsions and negatively charged cells of the ocular surface (Lallemand, Daull, Benita, Buggage, & Garrigue, 2012).

Ointments

Ophthalmic ointments are semisolid ophthalmic dosage forms that are comprised of solid or semisolid hydrocarbon bases (Patel, 2013). They have widely used medications for the treatment of inflammation and infection. After the application of the ointment, it converts into small droplets that reach the conjunctival sac and provide sustained release of the drug to the ocular tissues. Their main advantage over conventional eye drops includes sustained release, minimum tear dilution, reduced nasolacrimal drainage, and enhanced effective concentrations (Shell, 1984). Many ointments have been designed having improved bioavailability to achieve sustained drug release. The base of the ophthalmic ointment is highly compatible and does not cause any discomfort in the eye. Commonly used vehicles for the preparation of ointments are white petrolatum and liquid petrolatum (mineral oil). One of the attractive features of the utilization of ointments is their entrapment into the fornices, which acts as the reservoir of the drug (Rathore & Nema, 2009). Drugs having low water solubility can easily be incorporated into the formulations, so they can be delivered to the eye. Moreover, these ointments have a greater shelf life and the drug is highly stabilized in the matrix of the formulation. Ointments can be sustained up to 2–4 h and in some cases even up to 8 h after their application in the eye. Ointments having sorption promoters (penetration enhancers) have been developed which exhibited significantly higher drug release rates as compared to the ointments without these promoters. Despite the great and attractive advantages of the ointments, there are some issues associated with them. Ointments cause blurred vision due to the difference in the refractive index of non-aqueous ointment base and tears which is the main cause that they are recommended for bedtime installation (Dubald, Bourgeois, Andrieu, & Fessi, 2018). Sticking together eyelids after the application of ointments is also a common complaint from patients. Drug molecules can remain entrapped into the matrix of the ointment which results in the poor release of the drug toward the targeted site. To overcome these aforementioned issues, the utilization of water-soluble bases called gels has recently been increased due to their advantages over petroleum bases like better lubrication, stability, and low irritability.

Suspensions

Suspensions are considered to be one of the most efficient and non-invasive forms for ophthalmic topical drug administration. Suspensions are dispersions of the drug in an aqueous solvent consisting of appropriate suspending and dispersing agents (Patel, 2013). After the application of suspension into the eye, it retains in the precorneal pocket which results in longer contact time and higher bioavailability of drug as compared to drug solution (Patravale, Date, & Kulkarni, 2004). The critical parameter which ensures the optimum stability and bioavailability of the suspension is the particle size. Small size particles tend to have a larger amount of drug absorbed into ocular tissues from the precorneal pocket while larger particles have slower drug dissolution (Patravale et al., 2004). Therefore, optimum particle size has to be achieved for improved and sustained drug activity. Patient compliance is also dependent on the particle size of the suspension, since irregularly shaped and distorted particles can irritate the eye and cause discomfort. The average particle size of most ophthalmic suspensions is less than 10 mm. The development of effective and stable formulation always remains a challenging task due to the possibility of aggregation of particles and non-homogeneity of the sample (Krishna & Prabhakar, 2011). Incompatibility of the ingredients of suspension, cake formation, and settling are some of the major complications which have to be resolved and optimized for the development of a stable formulation having sustained release and enhanced bioavailability of the drug. The utilization of cyclodextrins for the improvement of bioavailability of suspension is an innovative approach (Muankaew, Jansook, Sigurcrossed, Signsson, & Loftsson, 2016). Cyclodextrins are cyclic sugars that have a hydrophilic exterior and a lipophilic cavity and the ability to form inclusion complexes with poorly water-soluble drugs. Various techniques have been used for the preparation of nanosuspensions including high-pressure homogenization, medical milling, and microfluidic nanoprecipitation which have shown to be highly promising for poorly water-soluble drugs like hydrocortisone.

Aqueous gels

To overcome the drawbacks of conventional eye drops, aqueous gels are a recent approach for ophthalmic treatment. These gels can be easily administered to the eye and provide a prolonged contact time of pharmacologically active molecules (Jothi, Harikumar, & Aggarwal, 2012). Aqueous gels are formulation based on hydrophilic polymers (hydrogels). The formulation matrix is comprised of water-soluble bases since they have greater advantages over petrolatum bases such as better stability, pH, spreadability, better lubrication, and lower chances of irritation to the ocular tissues. The polymers which are commonly used for the preparation of gels include carbopol, PEG 200, PEG 400, and hydroxypropyl methylcellulose (HPMC) (Patel et al., 2016). Aqueous gels allow the incorporation of ophthalmic pharmaceuticals to achieve sustained release of drugs at the ocular sites. Nanotechnology-based drug delivery systems have brought a revolution in the design and development of environmentally responsive systems to extend current treatment durations and patient compliance. A more advanced approach based on gels is the development of in-situ gels, which are based on stimuli-responsive polymers that allow incorporation of ophthalmic pharmaceuticals (Baranowski, Karolewicz, Gajda, & Pluta, 2014). Natural polymers like chitosan are highly biocompatible, biodegradable, and have a mucoadhesive character, which is being utilized for the preparation of in-situ gelling systems. The combination of pH-sensitive polymer, chitosan, as well as stimuli-responsive polymers, results in enhanced mechanical strength and improved therapeutic effect of the in-situ gelling system due to sustained release of drug and prolonged contact time. The development of gel integrated drug delivery devices including injections, intraocular pumps, and implants are novel applications that have great potential to reduce comorbidities generated by cataracts, glaucoma, diabetic retinopathies, and age-related macular degeneration (Al-Kinani et al., 2018).

Posterior ocular drug delivery

In the last two decades, the advancement of nanotechnology has led to the development of more effective and advanced nanocarrier-based drug delivery systems for the posterior segment of the eye. Apart from the conventional drug delivery methods, recent novel ophthalmic nanocarrier-based drug delivery systems are based on nanoparticles, liposomes, nanomicelles, dendrimers, contact lenses, and in situ thermoresponsive hydrogels (Thrimawithana, Young, Bunt, Green, & Alany, 2011). These systems provide targeted site action and sustained release over a longer period which are required for the improvement of the bioavailability of the drugs. Drug delivery to ocular tissues is restricted by precorneal, dynamic, and static ocular barriers which is the main reason that therapeutic drug levels are not maintained for a longer period in the ocular tissues and ultimately results in poor bioavailability of therapeutic agents. The main focus of ocular drug delivery research is toward the design and development of biocompatible, safe, and patient compliant formulations and drug delivery devices that overcome these barriers and enhance the bioavailability of drugs to the targeted site of action in the ocular tissues (Nayak & Misra, 2018). Posterior ocular delivery remains a challenging task due to ocular barriers and has been focused on the development of ophthalmic nanoformulations and drug-releasing devices for the treatment of vitreoretinal diseases (Wang, Xu, Gu, Cheng, & Cao, 2018). These formulations may help to overcome ocular barriers and reduce side effects associated with conventional methods currently available for the treatment of ocular diseases. The novel formulations are easy to formulate and have significant advantages like minimum irritation, high precorneal residence time, sustained drug release, and enhanced bioavailability of pharmacologically active compounds. The main focus for the optimization and development of nanocarrier-based ocular drug delivery systems is to achieve the following characteristics in the nanoformulation like increased residence time, bioavailability, stability of the formulation, extended retention time, and targeted delivery of drug at the desired site of action with minimum adverse effects.

Nanoparticles

Nanoparticles have at least one dimension on the scale of 1–100 nm. Nanoparticles have unique properties like ultra-small size and large surface area to the mass ratio which are different from their bulk counterparts. The diverse nature of nanoparticles is due to the great variety of their nature, morphology, and shapes in which they are available as well as their dispersion rate and medium in which they are present (Zhou, Hao, Wang, Zheng, & Zhang, 2013). Nanoparticles have significant biological importance for ocular drug delivery. For ocular delivery, nanoparticles are composed of biocompatible materials including proteins, lipids, and natural or synthetic polymers with improved sustainability, targeting efficiency with lower side effects. Albumin, chitosan, polylactic acid (PLA), poly (lactide-*co*-glycolide) (PLGA), and poly-caprolactone are commonly employed substances for the formation of nanoparticle-based ophthalmic formulations (Patel, Patel, Patel, Patel, & Prajapati, 2010). Due to their small size, nanoparticles do not cause irritation in the eye and provide sustained release thus, minimizing frequent drug administration.

In the past few decades, various scientists have developed drug-loaded nanoparticles for both anterior and posterior ocular drug delivery. Ocular drug delivery through nanoparticle formulation has many advantages as they surpass ocular barriers, can be easily formulated, have negligible irritation, provide sustained releases and enhance ocular bioavailability of therapeutics (Omerović & Vranić, 2020). Besides these advantages, aqueous suspensions of nanoparticles can be rapidly eliminated from the precorneal pocket. To overcome this drawback, nanoparticles having mucoadhesive properties have been developed to improve their sustainability and lead to prolonging precorneal residence time (Zhou et al., 2013). Chitosan, hyaluronic acid, and polyethylene glycol (PEG) are being widely used for the improvement of mucoadhesive character to achieve sustained release. Nanoparticles having a size in the range of 2–200 nm are most feasibly employed for therapeutic systemic administration. Nanoparticle-based drug delivery systems have been currently investigated for the treatment of ocular diseases, due to their unique properties and ultra-small size, nanoparticles can overcome the anatomical and physiological barriers of eyes, including cornea, conjunctiva, and blood-retinal barrier (BRB). Drug-loaded nanoparticles can be in the form of nanospheres or nanocapsules (Nagarwal, Kant, Singh, Maiti, & Pandit, 2009). In drug-loaded nanospheres, pharmacologically active molecules are evenly distributed throughout the polymeric matrix as shown in Fig. 10.4 while in nanocapsules, the drug is entrapped into the polymeric shell and is released at the targeted site due to the interaction of functional ligands with physiochemical metabolites. Targeted drug delivery is a unique feature for the localized delivery of pharmacologically active compounds to improve and ensure their effectiveness at the desired site of action (Barar, Aghanejad, Fathi, & Omidi, 2016). Various targeting ligands have been incorporated in nanoparticles-based drug delivery systems for active and passive targeting to the desired ocular tissues.

Nanoparticles-based ophthalmic formulations are still in the initial stage of development and further studies are required for the determination of their toxicity to the health and environmental conditions. Moreover, scaling up for their large-scale production is another challenging task for the commercialization of these formulations for ocular diseases.

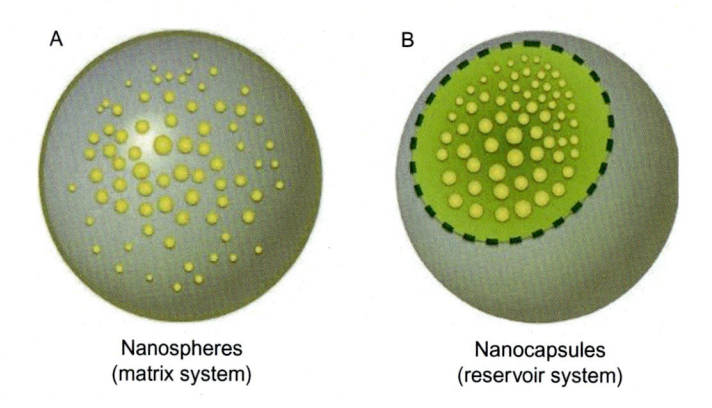

FIG. 10.4

Schematic illustration of nanospheres and nanocapsules. *From Bahamonde-Norambuena, D., Molina-Pereira, A., Cantin, M., Muñoz, M., Zepeda, K., & Vilos, C. (2015). Nanopartículas poliméricas en dermocosmética.* International Journal of Morphology, 33*(4), 1563–1568. https://doi.org/10.4067/S0717-95022015000400061.*

Liposomes

Liposomes are vesicular systems that are utilized as drug loading vehicles having a size range of 10 nm to 1 μm or greater (Fig. 10.5). These vesicular systems are comprised of phospholipid bilayers of natural or synthetic origin with an aqueous core (Meisner & Mezei, 1995). Types of liposomal carriers based on

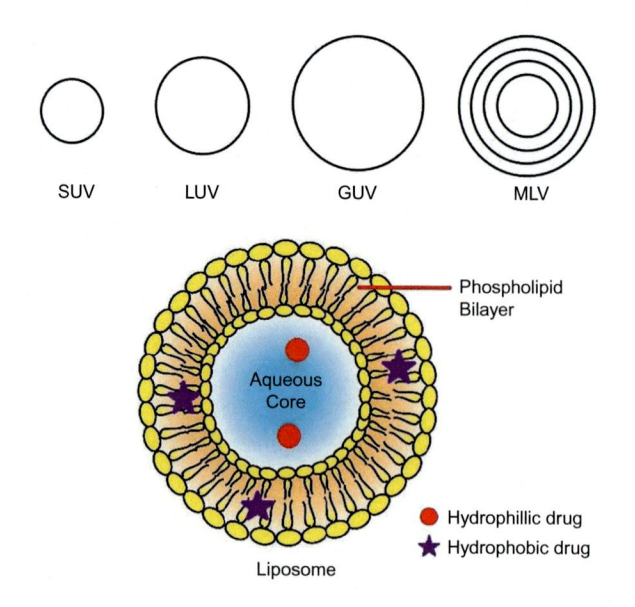

FIG. 10.5

Types and structure of liposome. *From Leung, A. W. Y., Amador, C., Wang, L. C., Mody, U. V., & Bally, M. B. (2019). What drives innovation: The Canadian touch on liposomal therapeutics.* Pharmaceutics, 11*(3). https://doi.org/10.3390/pharmaceutics11030124.*

Table 10.1 Size of different types of liposomes.

Vesicle type	Size
Small unilamellar vesicles (SUVs)	~20 to ~200 nm
Large unilamellar vesicles (LUVs)	~200 nm to ~1 μm
Multilamellar vesicles (MUVs)	>0.5 μm
Giant unilamellar vesicles (GUVs)	>1 μm

size are shown in Table 10.1 (Mishra, Bagui, Tamboli, & Mitra, 2011). Liposomal nanocarriers are being widely used for ocular drug delivery applications. Liposomal vesicles are mainly developed by using biocompatible materials including phosphatidylcholine (PC), lecithin, or phosphatidylglycerol with the addition of cholesterol (Agarwal et al., 2016). In the case of topical delivery, the main focus of the liposomal nanoformulations is to improve the bioavailability of the drug by increasing the corneal adhesion and permeation of the nanovesicles by using permeation enhancing and bioadhesive polymers. For posterior segment disorders, sustained drug release and targeted drug induction to the retina are achieved by designing the liposomal carrier having specific receptors or proteins bound with the surface of vesicles (Lai et al., 2019). The capacity of drug loading depends on many factors including size of liposomes, types of precursor lipid used for their preparation, nature of lipids used for its core composition, and physicochemical properties of therapeutic agent entrapped into the liposomal nanocarrier. Regarding the size of the liposomal nanocarriers, SUVs have the smallest size and their entrapment efficiency is poor as compared to MLVs, while LUVs provide optimum balance related to size and drug loading capacity (Mishra et al., 2011). One of the unique abilities of liposomes is their entrapment ability toward both hydrophobic and lipophilic pharmacologically active agents. Hydrophobic drugs are entrapped in lipid bilayers while hydrophilic drugs interact with aqueous layers. The loading efficiency of ionic molecules can be improved by the utilization of cationic or anionic lipids during the formation of liposomal nanocarriers (Akbarzadeh et al., 2013).

Nanomicelles

Micelles are self-assembly of amphiphilic molecules which are organized in various shapes and sizes, depending on the molecular weight of precursor molecules, surfactant, and characteristics of the medium (Fig. 10.6). Micelles having a size below 100 nm are referred to as nanomicelles. In the last 5 years, research has been mainly focused on the design and development of polymeric nanomicelles based on ophthalmic formulations to overcome some of the major challenges for the topical treatment of ocular diseases (Grimaudo et al., 2019).

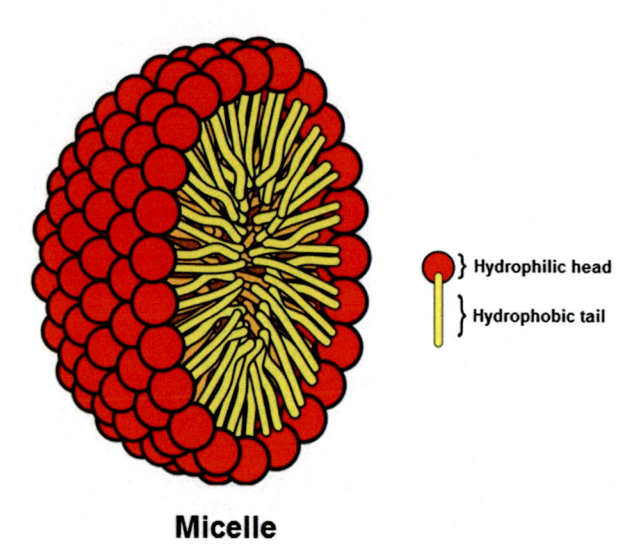

Micelle

FIG. 10.6

Structure of micelle. *No Permission Required.*

Nanomicelles acts as solubility enhancers for drugs having lower solubility and promote drug transportation across the cornea and sclera. Nanomicelles have higher drug loading efficiency, can be easily formulated and their smaller size improves drug availability in ocular tissues which results in better therapeutic outcomes (Vaishya, Khurana, Patel, & Mitra, 2014). Polymeric micelles are composed of biocompatible and biodegradable materials. Widely utilized polymers for the formation of nanomicelles are poly(propylene oxide) (PPO), poly(ethylene glycol) (PEG), poly(lactide), poly(glycolide), poly(ε-caprolactone) (PCL) and poly(lactide-*co*-glycolide) (Grimaudo et al., 2019). The methodology utilized for the preparation of micelles and physiochemical properties of the drug, both factors play a key role in the entrapment of pharmacologically active compounds into the nanomicelle formulation. The most frequently utilized methods for the preparation of nanomicelle are solvent or co-solvent evaporation, direct dissolution, film hydration, and dialysis. Drugs can be entrapped at various positions depending on their polarity. Hydrophobic drugs can adhere at the surface of the nanomicelle, drugs having intermediate hydrophobicity can attain intermediate positions within the nanomicelle while hydrophobic drugs are located in the core of the nanomicelles (Vaishya et al., 2014). Drug releasing behavior from polymeric micelles is dependent upon the nature of polymer and drug, method of preparation, and localization of drug into the nanoformulation. Nanomicelles are designed to have controlled and sustained release by derivatization and attachment of specific ligands in order to prevent burst release of the pharmacologically active compound upon the interaction of nanomicelles with the physiological fluids and to achieve the desired pharmacological effect (Trinh, Joseph, Cholkar, Mitra, & Mitra, 2017). For the increment in retention time and to improve the sustained release from the nanomicelles, the use of mucoadhesive polymers has also gained much attention. Chitosan, hyaluronate, or mucoadhesive gels provides adequate adhesion of nanomicelles to the pre-corneal conjunctival mucus layer through non-covalent bonds (Lynch et al., 2019).

Several studies have been performed to determine the potential of nanomicelles based formulations for the drug delivery of anterior and posterior segments of the eyes. Various nanomicelles based formulations have been patented and FDA has approved the first nanomicelle formulation containing cyclosporine A 0.09% (Cequa, Sun Pharmaceutical Industries Ltd) (Periman, Mah, & Karpecki, 2020). Despite the current advancement and progress in the development of nanomicelles based ophthalmic formulations, there is still a need to explore and study the underlying mechanisms and various properties of nanomicelles like size, shape, the rigidity of structure, and surface charge density for the development of efficient nanocarrier system for the delivery of pharmaceutically active compounds.

Dendrimers

Dendrimers are nanostructured polymers having tree-like structures as shown in Fig. 10.7 (Yavuz, Bozdağ Pehlivan, & Ünlü, 2013). Recent advancement in the field of ocular delivery has led to the development of dendrimers based drug delivery systems due to their ability to have multiple surface groups that

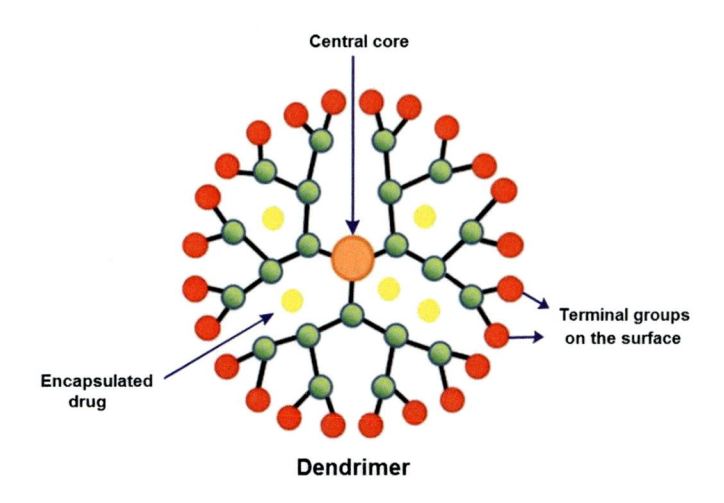

FIG. 10.7

The general structure of dendrimer. *Reproduced with permission from Sowinska, M., & Urbanczyk-Lipkowska, Z. (2014). Advances in the chemistry of dendrimers.* New Journal of Chemistry, 38*(6), 2168–2203. https://doi.org/10.1039/c3nj01239e;*

allow targeting and ease of functionalization (Sowinska & Urbanczyk-Lipkowska, 2014). Dendrimers have a variety of controlled terminal groups and their internal cavity allows them to encapsulate poorly water-soluble hydrophobic drugs. Dendrimers provide structure control as well as have unique biomimetic properties. Recent researchers have developed dendrimeric-based formulations that can enhance corneal residence time. The unique architecture of dendrimers has been utilized for the delivery of therapeutics due to their ability to enhance drug solubility, biocompatibility, and bioavailability and can be applied through various routes of drug administration (Yavuz et al., 2013). Dendrimers have also permeation enhancing properties (Abbasi et al., 2014) and an appropriate application route can allow the delivery of drug to the targeted site and allow to achieve desired pharmacokinetic parameters which makes them attractive ocular drug delivery systems. Dendrimeric drug delivery systems have been utilized in ophthalmology for drugs, peptide, and gene delivery purposes (Yavuz et al., 2013). Recent advancements in the development of improved dendrimeric drug delivery systems will not only enhance the drug delivery to the ocular surface but also will allow effective delivery of therapeutics to intraocular tissues using non-invasive delivery method. Dendrimeric formulations should be carefully designed with biocompatible compounds to reduce their toxicity. Dendrimers are not approved for clinical ophthalmic use yet but recent research reveals that they hold tremendous potential as drug delivery vehicles. Scientists around the world have been designing dendrimer-based materials that have shown improved efficacy for ophthalmic drug delivery.

Contact lenses

Ophthalmic formulations are mostly delivered through eye drops despite many deficiencies like poor bioavailability and less drug interaction with the ocular tissues. Contact lenses for ophthalmic drug delivery have become an ideal approach to achieve controlled and sustained release of pharmacologically active compounds to the ocular tissues (Ameeduzzafar et al., 2016). Contact lenses are highly suitable for the delivery of drugs to the anterior segment of the eye because they are directly placed over the cornea and due to extended wear, they enhance the bioavailability of drugs up to 50% (Fig. 10.8). Researchers have incorporated various nanosystems including nanoparticles, liposomes, micelles, and microemulsions with contact lenses to achieve sustained release and enhanced bioavailability of targeted ligand toward the ocular tissues (Maulvi, Soni, & Shah, 2016). Therapeutic contact lenses are being modified for the treatment of ocular diseases by various techniques including molecular imprinting, incorporation of drug-loaded colloidal nanoparticles, soaking method, drug film formation, and ion ligand polymeric systems (Maulvi et al., 2016). Besides the advantages achieved by the modification of contact lenses, critical parameters like water content, oxygen permeability, tensile strength, ion permeability, and transparency were disturbed, which limit

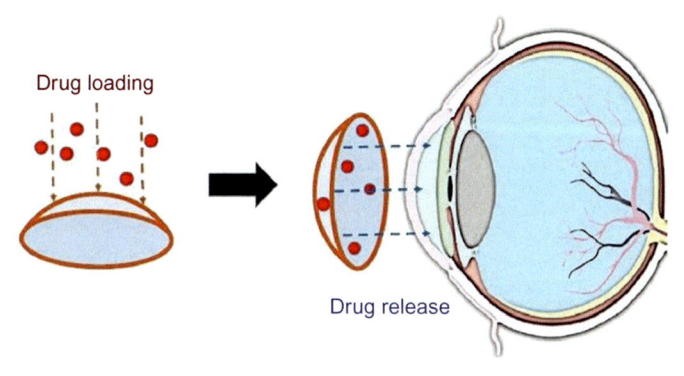

Drug eluting contact lenses

FIG. 10.8

Schematic representation of drug-eluting contact lenses. *Reproduced with permission from Fan, X., Torres-Luna, C., Azadi, M., Domszy, R., Hu, N., Yang, A., et al. (2020). Evaluation of commercial soft contact lenses for ocular drug delivery: A review.* Acta Biomaterialia, *115, 60–74. https://doi.org/10.1016/j.actbio.2020.08.025.*

the commercialization and mass production of these products. Furthermore, issues like drug stability throughout the fabrication process, burst release of the drug, protein adherence, drug release during storage, reduced shelf life are still needed to be assessed and addressed (Fan et al., 2020).

In situ thermoresponsive hydrogels

Hydrogels are a network of hydrophilic polymers having the ability to absorb and retain a huge amount of water (Soliman, Ullah, Shah, Jones, & Singh, 2019). Hydrogels are being used as vehicles for ocular drug delivery as they can prolong precorneal retention time and enhance the ocular bioavailability of pharmacologically active compounds. The design and development of advanced hydrogel drug delivery systems have improved the permeation and diffusion properties of gels; as a result, they can retain hydrophobic as well as hydrophilic molecules. In situ, gelling systems are solutions that convert into semisolid gels due to physical stimuli like physiological pH and body temperature (Fig. 10.9) (Wu et al., 2019). Advantages of hydrogels including rapid swelling, retention of drugs, and sustained release have made them an attractive vehicle for drug delivery (Cooper & Yang, 2019). The aqueous environment of hydrogels provides biocompatibility that can protect cells and sensitive drugs based on peptides and proteins. Due to the incorporation of various targeting ligands, scientists can develop various stimuli-responsive ocular drug delivery systems which are highly advantageous as they combine the merits of easy administration and accurate dosing of drugs with sustained release. In situ gelling systems are also promising vehicles for intraocular and periocular injections, where they develop depots into the vitreous humor or in periocular tissues for sustained drug release into the posterior segment of the eye

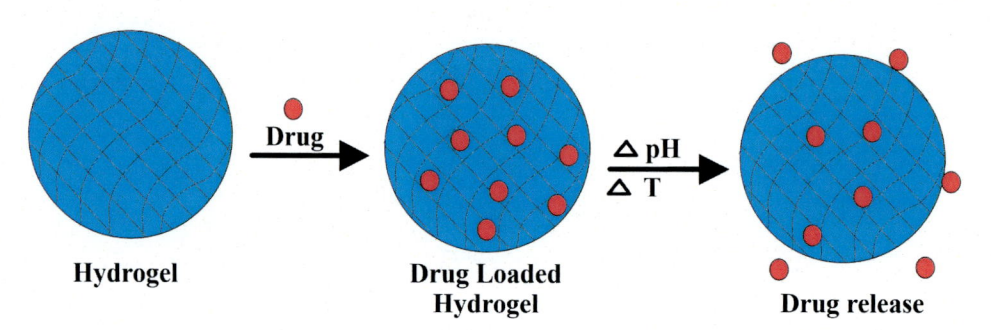

FIG. 10.9

Stimuli-responsive hydrogel for drug delivery. *No Permission Required.*

(Derwent & Mieler, 2008). Polymeric gelling systems that attain sol-gel transitions upon temperature changes are termed thermoresponsive gels (Dutta et al., 2020). Many natural and synthetic polymers show thermoresponsive gelling behavior, so they are being utilized as eye drops or injectable solutions for prolonging and sustained drug delivery. Aqueous solutions of methylcellulose (MC), hydroxypropylmethylcellulose (HPMC), and chitosan-based thermoresponsive hydrogels have been introduced as biocompatible and safe materials for ocular drug delivery (Fathi, Barar, Aghanejad, & Omidi, 2015). Polymers including poloxamers (Soliman et al., 2019), poly(N-isopropyl acrylamide) (pNIPAAm), poly(ethylene glycol) diacrylate (PEGDA), poly-(DL-lactic acid-co-glycolic acid)-PEG (PLGA-PEG) copolymer (Fathi et al., 2015) have also been utilized for the delivery of pharmacologically active compounds to the ocular tissues. Although in situ hydrogel-based drug delivery systems are being used to cure anterior eye diseases, their utilization for the treatment of posterior segmental conditions is limited due to the lack of studies on their biodegradation, retinal tolerability, biocompatibility, and long-term stability studies.

Conclusion

Ophthalmic research is mainly focused on the development of non-invasive methods for the delivery of therapeutically active compounds to both anterior and posterior segments of the eye. The advent of nanotechnology-based nanocarriers provides a unique dimension for the design of novel drug delivery systems that can surpass the ocular barriers and provide a patient-compliant route for drug administration. The development of smart nanocarriers has enabled targeted delivery of ophthalmic drugs with enhanced therapeutic efficacy and reduced treatment cost. Various methodologies are being adopted for the production of ophthalmic nanomedicines which are cost-effective and have higher efficacy toward ocular tissues. Most biocompatible materials include natural polymers are being explored for the development of effective nanoformulations but detailed toxicity and mechanistic studies are required for being considered by the drug regulatory department. Technological advances for large-scale production bring more hope for the commercialization of nanomedicines. Safety, reproducibility, and scale-up of nanomedicines are important parameters for commercial-scale production. Modification of nanocarriers by various parameters results in highly effective nanoformulations with superior therapeutic efficacy over conventional ocular drug delivery systems for both anterior and posterior eye segments.

References

Abbasi, E., Aval, S. F., Akbarzadeh, A., Milani, M., Nasrabadi, H. T., Joo, S. W., et al. (2014). Dendrimers: Synthesis, applications, and properties. *Nanoscale Research Letters, 9*(1), 1–10. https://doi.org/10.1186/1556-276X-9-247.

Agarwal, R., Iezhitsa, I., Agarwal, P., Abdul Nasir, N. A., Razali, N., Alyautdin, R., et al. (2016). Liposomes in topical ophthalmic drug delivery: An update. *Drug Delivery, 23*(4), 1075–1091. https://doi.org/10.3109/10717544.2014.943336.

Akbarzadeh, A., Rezaei-Sadabady, R., Davaran, S., Joo, S. W., Zarghami, N., Hanifehpour, Y., et al. (2013). Liposome: Classification, preparation, and applications. *Nanoscale Research Letters.* https://doi.org/10.1186/1556-276x-8-102.

Al-Kinani, A. A., Zidan, G., Elsaid, N., Seyfoddin, A., Alani, A. W. G., & Alany, R. G. (2018). Ophthalmic gels: Past, present and future. *Advanced Drug Delivery Reviews*, 113–126. https://doi.org/10.1016/j.addr.2017.12.017.

Ameeduzzafar, A., Ali, J., Fazil, M., Qumbar, M., Khan, N., & Ali, A. (2016). Colloidal drug delivery system: Amplify the ocular delivery. *Drug Delivery, 23*(3), 710–726. https://doi.org/10.3109/10717544.2014.923065.

Bachu, R. D., Chowdhury, P., Al-Saedi, Z. H. F., Karla, P. K., & Boddu, S. H. S. (2018). Ocular drug delivery barriers—Role of nanocarriers in the treatment of anterior segment ocular diseases. *Pharmaceutics, 10*(1). https://doi.org/10.3390/pharmaceutics10010028.

Baranowski, P., Karolewicz, B., Gajda, M., & Pluta, J. (2014). Ophthalmic drug dosage forms: Characterisation and research methods. *The Scientific World Journal, 2014.* https://doi.org/10.1155/2014/861904.

Barar, J., Aghanejad, A., Fathi, M., & Omidi, Y. (2016). Advanced drug delivery and targeting technologies for the ocular diseases. *BioImpacts: BI, 6*(1), 49–67. https://doi.org/10.15171/bi.2016.07.

Behar-Cohen, F. (2012). Drug delivery to the eye: Current trends and future perspectives. *Therapeutic Delivery, 3*(10), 1135–1137. https://doi.org/10.4155/tde.12.94.

Buech, G., Bertelmann, E., Pleyer, U., Siebenbrodt, I., & Borchert, H. H. (2007). Formulation of sirolimus eye drops and corneal permeation studies. *Journal of Ocular Pharmacology and Therapeutics, 23*(3), 292–303. https://doi.org/10.1089/jop.2006.130.

Cholkar, K., Patel, S. P., Vadlapudi, A. D., & Mitra, A. K. (2013). Novel strategies for anterior segment ocular drug delivery. *Journal of Ocular Pharmacology and Therapeutics, 29*(2), 106–123. https://doi.org/10.1089/jop.2012.0200.

Cooper, R. C., & Yang, H. (2019). Hydrogel-based ocular drug delivery systems: Emerging fabrication strategies, applications, and bench-to-bedside manufacturing considerations. *Journal of Controlled Release, 306*, 29–39. https://doi.org/10.1016/j.jconrel.2019.05.034.

Derwent, J. J. K., & Mieler, W. F. (2008). Thermoresponsive hydrogels as a new ocular drug delivery platform to the posterior segment of the eye. *Transactions of the American Ophthalmological Society*, 106.

Dubald, M., Bourgeois, S., Andrieu, V., & Fessi, H. (2018). Ophthalmic drug delivery systems for antibiotherapy—A review. *Pharmaceutics*, *10*(1). https://doi.org/10.3390/pharmaceutics10010010.

Dutta, K., Das, R., Ling, J., Monibas, R. M., Carballo-Jane, E., Kekec, A., et al. (2020). In situ forming injectable thermoresponsive hydrogels for controlled delivery of biomacromolecules. *ACS Omega*, *5*(28), 17531–17542. https://doi.org/10.1021/acsomega.0c02009.

Fan, X., Torres-Luna, C., Azadi, M., Domszy, R., Hu, N., Yang, A., et al. (2020). Evaluation of commercial soft contact lenses for ocular drug delivery: A review. *Acta Biomaterialia*, *115*, 60–74. https://doi.org/10.1016/j.actbio.2020.08.025.

Fathi, M., Barar, J., Aghanejad, A., & Omidi, Y. (2015). Hydrogels for ocular drug delivery and tissue engineering. *BioImpacts: BI*, *5*(4), 159–164. https://doi.org/10.15171/bi.2015.31.

Gautam, N., & Kesavan, K. (2017). Development of microemulsions for ocular delivery. *Therapeutic Delivery*, *8*(5), 313–330. https://doi.org/10.4155/tde-2016-0076.

Grimaudo, M. A., Pescina, S., Padula, C., Santi, P., Concheiro, A., Alvarez-Lorenzo, C., et al. (2019). Topical application of polymeric nanomicelles in ophthalmology: A review on research efforts for the noninvasive delivery of ocular therapeutics. *Expert Opinion on Drug Delivery*, *16*(4), 397–413. https://doi.org/10.1080/17425247.2019.1597848.

Jothi, M., Harikumar, S., & Aggarwal, G. (2012). In-situ ophthalmic gels for the treatment of eye diseases. *International Journal of Pharmaceutical Sciences and Research*, *3*(7), 1891–1904.

Krishna, K. B., & Prabhakar, C. (2011). A review on nanosuspensions in drug delivery. *International Journal of Pharma and Bio Sciences*, *2*(1), 549–558.

Lai, S., Wei, Y., Wu, Q., Zhou, K., Liu, T., Zhang, Y., et al. (2019). Liposomes for effective drug delivery to the ocular posterior chamber. *Journal of Nanobiotechnology*, *17*(1), 1711–1712. https://doi.org/10.1186/s12951-019-0498-7.

Lallemand, F., Daull, P., Benita, S., Buggage, R., & Garrigue, J.-S. (2012). Successfully improving ocular drug delivery using the cationic nanoemulsion, novasorb. *Journal of Drug Delivery*, 1–16. https://doi.org/10.1155/2012/604204.

Lynch, C., Kondiah, P. P. D., Choonara, Y. E., du Toit, L. C., Ally, N., & Pillay, V. (2019). Advances in biodegradable nano-sized polymer-based ocular drug delivery. *Polymers*, *11*(8), 1371. https://doi.org/10.3390/polym11081371.

Maulvi, F. A., Soni, T. G., & Shah, D. O. (2016). A review on therapeutic contact lenses for ocular drug delivery. *Drug Delivery*, *23*(8), 3017–3026. https://doi.org/10.3109/10717544.2016.1138342.

Meisner, D., & Mezei, M. (1995). Liposome ocular delivery systems. *Advanced Drug Delivery Reviews*, *16*(1), 75–93. https://doi.org/10.1016/0169-409X(95)00016-Z.

Mishra, G. P., Bagui, M., Tamboli, V., & Mitra, A. K. (2011). Recent applications of liposomes in ophthalmic drug delivery. *Journal of Drug Delivery*, 1–14. https://doi.org/10.1155/2011/863734.

Molokhia, S. A., Thomas, S. C., Garff, K. J., Mandell, K. J., & Wirostko, B. M. (2013). Anterior eye segment drug delivery systems: Current treatments and future challenges. *Journal of Ocular Pharmacology and Therapeutics*, *29*(2), 92–105. https://doi.org/10.1089/jop.2012.0241.

Muankaew, C., Jansook, P., Sigurcrossed, D., Signsson, H. H., & Loftsson, T. (2016). Cyclodextrin-based telmisartan ophthalmic suspension: Formulation development for water-insoluble drugs. *International Journal of Pharmaceutics*, *507*(1–2), 21–31. https://doi.org/10.1016/j.ijpharm.2016.04.071.

Nagarwal, R. C., Kant, S., Singh, P. N., Maiti, P., & Pandit, J. K. (2009). Polymeric nanoparticulate system: A potential approach for ocular drug delivery. *Journal of Controlled Release, 136*(1), 2–13. https://doi.org/10.1016/j.jconrel.2008.12.018.

Nayak, K., & Misra, M. (2018). A review on recent drug delivery systems for posterior segment of eye. *Biomedicine and Pharmacotherapy, 107*, 1564–1582. https://doi.org/10.1016/j.biopha.2018.08.138.

Omerović, N., & Vranić, E. (2020). Application of nanoparticles in ocular drug delivery systems. *Health and Technology, 10*(1), 61–78. https://doi.org/10.1007/s12553-019-00381-w.

Patel, A. (2013). Ocular drug delivery systems: An overview. *World Journal of Pharmacology, 47*. https://doi.org/10.5497/wjp.v2.i2.47.

Patel, C. M., Patel, M. A., Patel, N. P., Patel, C. N., & Prajapati, P. H. (2010). Poly lactic glycolic acid (PLGA) as biodegradable polymer. *Research Journal of Pharmacy and Technology, 3*(2), 353–360. https://rjptonline.org/HTML_Papers/Research%20Journal%20of%20Pharmacy%20and%20Technology__PID__2010-3-2-68.html.

Patel, N., Thakkar, V., Metalia, V., Baldaniya, L., Gandhi, T., & Gohel, M. (2016). Formulation and development of ophthalmic in situ gel for the treatment ocular inflammation and infection using application of quality by design concept. *Drug Development and Industrial Pharmacy, 42*(9), 1406–1423. https://doi.org/10.3109/03639045.2015.1137306.

Patravale, V. B., Date, A. A., & Kulkarni, R. M. (2004). Nanosuspensions: A promising drug delivery strategy. *Journal of Pharmacy and Pharmacology, 56*(7), 827–840. https://doi.org/10.1211/0022357023691.

Periman, L. M., Mah, F. S., & Karpecki, P. M. (2020). A review of the mechanism of action of cyclosporine A: The role of cyclosporine a in dry eye disease and recent formulation developments. *Clinical Ophthalmology, 14*, 4187–4200. https://doi.org/10.2147/OPTH.S279051.

Rathore, S. K., & Nema, R. K. (2009). An insight into ophthalmic drug delivery system. *International Journal of Pharmaceutical Sciences and Drug Research, 1*(1), 1–5. https://doi.org/10.25004/ijpsdr.2009.010101.

Shell, J. W. (1984). Ophthalmic drug delivery systems. *Survey of Ophthalmology, 29*(2), 117–128. https://doi.org/10.1016/0039-6257(84)90168-1.

Soliman, K. A., Ullah, K., Shah, A., Jones, D. S., & Singh, T. R. R. (2019). Poloxamer-based in situ gelling thermoresponsive systems for ocular drug delivery applications. *Drug Discovery Today, 24*(8), 1575–1586. https://doi.org/10.1016/j.drudis.2019.05.036.

Sowinska, M., & Urbanczyk-Lipkowska, Z. (2014). Advances in the chemistry of dendrimers. *New Journal of Chemistry, 38*(6), 2168–2203. https://doi.org/10.1039/c3nj01239e.

Thrimawithana, T. R., Young, S., Bunt, C. R., Green, C., & Alany, R. G. (2011). Drug delivery to the posterior segment of the eye. *Drug Discovery Today, 16*(5–6), 270–277. https://doi.org/10.1016/j.drudis.2010.12.004.

Trinh, H. M., Joseph, M., Cholkar, K., Mitra, R., & Mitra, A. K. (2017). Nanomicelles in diagnosis and drug delivery. In *Emerging nanotechnologies for diagnostics, drug delivery and medical devices* (pp. 45–58). Elsevier Inc. https://doi.org/10.1016/B978-0-323-42978-8.00003-6.

Tsai, C. H., Wang, P. Y., Lin, I. C., Huang, H., Liu, G. S., & Tseng, C. L. (2018). Ocular drug delivery: Role of degradable polymeric nanocarriers for ophthalmic application. *International Journal of Molecular Sciences, 19*(9). https://doi.org/10.3390/ijms19092830.

Vaishya, R. D., Khurana, V., Patel, S., & Mitra, A. K. (2014). Controlled ocular drug delivery with nanomicelles. *Wiley Interdisciplinary Reviews: Nanomedicine and Nanobiotechnology, 6*(5), 422–437. https://doi.org/10.1002/wnan.1272.

Vandamme, T. F. (2002). Microemulsions as ocular drug delivery systems: Recent developments and future challenges. *Progress in Retinal and Eye Research, 21*(1), 15–34. https://doi.org/10.1016/S1350-9462(01)00017-9.

Varela-Fernández, R., Díaz-Tomé, V., Luaces-Rodríguez, A., Conde-Penedo, A., García-Otero, X., Luzardo-álvarez, A., et al. (2020). Drug delivery to the posterior segment of the eye: Biopharmaceutic and pharmacokinetic considerations. *Pharmaceutics*, *12*(3). https://doi.org/10.3390/pharmaceutics12030269.

Wadhwa, S., Paliwal, R., & Vyas, S. P. (2009). Nanocarriers in ocular drug delivery: An update review. *Current Pharmaceutical Design*, *15*(23), 2724–2750. https://doi.org/10.2174/138161209788923886.

Wang, Y., Xu, X., Gu, Y., Cheng, Y., & Cao, F. (2018). Recent advance of nanoparticle-based topical drug delivery to the posterior segment of the eye. *Expert Opinion on Drug Delivery*, *15*(7), 687–701. https://doi.org/10.1080/17425247.2018.1496080.

Weng, Y., Liu, J., Jin, S., Guo, W., Liang, X., & Hu, Z. (2017). Nanotechnology-based strategies for treatment of ocular disease. *Acta Pharmaceutica Sinica B*, *7*(3), 281–291. https://doi.org/10.1016/j.apsb.2016.09.001.

Wu, Y., Liu, Y., Li, X., Kebebe, D., Zhang, B., Ren, J., et al. (2019). Research progress of in-situ gelling ophthalmic drug delivery system. *Asian Journal of Pharmaceutical Sciences*, *14*(1), 1–15. https://doi.org/10.1016/j.ajps.2018.04.008.

Xu, Q., Kambhampati, S. P., & Kannan, R. M. (2013). Nanotechnology approaches for ocular drug delivery. *Middle East African Journal of Ophthalmology*, *20*(1), 26–37. https://doi.org/10.4103/0974-9233.106384.

Yavuz, B., Bozdağ Pehlivan, S., & Ünlü, N. (2013). Dendrimeric systems and their applications in ocular drug delivery. *The Scientific World Journal*, *2013*. https://doi.org/10.1155/2013/732340.

Zhou, H. Y., Hao, J. L., Wang, S., Zheng, Y., & Zhang, W. S. (2013). Nanoparticles in the ocular drug delivery. *International Journal of Ophthalmology*, *6*(3), 390–396. https://doi.org/10.3980/j.issn.2222-3959.2013.03.25.

CHAPTER 11

Organ-specific toxicities of nanocarriers

Introduction

Nanotechnology is a multidisciplinary field of study that focuses on the development and use of materials with a diameter of less than 100 nm (Dusinska, Rundén-Pran, Schnekenburger, & Kanno, 2017; Radaic, Pugliese, Campese, Pessine, & De Jesus, 2016). In 2004, the Royal Society and the Royal Engineering Academy proposed this concept, which was associated with nanoscience as the branch of science that studies the phenomena of materials on the atomic, molecular, and macromolecular scales, whose properties differ significantly from those on larger scales (Ju-Nam & Lead, 2008; Paschoalino, Marcone, & Jardim, 2010). Nanoparticles (NPs) are ultrafine small particles with a diameter of 1–100 nm; however, this definition includes various systems that are not limited to small particles of a specific material, such as, nanospheres, nanotubes, and nanocapsules (Ju-Nam & Lead, 2008; Roberto & Christofoletti, 2019). Nanomaterials have unique properties that are being used in a variety of fields, from industry to medicine (Dusinska, Tulinska, et al., 2017; Laux et al., 2017). Nanomaterials are increasingly being used for commercial purposes as opacifiers, fillers, water filtration agents, semiconductors, cosmetic ingredients, electronic parts, and other applications, according to Arora et al. These same authors also reported that nanomaterials are being used in the medical field, primarily as drug delivery agents, imaging contrast agents and, biosensors, implying that human contact can occur both directly and indirectly, and that nanomaterials can be administered via ingestion or injection (Arora, Rajwade, & Paknikar, 2012).

Nanotechnology in drug delivery has the potential to revolutionize the treatment of a wide range of diseases, including diabetes, cancer, neurodegenerative and vascular diseases (Irvine & Dane, 2020). Nanoscale drug delivery systems are now routinely used in clinical trials, and Europe has been a leader in this field (Ledet & Mandal, 2012). Nanotechnology-based formulations are mostly parenteral on the market, though some are designed for oral administration (Hafner, Lovrić, Lakŏ, & Pepić, 2014). A large number of preclinical and clinical

245

trials are expected to advance the development of novel nanotherapeutics for non-parenteral delivery routes such as nasal, pulmonary, vaginal, dermal, and ocular delivery. The option of delivery and the obstacles to be overcome are of particular concern to drug delivery systems (European Commission/ETP) (Prasad et al., 2018). In recent years, novel drug delivery systems have steadily improved. The main goals of these enhancements are to attain more precise drug targeting, reduce drug toxicity without compromising efficacy, achieve biocompatibility with the development of safe new medications (De Jong & Borm, 2008). Proteins, dendrimers, polymers, micelles, emulsions, liposomes, NPs, and nanocapsules are some of the nanomedicines developed via nanotechnology-based engineered materials (Garnett & Kallinteri, 2006).

Nanotoxicology

Although nano-based drug delivery systems were developed to reduce the toxicity of the drug and increase biocompatibility (Bhoyar, Giri, Tripathi, & Alexander, 2012), the unique characteristics of these systems may present a risk. The current state of knowledge about the toxic effects of nanomaterials on human health is inadequate. But they may persist in the human body and environment and for a longer time than their larger counterparts due to their exceptionally small size and, in some cases, larger stability. Certain nanomaterials are also capable to enter the body via inhalation, ingestion, or dermal absorption. Therefore, there is a growing motivation to know more about their potential toxicity. Efforts have been made in recent years for the development of methods to assess the toxicity profile and health effects of these nanocarriers. These efforts have resulted in the emergence of a new branch of toxicology known as "nanotoxicology" (Oberdörster, 2010; Oberdörster, Oberdörster, & Oberdörster, 2005), an emerging and rapidly developing branch of modern toxicology concerned with the toxicity of nanomaterials in particular. The public is concerned about increased human exposure to nanomaterials in the environment, as well as their negative health effects. Nanomaterials' shape, size, morphology, distribution, composition, dispersion, surface chemistry, surface area, and reactivity are expected to have an impact on their toxicity (Oberdörster et al., 2005) making a complex evaluation of their toxicity (Dhawan & Sharma, 2010; Oberdörster et al., 2005). Chemically synthesized NPs are more toxic to human cells due to the presence of synthetic chemicals as capping agents and surface functional in comparison to biosynthesized NPs with biocompatible surface functional groups (Jeevanandam, Chan, & Danquah, 2016). On the other hand, certain biosynthesized NPs can also be toxic when they react with cells and disintegrate into simpler forms or accumulate (Naz, Gul, & Zia, 2020; Roy, Sadhukhan, Ghosh, & Das, 2015). Furthermore, the small size of NPs results in a large surface area per unit mass, which is often associated with higher biological reactivity. Additionally, a large surface area may increase the formation of free radicals such as hydroxyl radical and superoxide anion. Thus, oxidative stress, particularly in metal-based NPs, may play a key role in NP toxicity. The free radical formation, for example, can explain inflammatory responses to NPs (Donaldson et al., 2006).

The toxicological potential of nanomaterials is gaining interest due to the potential for exposure to a wide range of nanomaterials in the human environment. The goal of this chapter was to compile the organ-specific toxicity of nanomaterials that could be found in the environment while taking into account the wide range of toxicological endpoints to which humans could be exposed in their daily lives.

Organ-/tissue-specific nanotoxicity

The smaller size of NPs, as well as their high specificity for the tissue system, allow them to easily penetrate the tissue system and cause damage to body organs. NPs have been observed to move rapidly through the bloodstream and easily cross the blood-brain barrier, potentially causing toxicity and significantly harming the human organ systems (e.g., pulmonary systems, cardiovascular systems, hepatocellular systems, and renal systems).

Toxicity in the pulmonary system

When NPs are suspended in air, Brownian diffusion allows them to travel long distances. Therefore, inhalation is a common way for people to be exposed to nanomaterials in the air. Diffusion is the primary mechanism by which NPs are deposited in the respiratory tract (Semmler et al., 2004). Small NPs can easily penetrate the lungs, potentially cause lung injuries, and generate ROS (Warheit, Webb, Sayes, Colvin, & Reed, 2006). In rats, fine and ultrafine NPs such as nickel, carbon black, and TiO_2 particles were used to study their pulmonary toxicity. The ultrafine NPs, in particular, were found to cause increased pulmonary inflammation (Pettibone et al., 2008).

Only a few studies are showing the in vitro inhalation toxicity studies of nanomaterials. To study the toxicity of nanomaterials, researchers used the human lung cancer cell line A549 as an in vitro cell culture model. Foldbjerg et al. investigated the cytotoxic and genotoxic effects of silver NPs using this system (Foldbjerg, Dang, & Autrup, 2011). The researchers used atomic absorption spectroscopy and flow cytometry to determine the amount of NPs that were taken up by cells. MTT and annexin V/propidium iodide assays were used to determine the dose-dependent cytotoxicity of silver NPs. Pretreatment of silver NPs with the N-acetyl-cysteine (antioxidant) significantly reduced their cytotoxicity. They discovered a strong relation between ROS levels and mitochondrial damage, as well as early apoptosis. They used P-post labeling to assess ROS-induced DNA damage caused by increased DNA adducts after NP exposure (Meek & Doull, 2009). The amount of DNA adducts was highly dependent on the amount of cellular ROS and was inhibited by the pretreatment with antioxidants. According to their findings, silver NPs act as a mediator of ROS-induced genotoxicity and cytotoxicity. Jugan et al. recently investigated the genotoxic and cytotoxic effects of titanium oxide NPs on the human lung cell line A549 alveolar epithelial cells. According to reports, the toxic effects of NPs on the lungs include exaggerated lung responses, cellular proliferation, a high pulmonary inflammation rate, poor clearance, fibroproliferative effects, and inflammatory-derived mutagenesis, ultimately leading to chronic effects such as tumor development in the lungs. The factors that have the greatest influence on nanotoxicity in the lungs are the particulate characteristics of NPs, such as surface area, particle size, surface dose, degree of aggregation, surface modifications, and particle synthesis method (Nemmar, Hoylaerts, Hoet, Vermylen, & Nemery, 2003).

Toxicity in cardiovascular systems

Positively charged NPs such as gold and polystyrene have been shown to cause blood clotting and hemolysis, whereas negatively charged NPs be nontoxic. In hypertensive rats, increased inhalation exposure to diesel-exposed particles (DEP) resulted in an altered heart rate, as measured by the pacemaker (Hansen et al., 2007). In the hepatic microvasculature of normal rats, platelet accumulation occurred after being injected ultrafine carbon black NPs into their blood, as they did prothrombotic changes on the endothelial surface of the hepatic microvessels (Simeonova & Erdely, 2009). Du et al. used intratracheal instillation to study the cardiovascular toxicity of silica NPs in rats (Du et al., 2013). Hematologic parameters, oxidative stress, inflammatory responses, endothelial dysfunction, and myocardial enzymes were all measured in serum. Silica NPs were found to pass through the alveolar-capillary barrier and enter into the systemic circulation. The toxicity of silica NPs on the cardiovascular system was strongly dependent on particle size and dosage, according to their findings. Inflammatory reactions and endothelial dysfunction were associated with oxidative stress. Zebrafish model in vivo and endothelial cells in vitro were used by Duan et al. to assess the effect of silica NPs on the cardiovascular system (Duan, Yu, Li, Yu, & Sun, 2013). Their toxicological biomarkers include oxidative stress, cytotoxicity, and apoptosis. They discovered that apoptosis and oxidative stress were the most important factors in endothelial cell dysfunction. They concluded that silica NP exposure could be a risk factor for the failure of the cardiovascular system.

Yang et al. found that mice exposed to gold NPs of different sizes through the tail vein developed chronic cardiac toxicity (Yang, Tian, & Li, 2016). They examined the effect of NPs accumulation in the mouse heart on cardiac structure, function, inflammation, and fibrosis. The NPs had almost no effect on the systolic function of the heart. However, after only 2 weeks of exposure, mice receiving the smallest (10 nm) NPs had significantly higher left ventricular end-diastolic inner dimension, left ventricular mass, and heart weight/body weight. They concluded that gold NPs caused cardiac hypertrophy.

Toxicity in hepatocellular system

The liver is considered the primary organ involved in the metabolism of xenobiotics and detoxification. Liver filters out the blood-carrying toxicants before their distribution in the other body parts. Because of the high blood flow rate to the liver, the toxicant is delivered in high concentrations to this organ. The liver is a major target organ for toxicants due to its high levels of exposure and metabolic activity. After being ingested, absorbed through the skin, inhaled, or administered through medical devices and intravenous injections, the nanomaterial enters the circulation and is translocated to the liver. According to some studies, nanomaterials are entrapped by the reticuloendothelial system, with both the liver and spleen being the primary targets (Hillyer & Albrecht, 2001). Therefore, nanomaterials may be hepatotoxicants, and hepatotoxicity testing is an important testing strategy for assessing the safety of nanomaterials when necessary.

The number of in vitro hepatotoxicity studies is limited. Several in vitro studies, however, have suggested the hypotonicity of certain nanomaterials. Hussain et al. assessed the toxicity of silver and TiO_2 NPs in vitro using the rat liver cell line BRL 3A (Hussain, Hess, Gearhart, Geiss, & Schlager, 2005). They defined hepatotoxicity as cytotoxicity, mitochondrial dysfunction, and oxidative stress. The study concluded that silver was far more cytotoxic than TiO_2 NPs. Sahu et al. discovered nanosilver cytotoxicity in human liver HepG2 and colon Caco-2 cells (Sahu et al., 2014).

Jeon et al. searched for proteins that were differentially expressed in the mouse liver as a result of titanium NP toxicity (Jeon, Park, & Lee, 2011). Using liquid chromatography-mass spectrometry, they discovered 15 proteins that showed a 2-fold change in expression in response to TiO_2 NPs. Twelve proteins were downregulated and three proteins were upregulated after being exposed to TiO_2 NPs. The 15 differentially expressed proteins could be used in the treatment of acute hepatic damage with TiO_2 NPs to detect inflammation, apoptosis, and antioxidative reactions.

Toxicity in the renal system

The kidney is one of the common target organs for nanomaterial toxicity. The kidney has been identified as an important target organ for toxicity and the primary organ for nanomaterial clearance (Wang et al., 2009; Yan et al., 2012). Yan et al. studied the nephrotoxicity of ZnO NPs in rats and observed that the ZnO NPs caused severe damage in mitochondria and cell membrane of the rat kidney, resulting in nephrotoxicity (Yan et al., 2012). In rats exposed to copper NPs, Liao and Liu investigated the mechanism of nanomaterial-induced nephrotoxicity (Liao & Liu, 2012). They discovered nano copper-induced renal proximal tubule necrosis using renal gene expression profiles. On the other hand, carbon nanotubes (CNTs) that are both pristine and functionalized have the potential to remain in the lungs for up to a month or even a year (Jacobsen et al., 2017). After 20 days of injection (8 nm), MWCNT functionalized with carboxylic group caused significant changes in the kidney of the Wistar rat, even at a low dose (2.5 mg/kg) in a study. It was found that the lesions included inflammation in the medulla and cortex, glomerular degeneration, accumulation of hyaline-like substances, and proximal tubular necrosis, all of which were dose-dependent (Poormohammad Matouri & Noori, 2018). In rats, pristine MWCNT (1 mg/kg for 30 days) injected intratracheally caused nephrotoxicity, whereas no toxic effect was observed in those treated with MWCNT and PEG. Collapsed glomeruli, endothelial cells, packed mesangial, and apoptosis were all signs of renal tissue damage (Abu Gazia & El-Magd, 2019). Thus it is possible that vectors have an impact on CNTs' toxic endpoints.

There are few studies on the nephrotoxicity of nanomaterials in the literature. According to the findings, certain nanomaterials are responsible to cause nephrotoxicity. However, it should be the main concern regarding these studies because they used different nanomaterial testing models, different sources of test nanomaterial, different characterization methods, and different experimental conditions. Therefore, the findings of these studies might not be comparable.

Conclusion

Nanoscaled materials have revolutionized numerous fields, including medicine and health, particularly in terms of the development of various types of targeted drug delivery devices for the early detection and successful treatment of a variety of diseases. Despite their enormous contribution to improving prognostic and therapeutic modalities, biological interactions between nanomaterials and various body tissues have the potential to cause severe nanotoxicity in a variety of organs, including the heart, liver, kidney, lungs, gastrointestinal system, skin, and nervous system. As a result of the contents of this chapter, it can be inferred that various parameters of nanomaterials result in a variety of toxicities in essential organs. It is critical to examine these parameters throughout the creation of nanomaterials in order to control their associated toxicities.

References

Abu Gazia, M., & El-Magd, M. A. (2019). Effect of pristine and functionalized multiwalled carbon nanotubes on rat renal cortex. *Acta Histochemica, 121*(2), 207–217. https://doi.org/10.1016/j.acthis.2018.12.005.

Arora, S., Rajwade, J. M., & Paknikar, K. M. (2012). Nanotoxicology and in vitro studies: The need of the hour. *Toxicology and Applied Pharmacology, 258*(2), 151–165. https://doi.org/10.1016/j.taap.2011.11.010.

Bhoyar, N., Giri, T. K., Tripathi, D. K., & Alexander, A. (2012). Recent advances in novel drug delivery system through gels: Review. *Journal of Pharmacy and Allied Health Sciences*, 21–39. https://doi.org/10.3923/jpahs.2012.21.39.

De Jong, W. H., & Borm, P. J. A. (2008). Drug delivery and nanoparticles: Applications and hazards. *International Journal of Nanomedicine, 3*(2), 133–149.

Dhawan, A., & Sharma, V. (2010). Toxicity assessment of nanomaterials: Methods and challenges. *Analytical and Bioanalytical Chemistry, 398*(2), 589–605. https://doi.org/10.1007/s00216-010-3996-x.

Donaldson, K., Aitken, R., Tran, L., Stone, V., Duffin, R., Forrest, G., et al. (2006). Carbon nanotubes: A review of their properties in relation to pulmonary toxicology and workplace safety. *Toxicological Sciences, 92*(1), 5–22. https://doi.org/10.1093/toxsci/kfj130.

Du, Z., Zhao, D., Jing, L., Cui, G., Jin, M., Li, Y., et al. (2013). Cardiovascular toxicity of different sizes amorphous silica nanoparticles in rats after intratracheal instillation. *Cardiovascular Toxicology, 13*(3), 194–207. https://doi.org/10.1007/s12012-013-9198-y.

Duan, J., Yu, Y., Li, Y., Yu, Y., & Sun, Z. (2013). Cardiovascular toxicity evaluation of silica nanoparticles in endothelial cells and zebrafish model. *Biomaterials, 34*(23), 5853–5862. https://doi.org/10.1016/j.biomaterials.2013.04.032.

Dusinska, M., Rundén-Pran, E., Schnekenburger, J., & Kanno, J. (2017). Toxicity tests: In vitro and in vivo. In *Adverse effects of engineered nanomaterials: Exposure, toxicology, and impact on human health* (2nd ed., pp. 51–82). Elsevier Inc. https://doi.org/10.1016/B978-0-12-809199-9.00003-3.

Dusinska, M., Tulinska, J., El Yamani, N., Kuricova, M., Liskova, A., Rollerova, E., et al. (2017). Immunotoxicity, genotoxicity and epigenetic toxicity of nanomaterials: New strategies for toxicity testing? *Food and Chemical Toxicology, 109*, 797–811. https://doi.org/10.1016/j.fct.2017.08.030.

Foldbjerg, R., Dang, D. A., & Autrup, H. (2011). Cytotoxicity and genotoxicity of silver nanoparticles in the human lung cancer cell line, A549. *Archives of Toxicology, 85*(7), 743–750. https://doi.org/10.1007/s00204-010-0545-5.

Garnett, M. C., & Kallinteri, P. (2006). Nanomedicines and nanotoxicology: Some physiological principles. *Occupational Medicine, 56*(5), 307–311. https://doi.org/10.1093/occmed/kql052.

Hafner, A., Lovrić, J., Lakǒ, G. P., & Pepić, I. (2014). Nanotherapeutics in the EU: An overview on current state and future directions. *International Journal of Nanomedicine, 9*(1), 1005–1023. https://doi.org/10.2147/IJN.S55359.

Hansen, C. S., Sheykhzade, M., Møller, P., Folkmann, J. K., Amtorp, O., Jonassen, T., et al. (2007). Diesel exhaust particles induce endothelial dysfunction in apoE-/- mice. *Toxicology and Applied Pharmacology, 219*(1), 24–32. https://doi.org/10.1016/j.taap.2006.10.032.

Hillyer, J. F., & Albrecht, R. M. (2001). Gastrointestinal persorption and tissue distribution of differently sized colloidal gold nanoparticles. *Journal of Pharmaceutical Sciences, 90*(12), 1927–1936. https://doi.org/10.1002/jps.1143.

Hussain, S. M., Hess, K. L., Gearhart, J. M., Geiss, K. T., & Schlager, J. J. (2005). In vitro toxicity of nanoparticles in BRL 3A rat liver cells. *Toxicology in Vitro, 19*(7), 975–983. https://doi.org/10.1016/j.tiv.2005.06.034.

Irvine, D. J., & Dane, E. L. (2020). Enhancing cancer immunotherapy with nanomedicine. *Nature Reviews Immunology, 20*(5), 321–334. https://doi.org/10.1038/s41577-019-0269-6.

Jacobsen, N. R., Møller, P., Clausen, P. A., Saber, A. T., Micheletti, C., Jensen, K. A., et al. (2017). Biodistribution of carbon nanotubes in animal models. *Basic & Clinical Pharmacology & Toxicology*, *121*, 30–43. https://doi.org/10.1111/bcpt.12705.

Jeevanandam, J., Chan, Y. S., & Danquah, M. K. (2016). Biosynthesis of metal and metal oxide nanoparticles. *ChemBioEng Reviews*, *3*(2), 55–67. https://doi.org/10.1002/cben.201500018.

Jeon, Y.-M., Park, S.-K., & Lee, M.-Y. (2011). Proteomic analysis of hepatotoxicity induced by titanium nanoparticles in mouse liver. *Journal of Korean Society for Applied Biological Chemistry*, 852–859. https://doi.org/10.1007/bf03253172.

Ju-Nam, Y., & Lead, J. R. (2008). Manufactured nanoparticles: An overview of their chemistry, interactions and potential environmental implications. *Science of the Total Environment*, *400*(1–3), 396–414. https://doi.org/10.1016/j.scitotenv.2008.06.042.

Laux, P., Riebeling, C., Booth, A. M., Brain, J. D., Brunner, J., Cerrillo, C., et al. (2017). Biokinetics of nanomaterials: The role of biopersistence. *NanoImpact*, *6*, 69–80. https://doi.org/10.1016/j.impact.2017.03.003.

Ledet, G., & Mandal, T. K. (2012). Nanomedicine: Emerging therapeutics for the 21st century. *US Pharm*, *37*(3), 7–11.

Liao, M. Y., & Liu, H. G. (2012). Gene expression profiling of nephrotoxicity from copper nanoparticles in rats after repeated oral administration. *Environmental Toxicology and Pharmacology*, *34*(1), 67–80. https://doi.org/10.1016/j.etap.2011.05.014.

Meek, B., & Doull, J. (2009). Pragmatic challenges for the vision of toxicity testing in the 21st century in a regulatory context: Another ames test? .. or a new edition of "the red book"? *Toxicological Sciences*, *108*(1), 19–21. https://doi.org/10.1093/toxsci/kfp008.

Naz, S., Gul, A., & Zia, M. (2020). Toxicity of copper oxide nanoparticles: A review study. *IET Nanobiotechnology*, *14*(1), 1–13. https://doi.org/10.1049/iet-nbt.2019.0176.

Nemmar, A., Hoylaerts, M. F., Hoet, P. H. M., Vermylen, J., & Nemery, B. (2003). Size effect of intratracheally instilled particles on pulmonary inflammation and vascular thrombosis. *Toxicology and Applied Pharmacology*, *186*(1), 38–45. https://doi.org/10.1016/S0041-008X(02)00024-8.

Oberdörster, G. (2010). Safety assessment for nanotechnology and nanomedicine: Concepts of nanotoxicology. *Journal of Internal Medicine*, *267*(1), 89–105. https://doi.org/10.1111/j.1365-2796.2009.02187.x.

Oberdörster, G., Oberdörster, E., & Oberdörster, J. (2005). Nanotoxicology: An emerging discipline evolving from studies of ultrafine particles. *Environmental Health Perspectives*, *113*(7), 823–839. https://doi.org/10.1289/ehp.7339.

Paschoalino, M. P., Marcone, G. P. S., & Jardim, W. F. (2010). Os nanomateriais e a questão ambiental. *Quimica Nova*, *33*(2), 421–430. https://doi.org/10.1590/s0100-40422010000200033.

Pettibone, J. M., Adamcakova-Dodd, A., Thorne, P. S., O'Shaughnessy, P. T., Weydert, J. A., & Grassian, V. H. (2008). Inflammatory response of mice following inhalation exposure to iron and copper nanoparticles. *Nanotoxicology*, *2*(4), 189–204. https://doi.org/10.1080/17435390802398291.

Poormohammad Matouri, Z., & Noori, A. (2018). Effect of multi-wall carbon nanotubes toxicity on kidney function and tissue in rats. *Journal of Gorgan University of Medical Sciences*, *20*(1), 22–28. http://goums.ac.ir/journal/article-1-3273-en.html&sw=.

Prasad, M., Lambe, U. P., Brar, B., Shah, I., Manimegalai, J., Ranjan, K., et al. (2018). Nanotherapeutics: An insight into healthcare and multi-dimensional applications in medical sector of the modern world. *Biomedicine and Pharmacotherapy*, *97*, 1521–1537. https://doi.org/10.1016/j.biopha.2017.11.026.

Radaic, A., Pugliese, G. O., Campese, G. C., Pessine, F. B. T., & De Jesus, M. B. (2016). Como estudar interações entre nanopartículas e sistemas biológicos. *Quimica Nova*, *39*(10), 1236–1244. https://doi.org/10.21577/0100-4042.20160146.

Roberto, M. M., & Christofoletti, C. A. (2019). How to assess nanomaterial toxicity? An environmental and human health approach. In *Nanomaterials: Toxicity, human health and environment* IntechOpen.

Roy, S., Sadhukhan, R., Ghosh, U., & Das, T. K. (2015). Interaction studies between biosynthesized silver nanoparticle with calf thymus DNA and cytotoxicity of silver nanoparticles. *Spectrochimica Acta—Part A: Molecular and Biomolecular Spectroscopy*, *141*, 176–184. https://doi.org/10.1016/j.saa.2015.01.041.

Sahu, S. C., Zheng, J., Graham, L., Chen, L., Ihrie, J., Yourick, J. J., et al. (2014). Comparative cytotoxicity of nanosilver in human liver HepG2 and colon Caco2 cells in culture. *Journal of Applied Toxicology*, *34*(11), 1155–1166. https://doi.org/10.1002/jat.2994.

Semmler, M., Seitz, J., Erbe, F., Mayer, P., Heyder, J., Oberdörster, G., et al. (2004). Long-term clearance kinetics of inhaled ultrafine insoluble iridium particles from the rat lung, including transient translocation into secondary organs. *Inhalation Toxicology*, *16*(6–7), 453–459. https://doi.org/10.1080/08958370490439650.

Simeonova, P. P., & Erdely, A. (2009). Engineered nanoparticle respiratory exposure and potential risks for cardiovascular toxicity: Predictive tests and biomarkers. *Inhalation Toxicology*, *21*(1), 68–73. https://doi.org/10.1080/08958370902942566.

Wang, F., Gao, F., Lan, M., Yuan, H., Huang, Y., & Liu, J. (2009). Oxidative stress contributes to silica nanoparticle-induced cytotoxicity in human embryonic kidney cells. *Toxicology in Vitro*, *23*(5), 808–815. https://doi.org/10.1016/j.tiv.2009.04.009.

Warheit, D. B., Webb, T. R., Sayes, C. M., Colvin, V. L., & Reed, K. L. (2006). Pulmonary instillation studies with nanoscale TiO2 rods and dots in rats: Toxicity is not dependent upon particle size and surface area. *Toxicological Sciences*, *91*(1), 227–236. https://doi.org/10.1093/toxsci/kfj140.

Yan, G., Huang, Y., Bu, Q., Lv, L., Deng, P., Zhou, J., et al. (2012). Zinc oxide nanoparticles cause nephrotoxicity and kidney metabolism alterations in rats. *Journal of Environmental Science and Health, Part A*, *47*(4), 577–588. https://doi.org/10.1080/10934529.2012.650576.

Yang, C., Tian, A., & Li, Z. (2016). Reversible cardiac hypertrophy induced by PEG-coated gold nanoparticles in mice. *Scientific Reports*, *6*. https://doi.org/10.1038/srep20203.

Index

Note: Page numbers followed by *f* indicate figures and *t* indicate tables.

Printed in the United States
by Baker & Taylor Publisher Services